Green Sustainable Process for
CHEMICAL AND ENVIRONMENTAL ENGINEERING AND SCIENCE

Green Sustainable Process for
CHEMICAL AND ENVIRONMENTAL ENGINEERING AND SCIENCE

Methods for Producing Smart Packaging

Edited by

INAMUDDIN
Department of Applied Chemistry, Zakir Husain College of Engineering and Technology, Faculty of Engineering and Technology, Aligarh Muslim University, Aligarh, Uttar Pradesh, India

TARIQ ALTALHI
Department of Chemistry, College of Science, Taif University, Taif, Saudi Arabia

JORDDY NEVES CRUZ
Adolpho Ducke Laboratory, Botany Coordination, Museu Paraense Emílio Goeldi, Belém, PA, Brazil

ELSEVIER

Elsevier
Radarweg 29, PO Box 211, 1000 AE Amsterdam, Netherlands
The Boulevard, Langford Lane, Kidlington, Oxford OX5 1GB, United Kingdom
50 Hampshire Street, 5th Floor, Cambridge, MA 02139, United States

Copyright © 2023 Elsevier Inc. All rights reserved.

No part of this publication may be reproduced or transmitted in any form or by any means, electronic or mechanical, including photocopying, recording, or any information storage and retrieval system, without permission in writing from the publisher. Details on how to seek permission, further information about the Publisher's permissions policies and our arrangements with organizations such as the Copyright Clearance Center and the Copyright Licensing Agency, can be found at our website: www.elsevier.com/permissions.

This book and the individual contributions contained in it are protected under copyright by the Publisher (other than as may be noted herein).

Notices

Knowledge and best practice in this field are constantly changing. As new research and experience broaden our understanding, changes in research methods, professional practices, or medical treatment may become necessary.

Practitioners and researchers must always rely on their own experience and knowledge in evaluating and using any information, methods, compounds, or experiments described herein. In using such information or methods they should be mindful of their own safety and the safety of others, including parties for whom they have a professional responsibility.

To the fullest extent of the law, neither the Publisher nor the authors, contributors, or editors, assume any liability for any injury and/or damage to persons or property as a matter of products liability, negligence or otherwise, or from any use or operation of any methods, products, instructions, or ideas contained in the material herein.

ISBN: 978-0-323-95644-4

For information on all Elsevier publications visit our website at https://www.elsevier.com/books-and-journals

Publisher: Joseph Hayton
Editorial Project Manager: Emerald Li
Production Project Manager: Sruthi Satheesh
Cover Designer: Christian Bilbow

Typeset by TNQ Technologies

Contents

List of contributors   xi

1. Antioxidant packaging   1
Olufunmilola Adunni Abiodun, Shalom Olamide Abiodun and Abimbola Kemisola Arise

     1. Introduction   1
     2. Types of antioxidant   1
     3. Synthetic antioxidants   5
     4. Sources of antioxidants   5
     5. Food packaging material   8
     6. Oxidation in food products   10
     7. Mechanism of peroxidation   11
     8. Types and features of smart packaging   12
     9. Production of films for antioxidant packaging on food product   13
     10. Mechanism of antioxidant release from packaging film to food products   14
     11. Application of antioxidant in food   15
     12. Conclusions   16
     References   17

2. Challenges and perspectives in application of smart packaging   25
Nadia Akram, Khawaja Taimoor Rashid, Tanzeel Munawar, Muhammad Usman and Fozia Anjum

     1. Introduction   25
     2. Classification of packaging systems   26
     3. The applications and market opportunities   30
     4. The challenges and prospects of research   31
     5. Conclusion   34
     References   34

3. Innovations in smart packaging technologies for monitoring of food quality and safety   39
Biplab Roy, Deepanka Saikia, Prakash Kumar Nayak, Suresh Chandra Biswas, Tarun Kanti Bandyopadhyay, Biswanath Bhunia and Pinku Chandra Nath

     Abbreviations   39
     1. Introduction   39
     2. Overview of smart packaging technology   40
     3. Conclusion   55
     References   56

4. Food safety guidelines for food packaging — 59
Kashish Gupta

1. Introduction to food packaging — 59
2. Food packaging role in food safety — 59
3. International legislation related to food safety — 60
4. Food safety laws — 61
5. Food contact legislation — 62
6. Condition for sale and license — 62
7. Packaging label requirements of oils — 64
8. Packaging labeling regulations in different countries — 65

References — 68

5. Industrial barriers for the application of active and intelligent packaging — 71
Partha Pratim Sarma, Kailash Barman and Pranjal K. Baruah

1. Introduction — 71
2. Need of packaging — 72
3. Traditional packaging — 73
4. Advantages and disadvantages of different types of packing materials — 74
5. Smart packaging — 77
6. Active packaging — 77
7. Intelligent packaging — 83
8. Differences between active and intelligent packaging — 85
9. Advantages of active and intelligent packaging materials — 86
10. Industrial barriers for the application of active and intelligent packaging — 86
11. Conclusion — 91

References — 92

6. Legislation on active and intelligent packaging — 97
Shashi Kiran Misra and Kamla Pathak

1. Introduction — 97
2. Regulatory aspects in European Union — 100
3. Regulatory aspects in India and South East Asian countries (Thailand, Malaysia and Singapore) — 104
4. Bureau of Indian standards (BIS) and food safety and standards authority (FSSAI) — 105
5. Regulatory aspects in US — 107
6. FDA regulations — 108
7. Threshold of regulation rule (1995) — 109
8. Regulatory aspects in Brazil, South and Central America — 109
9. Conclusion — 111

References — 112

7. Market demand for smart packaging versus consumer perceptions — 115
Madhur Babu Singh, Prashant Singh and Pallavi Jain

1. Introduction — 115
2. Smart packaging — 117
3. Active packaging (AP) — 118
4. Intelligent packaging (IP) — 123
5. Consumer perception — 124
6. Conclusion — 126
References — 126

8. Metal packaging for food items advantages, disadvantages and applications — 129
Nadia Akram, Muhammad Saeed, Asim Mansha, Tanveer Hussain Bokhari and Akbar Ali

1. Introduction — 129
References — 140

9. An approach of smart packaging for home meals — 143
Uzma Hira and Muhammad Husnain

1. Smart packaging approach — 143
2. Current need of smart packaging for home meal — 145
3. Different HFR goods and essential smart packaging features — 146
4. Impact of smart packaging in HFR technology — 149
5. Smart packaging technologies available in HFR industries — 150
6. Different smart packaging approaches for different products — 162
7. Smart packaging trends for home meals — 163
8. Consumer benefits of smart packaging — 166
9. Issues related to the smart packaging — 166
10. Conclusion — 167
References — 167

10. Perspective and challenges: intelligent to smart packaging for future generations — 171
Sri Bharti, Shambhavi Jaiswal, V.P. Sharma and Inamuddin

1. Introduction — 171
2. Connected and smart packaging: past, present, future … — 172
3. Bio-based plastics — 174
4. Cutting edge advancement — 174
5. Packaging types — 176
References — 182

11. Production of smart packaging from sustainable materials 185
Adeshina Fadeyibi

1. Introduction	185
2. Mechanical production of SPMs	186
3. Biochemical production of SPMs	187
4. Prospects and conclusions	193
References	193

12. Smart packaging for commercial food products 197
Pinku Chandra Nath, Nishithendu Bikash Nandi, Shamim Ahmed Khan, Biswanath Bhunia, Tarun Kanti Bandyopadhyay and Biplab Roy

Abbreviations	197
1. Introduction	197
2. Smart packaging technologies for food products	198
3. Consumer advantages and comfort factors of smart packaging	205
4. Conclusion	207
References	207

13. Smart applications for fish and seafood packaging systems 211
Oya Irmak Sahin, Furkan Turker Saricaoglu, Ayse Neslihan Dundar and Adnan Fatih Dagdelen

1. Introduction	211
2. Sensors	212
3. Indicators	216
4. Blockchain systems	219
5. Electronic sensing systems	220
6. Conclusions and potential trends	222
References	223

14. Smart packaging for medicinal food supplements 229
Vipul Prajapati and Salona Roy

1. Introduction	229
2. Medicinal food products and its types	230
3. Concept of traditional and smart packaging	230
4. Necessity of the smart packaging	230
5. Smart packages	234
6. Patented products of smart packaging	257
7. Future challenges and scope of smart packaging	257
8. Conclusion	262
References	262

15. Smart packaging to preserve fruit quality — 267
Pinku Chandra Nath, Biswanath Bhunia, Tarun Kanti Bandyopadhyay and Biplab Roy

Abbreviations — 267
1. Introduction — 267
2. Fruit freshness is related to classification, stages, and harvesting — 269
3. Smart packaging technologies for fruit preservation and freshness — 269
4. Conclusion — 279
References — 280

16. Evaluating the sustainable metal packaging for cooked foods among food packaging materials — 283
Figen Balo and Lutfu S. Sua

1. Introduction — 283
2. Major food packaging materials and metal packing — 286
3. Results and discussion — 297
4. Conclusions — 301
References — 301

17. Smart packaging products and smart showcase design — 303
Mustafa Kucuktuvek and Caglar Altay

1. Introduction — 303
2. Smart packaging — 305
3. Smart showcase design — 306
4. Showcase design principles — 308
5. Showcase accessories — 310
6. Showcase materials — 310
7. Smart showcase design proposal — 313
References — 314

18. Biodegradable polymers- a greener approach for food packaging — 317
Bably Khatun, Jonali Das, Shagufta Rizwana and T.K. Maji

1. Introduction — 317
2. History of food packaging — 318
3. Characteristics and criteria of packaging materials — 319
4. Types of food packaging — 322
5. Why is biodegradable packaging important? — 324
6. Biopolymers used for food packaging — 324
7. Biodegradable polymers' limitations in food packaging — 344
8. Role or impact of different ingredients in biopolymer modification — 345
9. Advances in food packaging — 353
10. Characterization of biodegradable packaging material — 355
11. Polymer fabrication technology for packaging food — 356

12.	Conclusion and future scope	360
	References	362

19. Application of environmentally benign biodegradable composite in intelligent and active packaging   371

Mira chares Subash and Muthiah Perumalsamy

1.	Introduction	371
2.	Biodegradable composite	373
3.	Smart packaging	377
4.	Benefits of smart packaging	381
5.	Advanced techniques in smart packaging	385
6.	Future scope and challenges of smart packaging	386
7.	Conclusion	387
	References	387

Index   *391*

List of contributors

Olufunmilola Adunni Abiodun
Department of Home Economics and Food Science, University of Ilorin, Ilorin, Kwara State, Nigeria

Shalom Olamide Abiodun
Department of Biochemistry, Landmark University, Omu Aran, Kwara State, Nigeria

Nadia Akram
Department of Chemistry, Government College University Faisalabad, Faisalabad, Punjab, Pakistan

Akbar Ali
Department of Chemistry, Government College University Faisalabad, Faisalabad, Punjab, Pakistan

Caglar Altay
Department of Interior Design, Aydın Vocational School, Aydın Adnan Menderes University, Aydın, Turkey

Fozia Anjum
Department of Chemistry, Government College University Faisalabad, Faisalabad, Punjab, Pakistan

Abimbola Kemisola Arise
Department of Home Economics and Food Science, University of Ilorin, Ilorin, Kwara State, Nigeria

Figen Balo
Department of METE, Firat University, Elazig, Turkey

Tarun Kanti Bandyopadhyay
Department of Chemical Engineering, National Institute of Technology Agartala, Jirania, Tripura, India

Kailash Barman
Department of Food Engineering and Technology, Central Institute of Technology, Kokrajhar, Assam, India

Pranjal K. Baruah
Department of Applied Sciences, GUIST, Gauhati University, Guwahati, Assam, India

Sri Bharti
CSIR–IITR, Lucknow, Uttar Pradesh, India

Biswanath Bhunia
Department of Bio Engineering, National Institute of Technology Agartala, Jirania, Tripura, India

Tanveer Hussain Bokhari
Department of Chemistry, Government College University Faisalabad, Faisalabad, Punjab, Pakistan

Suresh Chandra Biswas
Department of Home Science, Khowai, Tripura, India

Pinku Chandra Nath
Department of Bio Engineering, National Institute of Technology Agartala, Jirania, Tripura, India

Adnan Fatih Dagdelen
Department of Food Engineering, Faculty of Engineering and Natural Science, Bursa Technical University, Bursa, Turkey

Jonali Das
Department of Chemical Sciences, Tezpur University, Tezpur, Assam, India

Ayse Neslihan Dundar
Department of Food Engineering, Faculty of Engineering and Natural Science, Bursa Technical University, Bursa, Turkey

Adeshina Fadeyibi
Department of Food and Agricultural Engineering, Faculty of Engineering and Technology, Kwara State University, Ilorin, Kwara, Nigeria

Kashish Gupta
Noida International University, Greater Noida, Uttar Pradesh

Uzma Hira
School of Physical Sciences (SPS), University of the Punjab, New Campus, Lahore, Punjab, Pakistan

Muhammad Husnain
School of Physical Sciences (SPS), University of the Punjab, New Campus, Lahore, Punjab, Pakistan

Inamuddin
Department of Applied Chemistry, Zakir Husain College of Engineering and Technology, Faculty of Engineering and Technology, Aligarh Muslim University, Aligarh, Uttar Pradesh, India

Pallavi Jain
Department of Chemistry, SRM Institute of Science and Technology, Delhi NCR Campus, Ghaziabad, Uttar Pradesh, India

Shambhavi Jaiswal
CSIR-IITR, Lucknow, Uttar Pradesh, India

Shamim Ahmed Khan
Department of Chemistry, National Institute of Technology Agartala, Jirania, Tripura, India

Bably Khatun
Department of Chemical Sciences, Tezpur University, Tezpur, Assam, India

Mustafa Kucuktuvek
Department of Interior Architecture, Faculty of Architecture, İskenderun Technical University, İskenderun, Hatay, Turkey

T.K. Maji
Department of Chemical Sciences, Tezpur University, Tezpur, Assam, India

Asim Mansha
Department of Chemistry, Government College University Faisalabad, Faisalabad, Punjab, Pakistan

Shashi Kiran Misra
School of Pharmaceutical Sciences, CSJM University, Kanpur, Uttar Pradesh, India

Tanzeel Munawar
Department of Chemistry, Government College University Faisalabad, Faisalabad, Punjab, Pakistan

Nishithendu Bikash Nandi
Department of Chemistry, National Institute of Technology Agartala, Jirania, Tripura, India

Prakash Kumar Nayak
Department of Food Engineering and Technology, Central Institute of Technology, Kokrajhar, Assam, India

Kamla Pathak
Faculty of Pharmacy, Uttar Pradesh University of Medical Sciences Saifai, Etawah, Uttar Pradesh, India

Muthiah Perumalsamy
Department of Chemical Engineering, National Institute of Technology, Tiruchirappalli, Tamilnadu, India

Vipul Prajapati
Department of Pharmaceutics, SSR College of Pharmacy, Affiliated to Savitribai Pule Pune University, Silvassa, Union Territory of Dadra Nagar Haveli and Daman Diu, India

Khawaja Taimoor Rashid
Department of Chemistry, Government College University Faisalabad, Faisalabad, Punjab, Pakistan

Shagufta Rizwana
Department of Food Engineering and Technology, Tezpur University, Tezpur, Assam, India

Biplab Roy
Department of Chemical Engineering, National Institute of Technology Agartala, Jirania, Tripura, India

Salona Roy
Department of Pharmaceutics, SSR College of Pharmacy, Affiliated to Savitribai Pule Pune University, Silvassa, Union Territory of Dadra Nagar Haveli and Daman Diu, India

Muhammad Saeed
Department of Chemistry, Government College University Faisalabad, Faisalabad, Punjab, Pakistan

Oya Irmak Sahin
Department of Chemical Engineering, Faculty of Engineering, Yalova University, Yalova, Turkey

Deepanka Saikia
Department of Food Engineering and Technology, Central Institute of Technology, Kokrajhar, Assam, India

Furkan Turker Saricaoglu
Department of Food Engineering, Faculty of Engineering and Natural Science, Bursa Technical University, Bursa, Turkey

Partha Pratim Sarma
Department of Applied Sciences, GUIST, Gauhati University, Guwahati, Assam, India

V.P. Sharma
CSIR-IITR, Lucknow, Uttar Pradesh, India

Madhur Babu Singh
Department of Chemistry, SRM Institute of Science and Technology, Delhi NCR Campus, Ghaziabad, Uttar Pradesh, India; Department of Chemistry, Atma Ram Sanatan Dharma College, University of Delhi, New Delhi, India

Prashant Singh
Department of Chemistry, Atma Ram Sanatan Dharma College, University of Delhi, New Delhi, India

Lutfu S. Sua
Department of Management and Marketing, Southern University and A&M College, Baton Rouge, LA, United States

Mira chares Subash
Department of Chemical Engineering, National Institute of Technology, Tiruchirappalli, Tamilnadu, India

Muhammad Usman
Department of Chemistry, Government College University Faisalabad, Faisalabad, Punjab, Pakistan

CHAPTER 1

Antioxidant packaging

Olufunmilola Adunni Abiodun[a], Shalom Olamide Abiodun[b] and Abimbola Kemisola Arise[a]

[a]Department of Home Economics and Food Science, University of Ilorin, Ilorin, Kwara State, Nigeria; [b]Department of Biochemistry, Landmark University, Omu Aran, Kwara State, Nigeria

1. Introduction

Antioxidants are compounds that inhibit lipid oxidation which is a chemical reaction that can release free radicals and chain reactions that cause damage the cells of an organisms. Antioxidants are considered as free radical scavengers and is essential in maintaining cell functions and integrity [1,2]. Similarly, antioxidants react with the reactive oxygen species (ROS) and support the normal cell activities. Antioxidants prevents the configuration of free radical from damaging tissues by delaying the production of radicals or causing degradation of free radical [3]. Antioxidants slow down oxidation processes and thus prevent oxidative stress [4]. Antioxidants can either be classified as synthetic or natural in nature. The natural antioxidants are obtained from fruits and vegetables, grain seeds, herbs and spices etc.

2. Types of antioxidant

Antioxidant can be divided in two based on their sources:
1. Endogenous Antioxidant
2. Exogenous Antioxidant

2.1 Endogenous antioxidant

Endogenous antioxidants are products of the body's metabolism. Because they are produced by our own bodies and not obtained from food sources, endogenous antioxidants are much more powerful than exogenous antioxidants. Endogenous antioxidants repair damage caused by free radicals by initiating cell regeneration from the inside on out, while exogenous antioxidants repair only part of the damage caused by free radicals from the outside out inside by stimulating (not initiating) cell regeneration. Endogenous antioxidants are those that ones that can produce their own antioxidant substances. The endogenous antioxidants are divided into enzyme and non-enzyme antioxidants based on their structural characteristics. Antioxidant enzyme is an important enzyme that defend the body against oxidative stress. These enzymes include: catalase, superoxide dismutase, glutathione reductase and glutathi-one peroxidase [5]. Enzymatic antioxidants breakdown the free radicals and convert dangerous oxidant products into hydrogen peroxide and water especially in the presence of metallic catalyst such as zinc, copper, iron and manganese. The non-enzyme antioxidants includes: Glutathione GSH, Alpha-lipoic acid (LA), uric acid, Coenzyme Q and Ferritin.

2.2 Exogenous antioxidant

Exogenous antioxidants are the compounds which are derived outside the body with significant roles as preservatives for products such as cosmetics, pharmaceuticals, food and human health. Among the exogenous antioxidants are vitamins (A, C, E), selenium, carotenoids, catechins, polyphenols and phytoestrogens. These antioxidants and phytonutrients are found in plant-based foods.

2.2.1 Ascorbic acid

Ascorbic acid (vitamin C) is a natural component in plant tissues and is water-soluble therefore diffuses easily in food. The nutrient is majorly found in fruits, vegetables and many other crops such as pineapple, citrus, papaya, kiwifruits etc. [6]. Ascorbic acid has antioxidant property which makes it useful and important in the food industry [7]. Ascorbic acid acts as electron donor in food thereby releasing hydrogen atoms leading to termination of the chain reactions. According to Cort (1982), the presence of hydroxyl groups in the ascorbic acid indicated its role as antioxidant agent. Ascorbic acid scavenges oxygen and reduced the oxidation state of the metals thereby reducing the oxidation catalysis in food [8]. It has strong oxidizing ability against the free radicals and arrests the chain reactions that may cause undesirable effect on consumers [7]. Taniguchi et al. [9] stated that the level of lipid oxidation could be reduced in the presence of ascorbic acid. Apart from the antioxidant properties, ascorbic acid prevents browning reactions in food by decreasing the o-quinone synthesized by polyphenol oxidase. Ascorbic acid decreases the pH and is used as natural agent for stabilizing the color of food products such as meat and pork as reported by Yin et al. [10]. Ascorbic acid had been applied to food products such as meat, beer, wine bread etc due the strong oxygen scavenging ability [8]. It is affected by pH, concentration, heat and light.

2.2.2 Carotenoid compounds

Carotenoids are isoprenoid compounds with eight isoprene units. The major two classes of carotenoids are the carotenes (comprises of lycopene and β-carotene) and oxycarotenoids (xanthophylls and lutein) [11]. The major carotenoids which have antioxidants properties in the diets include lycopene, β-carotene, lutein and β-cryptoxanthin. The ability of carotenoids to eliminate the ROS depends on the bonds (single or double) of the polyene backbone. Carotenoids generally are affected by light, heat and oxygen during storage and processing into product. Therefore, exposure to high temperature during processing could cause isomerization of the double bond which changes the trans to cis form in the carotenoids [12,13]. β-Carotene belong to carotenoid group and is regarded as a secondary metabolite which is synthesized by plants. It is a polyene compound with conjugated double bonds [14]. Carotene are vitamin A precursor or provitamin A which is useful in vision and eyesight. It acts as antioxidant in food by scavenging free radicals which can cause damaging effect on consumers [15]. Lycopene also belong to carotenoids family known as the tetraterpenoids with 11 conjugated double bonds which naturally occurs in trans form [16,17]. It consists of red pigment present in fruits and vegetables such as in watermelon, grape fruits, carrot, tomato etc. Lycopene is affected by the degree of ripeness, environmental factors, storage, varieties of fruits and processing condition. Apart from adding color to food, lycopene act as antioxidant agent which protect against the oxidative stress in plant. The antioxidant activities protect against cancer especially prostrate and skin cancer but also reduce cardiovascular and related diseases [17]. Xanthophylls belong to the carotenoid group and are yellow pigments containing oxygen atoms in the form of epoxides and hydroxyl group. Xanthophyll consists of the following lutein, astaxanthin, zeaxanthin, fucoxanthin, β-cryptoxanthin and capsanthin [18]. The three major types of xanthophylls which are valued for their roles in the food and human vision are crytoxanthanin, lutein and zeaxanthin (3,3′-dihydroxy-β-carotene) [19]. Lutein is found in fruits and vegetables such as spinach, yellow orange, guava, melon, while zeaxanthin are present in leafy vegetables and yellow/orange fruits. Both lutein and zeaxanthin are found in maize and egg yolk [20]. They act as antioxidant agent in food.

2.2.3 Polyphenol compounds

Phenolic compounds are chemical substances and metabolites occurring largely in plant derived foods. They have aromatic rings with a single or multiple hydroxyl compounds and functional derivatives such as esters, glycosides, methyl esters etc. [21,22]. Classification of polyphenols depends on their chemical structures, biological functions and origin. Most polyphenolic compounds in plants are in form of glycosides coupled with different sugar units and acylated sugars in the polyphenol skeletons [23]. Classifications of polyphenols are flavonoids, tannins, phenolic acids and stilbenes. Phenolic acids include hydroxycinnamic acid (ferulic, caffeic, sinapic acids and *p*-coumaric) and hydroxybenzoic acid (*p*-hydroxybenzoic, protocatechuic, vanillic, syringic acid and gallic) derivatives. Examples of food rich in phenolic acids include berries (cranberry, blueberry), citrus fruits, coffee, tea, potato, apple, lettuce, pear [22]. The two types of flavonoids are anthocyanins (anthocyanidin) and anthoxanthins (flavans, flavones (luteolin and Apigenin), isoflavones, flavonols (quercetin and kaempferol) and their glycosides). Tannin are water soluble polyphenols which are hydrolysable (esters of gallic acids), condensed tannins (proanthocyanidins) and phlorotannins (phloroglucinol). Examples of products containing tannin include olive, peach, peas, cocoa, tea, lentils etc. Stilbenes contains 1,2-diphenylethylene nucleus and hydroxyl groups substituted in the aromatic rings with the form of monomers or oligomers such as trans-resveratrol with a trihydroxystilbene skelelton. Examples food containing stilbenes are peanuts, grapes, wine, soy and peanut products. Polyphenolic compounds terminate free radicals and chelate metal ions which can catalyze lipid oxidation [22]. Polyphenol antioxidant activities depend on the hydroxylation status of the aromatic rings which include chelating and stabilizing the divalent cations, modulation of antioxidant endogenous enzymes and free radicals scavenging activities [22]. Sources of polyphenols in food include the vegetables, fruits and whole grains. Polyphenol compounds have strong antioxidant activities which scavenge free radicals by releasing or donating an electron. The hydroxylation and conjugated patterns shown by the hydroxy group are responsible for the antioxidant activities of the flavonol [23]. Polyphenolic compounds prevent free radicals generation thereby reducing the rate of lipid oxidation. Moreover, polyphenols are referred to as metal chelators by binding with free metals that can speed up the oxidation reaction process and chain breakers by releasing an electron, neutralizing the free radicals and forming stable radicals thus terminating the chain reactions [23,24]. Du et al. [25] also observed that polyphenols could cause enzymes like superoxide dismutase, glutathione peroxidase and catalase to decompose hydroperoxides, hydrogen peroxide and superoxide anions respectively. The major factors affecting polyphenol contents of food according to Manach et al. [26] are sun exposure, rainfall, soil type, degree of ripening and agronomic practices. Presence of polyphenols help plant to respond to stress of different types and healing of damaged plant parts through lignifications process.

2.2.4 Tocopherol

Vitamin E refers to all tocol and tocotrienol compounds which are present in plants. α-tocopherols are the most common types, biologically active compound and is located in the chloroplast whereas others, are formed outside the chloroplast [27]. The tocopherols have 6-chromanol ring structure at the fifth seventh and eighth positions with C_{16} saturated chain at the 2' position. However, tocotrienols are unsaturated at third, seventh and eleventh positions of the side chain. Therefore, the number and positions of the methyl groups in the structures are the major differences in tocopherols and tocotrienols [28]. Tocopherols are antioxidant compounds and also help in stabilization of the membrane. It is involved in scavenging free radicals by donating hydrogen atom thereby inhibiting lipid oxidation [29]. α-Tocopherol is situated in the cell membrane for lipoprotein protection and is the main antioxidant in animal and plant cells. It is regarded as chain breaker due to its peroxy radical scavenging ability and prevention of oxidation in food products [28]. Sources of vitamin E are oil, soybeans, cereal, peanut, eggs, margarine, tomato products etc.

3. Synthetic antioxidants

The popular and well known synthetic antioxidants are the butylated hydroxytoluene (BHT), butylated hydroxyanisole (BHA), propyl gallate (PG) and tert-butyl hydroquinone (TBHQ). Major synthetic antioxidants which are capable of preventing free radicals and chain reactions are the BHA (E-320) and BHT (E-321). The antioxidants stabilize the color, flavor, nutritive value and freshness of food. They are both effective in animal fat than vegetable fat. They are not stable at high temperature. BHT and BHA are food additives for fat and oil. The permissible level in food is 100 mg/kg fat as reported by Husøy et al. [30]. BHA is used in food products such as snacks, flavoring agents, beer, lard, meat, butter, baked goods, sweets, vegetable oil, chewing gum etc [31]. TBHQ (E-319) is a synthetic antioxidant commonly used in both vegetable and animal fat. It is more effective and heat stable than BHA and BHT. Propyl Galate (E-310) is not stable at high temperature and used as liposoluble synthetic antioxidants. Application of chemical or synthetic antioxidant is generally reducing due to its toxicity effect on the consumers. According to EFSA [32], the use of high doses of BHT results in tumor promoting effects in animal studies. SCCS [33] suggested the maximum limits of BHT in products in order to avoid its toxicity. The use of BHT at concentration of 0.0002–0.8% in cosmetics was recommended.

4. Sources of antioxidants

Lots of natural antioxidants are derived from animal products, herbs, cereals, spices, oilseeds, nuts, legumes, vegetables, fruits and microbial products had been reported [34–37]. The natural sources are endowed with essential antioxidant properties such as the phenols, ascorbic acids, vitamins and volatile compounds which chelate metals and scavenge the free radicals in food thereby prolonging the quality of food products.

4.1 Herbs and spices

Herbs and spices are annual, biennial or perennial plants and are major natural food ingredients used in beverages, perfumery, medicine, cosmetics, coloring and in other food products. They are useful due to their antimicrobial, antioxidants, nutritional and pharmaceutical properties. Herbs and spices are used as flavoring agent (cinnamon, mustard, mint, dill, parsley, bay leaves, clove, nutmeg, coriander, vanilla, onion, ginger, oregano, garlic, etc), deodorizing agent (rosemary, onion, thyme, garlic, clove, sage, bay leaves etc), pungency (clove, allspice, onion, coriander, cumin, mace, fennel, cardamom, mustard, pepper, nutmeg, sesame, thyme, rosemary, cinnamon etc) and coloring agent (Saffron, paprika and turmeric) [38]. The coloring compounds in herbs and spices include the carotenoids (β-carotene, cryptoxanthin, lutin, neoxanthin, crocetin, capsanthin, zeaxanthin, violaxanthin) which ranged from reddish orange, purple, yellow and dark red in color. Ravindran et al. [39] observed the existence of crocin, curcumin and chlorophylls in green herbs. Apart from the above functions, herbs and spices have the ability to preserve food by hindering or delaying deterioration thereby increasing the keeping quality of the food products. Herbs and spices use are not limited to food industries alone but finds its way into cosmetics industries where they are incorporated into toiletries manufacturing, cream, and so on. Major active ingredients in herbs and spices are phenolic compounds, alkaloids, coumarines, resins, volatile oils, anthraquinones, flavones, saponnins, tannins, glycosides etc. Which are characterized as antiseptic, fungicidal, stimulant, anti-inflammatory, soothing, antibacterial, anticoagulant, and so on [38]. Phenolic compounds, aqueous extract and essential oils (oleoresins) contributed to the antioxidative properties of herbs and spices [40]. The dominant chemical compounds in the herbs and spices essential oils are the terpene compounds consisting of mono, di, tri and sesquiterpenes. The essential oil depends on the extraction methods, nature of volatiles and types of herbs/spices. The essential oil is made up of terpenes (camphene, α-terpinene, limonene, α-pinene, phellandrene, sabinene and myrcene), oxygenated hydrocarbons (geraniol, borneol, linalool, carveol, citronellol, menthol, nerol, fenchone and tumerone), benzene compounds (acids, alcohols, phenols, lactones and esters) and nitrogen or sulfur-containing compounds (hydrogen sulfide, indole, sinapine hydrogen sulfate and methyl propyl disulfide) [41].

4.2 Tea

Tea is a shrub belonging to the species *Camelia* and family *Theaceae*. The two popular varieties are the *Camelia sinensis and Camelia assamica* [42]. Green tea (***Camellia sinensis*, Theaceae**) has the highest phenol concentration. The phenols include **p**-coumaroylquinic acid, −epicatechin, -gallocatechin, gallic acid, -gallocatechin, -epigallocatechin gallate and -gallocatechin-3-gallate. Green tea had been reported to contain hydrolysable and condensed tannins [43]. Green tea polyphenols have cancer preventive effects and reactive oxygen species scavenger by stabilizing the phenolic radicals [44]. Polyphenolic components of green tea extract have high antioxidant properties of which the major active compounds are the flavonol monomers called catechins. According to Senanayake [43], the two flavonol with the most effective antioxidant compounds are epicatechin-3-gallate and epigallocatechin-3-gallate. Catechins (epigallocatechin and epicatechin) are among the active ingredients present in tea. Green tea polyphenol produces distinctive taste, aroma and color. The tea extracts are used in fatty foods so as to prevent lipid oxidation and improve the keeping qualities of the food products [43]. Tea, extracts from grape seeds and skins contain antioxidative activities due to the presence of phenolic acids, catechins, resveratrol, epicatechins and proanthocyanidins [45].

4.3 Fruits and vegetables

Fruits and vegetables are important in our diets because they are good sources of fibers and essential micronutrients. They also contain phytochemicals which are beneficial to health when taken singly or in combination with other food [46,47]. Fruits and vegetables are referred to as functional foods due to their protective and antioxidants properties. They contain essential vitamins, mineral and dietary fiber. The fruits and vegetables contain appreciable amount of vitamin C, carotenoids, polyphenols and lycopene. Examples of fruits rich in antioxidants include berry fruits, pineapple, blackcurrant, strawberry, pomace from vine production, red pepper, tomatoes, beetroot, garlic, pumpkin, onion etc. [48]. The antioxidants compounds in the fruits and vegetables have successfully been demonstrated by authors [49,50]. The antioxidant compounds present in the fruits are regarded as chemopreventers because they have the ability to chelate free radicals, prevent oxidative stress in cells and tissues thereby prevent diseases such as cancer [47]. The polyphenols found in grape seed contain catechins, gallic acid and epicatechins but the peels contain myricetin, quercetin, kaempferol, trans-resveratrol and ellagic acid [51,52]. Lourenco et al. [53] stated that the quality of the extract and their antioxidant properties depends on the source and technologies for extraction. Antioxidants are extracted from roots, leaves, fruits, stems, peels and seeds [35]. Citrus fruit peels such as lemon, grape and oranges was found to have high phenolic compounds.

4.4 Cereals, legumes and nut

Cereals such as rice, wheat, rye, maize, barley and oat had been established to have antioxidant properties which were linked to the carotenoids, tocopherol, tocotrienols and polyphenolic compounds present in the crops [54,55]. Angioloni and Collar [56] observed that mixtures of different cereal crops increased the antioxidant activities of the food. Cereals also contain other compounds such as polysaccharides, phytate, protein, lignans which help in minimizing the oxidation reaction effects. Major phenolic and other compounds found in cereals are ferulic, vanillic, p-coumaric acids, sinapic acids, procyanidins, galloylated propelargonidins salicylic, p-hydroxybenzoic, protocatechuic, syringic, quercetin, selenium, caffeic, isoamericanol A, Proanthocyanidins, rutin, avenanthramides etc. These compounds are extracted and derived from the bran, husk, leaves, germ and oil. The compounds scavenge the reactive nitrogen oxide, decrease the glutathione peroxidase and superoxide dismutase activities. Legumes are member of family Leguminosae which are grouped into three namely the forage, pulses and oilseeds. Peanut, soybeans are examples of oil seeds while pulses include the lentils. Examples of legumes also include kidney beans, peas, faba beans, chickpeas, lentils, and lupins [57]. Grain legumes seeds had been reported to contain antioxidant compounds which retard oxidation process in food by giving electrons to free radicals [58]. The antioxidant compounds in legumes include tocopherols, vitamin C, flavonoids, saponnin, phenolic acids etc. [58] while the major antioxidant in oilseed oil is sterol. The common oilseeds with high antioxidant activities are the cottonseed, sun flower seeds, flaxseed, canola, and soybean. Sterols are the major antioxidants in oilseeds reported to prevent oxidative degradation of the oil during thermal processing [34]. Other antioxidant compounds include tocopherols, tocotrienols with samin, sesamin and sesangolin in sesame, lignan, diglucoside and secoisolariciresinol in flatseed, isoflavonol and phenolic compounds in soy extract. Nuts and their oils are rich source of antioxidants with the highest value in the seed coat. Examples of nut are walnut, hazelnut, peanuts, cashew nut and they possess high unsaturated fatty acids, squalene, phytosterols and tocopherols [59].

5. Food packaging material

Packaging materials are important items for preserving the product from external barriers. Food packaging helps to reduce food loss/waste and increase the keeping quality of food products. Pongrácz [60] discussed the three categories of packaging which are categorized as primary, secondary and tertiary packaging depending on their functions. The major functions of packaging materials are for preservation, protection, containment, convenience, traceability communication and identification [61,62]. The selection of packaging materials depends largely on the permeability to oxygen, moisture, carbon dioxide, light, aroma and odor [63]. The materials used for packaging can be classified as traditional and industrial packaging materials. The traditional packaging materials comprises of the wood, leaves, jute bags, fibers, earthen wares, ceramic, leather etc. While the packaging materials used industrially include the plastic, glass, paper, cardboard and metals (Table 1.1).

5.1 Plastic

Plastic materials have lots of advantages over other packaging materials. Plastics are resistant, flexible, inexpensive, lightweight, heat sealable and can easily be incorporated into production processes. EPA [65] categorized plastics materials into two namely thermosets and thermoplastics depending on the moldability and solidification. Thermosets plastic when solidify and heated cannot be remolded. This type of plastic is used in construction and automobiles whereas thermoplastics materials are used in food industry. Thermoplastics materials are easily molded, remolded, reused/recyclable when heat is applied [61]. Examples of plastic materials are shown in Table 1.1. The choice of plastic packaging material used depends on the permeability which varies from material to material. Migration occurs in plastic packaging material overtime between the food and the plastic. This characteristics make plastic important in smart packaging technology.

Table 1.1 Types of packaging materials.

Packaging materials	Examples	Author
Plastic	Polyolefines, Polystyrene (PS), Polyvinyl-chloride (PVC) Polyamide (PA), Polyesters (PE), Polyethylene terephthalate (PET) Polyethylene (Low-density Polyethylene (LDPE) and High-density polyethylene (HDPE)) Polypropylene Films Degradable plastics	Sharma [64] Robertson [62] Marsh and Bugusu [61]
Metals	Aluminum Iron Steel Tin Chromium	Sharma [64] Robertson [62]
Paper Paperboard	Kraft paper, Grease proof, Bleached paper, Glassine paper Vegetable parchment paper Tissue paper, Waxed paper White board, Solid board, Paper laminates, Chipboard, Fiberbaord	Sharma [64] Robertson [62] Marsh and Bugusu [61],
Glass	White Flint glass Pale green glass Dark green glass Amber glass Blue glass	Sharma [64]

5.2 Glass

Glass is one of the most valuable materials used in food and non-food industries due to their functions. Glass is solid, translucent, transparent, reusable, recyclable, used for decorative and technological purposes. Glass is durable and resistant to internal and external forces. Glass has no permeability to water vapor, gas, liquid and odor. The smooth and shiny structure material does not erode nor react with the food components [63]. Glass packaging are commonly in cylindrical shape but are produced in different shapes with narrow rims, wide mouths (jars), vials and ampoules.

5.3 Paper and paper

Paper and paperboard are materials which are made from cellulose fibers obtained from wood. These fibers are bleached and chemically treated to produce paper. These materials are used in boxes, cartons, bags, wrapping paper, sacks. Examples of paper and paperboards products are paper plates, cups, tissue paper etc. Paper is used to package food for short period because of its unfavorable barrier properties and poor heat sealability. Plain paper properties are improved when they are laminated, coated and incorporated with other materials like resins, lacquers and waxes. Treated papers are best used as primary packaging material where the paper gets in contact with the food product. On the other hand, paperboard are rarely used as primary packaging material but used in shipping containers because they are thicker than paper [61].

5.4 Metals

Metal cans are the most available metal packaging material in food industry. According to Debeaufort [66], metallic material are classified into three namely; aluminum, coated plates and stainless steel plates. Metal packaging materials are used widely in food industries as a result of their ability to prevent movement of matter (gas, water) in and out of the of the packaging. The problem with metal packaging materials is the migration of metals such as lead, bisphenol A, cadmium, aluminum, mercury, iron, tin and nickel into the food. Also, corrosion and bulging of cans have been noticed in metal cans. Metals are coated with protective lacquers in order to hinder the reactions between the food products and the metals. Metal can be melted, remolded and recycled for other purposes [67].

6. Oxidation in food products

The major quality deterioration in food and food product is caused by lipid oxidation [68]. Lipid oxidation is initiated by several molecular reactions resulting in production of free radicals. This led to interactions in the food components resulting in generation of desirable and undesirable products. Fatty foods are more prone to oxidation especially during processing, storage and distribution. Lipid oxidation causes rancidity by altering the nutrient composition of food, change in color, giving off flavor etc [68]. Chelatin/antioxidant agents (inhibitors) are added to food products for preservation against oxidation. Lipid peroxidation is an oxidative damage or breakdown of double or multiple bonds from unsaturated fatty acids, cholesterol, glycolipids and their esters. This is regarded as a chain reaction caused by addition or removal of radical responsible for damage in unsaturated fatty acids [69]. Reactive species are the collective name for oxygen and nitrogen species known as reactive oxygen nitrogen species (RONS). ROS is a by-product of oxygen metabolic processes which is used in the homeostasis and cell signaling. ROS reacts with nutrients (lipids, proteins, carbohydrates) and nucleic acids at high concentration leading to cell structure damage and oxidative stress [70,71]. ROS are molecules emanated from oxygen free radicals (e.g superoxide, hydroxyl, peroxyl, alkoxyl) while reactive nitrogen species (RNS) are nitrogen-containing oxidants (e.g nitric oxide, peroxynitrite) [72]. Oxidation of lipids have been related to development of diseases like diabetes, organ injury, premature birth disorders, parkinson disease, fibrosis, alzheimer disease, silicosis, inflammation and cancer. Lipid peroxidation by-products may react with the enzyme superoxide dismutase causing oxidative modification which led to reduction in enzyme activity. The peroxidation chain reactions begin with the reactive oxygen species (ROS) which attack the multiple bonds unsaturated fatty acids with hydrogen atoms [73].

7. Mechanism of peroxidation

In the initiation stage, lipid radical L• is formed. This is activated in the presence of exogenous agents such as radiation, UV light, etc while the endogenous enzymes (cytochrome P450, xanthine oxidase, NADPH oxidase) can produce free radicals. Other ways by which free radicals can be produced were highlighted by Yin et al. [74]. These include the electron transport chain (ETC) in mitochondria to produce superoxide and in-vitro oxidation of unsaturated fatty acid-containing lipids which is activated by the use of cofactor like metals (eg iron), enzymes, irradiation and hydroxyl radical to produce ROS or RNS [74]. When the lipid is exposed to light, metal irons and heat, the hydrogen atom in the multiple bonds are free or form alkyl radical. In the propagation stage, oxygen is applied to the L• radical to produce a LOO• (peroxyl radical). The free radicals formed through peroxyl radical (major chain carrying species) are transfer to the hydrogen atom from the substrate. Fatty acids especially the polyunsaturated fatty acids and their esters (arachidonate and linoleate) are easily subjected to autooxidation. In the termination stage, hydrogen atom is released to the peroxyl radical species by antioxidant therefore forming non radical products and oxygen [75].

$$\text{Initiation: } \text{ln-ln} \rightarrow \text{ln}^{\cdot} + \text{ln}^{\cdot}$$

$$\text{ln}^{\cdot} + \text{L-H} \xrightarrow{k_{iLH}} \text{ln-H} + \text{L}^{\cdot}$$

$$\text{Propagation: } \text{L}^{\cdot} + O_2 \xrightarrow{k_{perox}} \text{L-OO}^{\cdot}$$

$$\text{L-OO}^{\cdot} + \text{L-H} \xrightarrow{k_p} \text{L-OOH} + \text{L}^{\cdot}$$

$$\text{Termination: } 2\text{L-OO}^{\cdot} \xrightarrow{k_t} [\text{L-OO-OO-L}]$$

$$[\text{L-OO-OO-L}] \rightarrow \text{Non radical products (NRP)} + O_2$$

$k_{iLH} = 6 \times 10^1 \text{ M}^{-1}\text{s}^{-1}$, $k_{perox} = 10^9 \text{ M}^{-1}\text{s}^{-1}$, $k_p = 6 \times 10^1 \text{ M}^{-1}\text{s}^{-1}$ for linoleate, $k_t = 1 \times 10^5$ to $10^7 \text{M}^{-1}\text{s}^{-1}$, ln = initiator, L = lipid

Yin et al. [74].

8. Types and features of smart packaging

Fig. 1.1 shows the classification of smart packaging according to Salgado et al. [76] and Biji et al. [77]. Smart packaging is the latest technology, efficient, quickest and cheapest ways of monitoring the surrounding conditions of food products [78]. Smart packaging is a total packaging systems containing the sensor technology used for monitoring the activities in and out of the packaging materials, displaying information, extending the shelf life, improving products and consumer safety. Smart packaging methods are used in the pharmaceuticals, food and other products [79]. Smart packaging are divided into two parts; namely the intelligent and active packaging. Intelligent packaging do not interfere with the food products but monitor, trace, record, send signals and information about the food quality [76,79,80]. Dobrucka and Przekop [81] established that intelligent packaging could be used in making decisions relating to shelf life, safety and quality of food products. Intelligent packagings are grouped into three categories namely, the

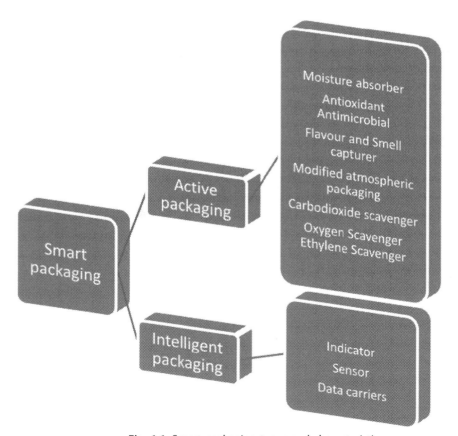

Fig. 1.1 Smart packaging types and characteristics.

indicator, sensor and data carriers. The indicator supplies signals about the changes in the product and its environment. Secondly, the sensors are used to quantify, locate and detect energy. It measures the properties captured by the device and detect smaller molecules in the food matrixes. Lastly, data carriers check the flow of the products and use the data for automation, traceability, forgery protection, identification, anti-theft prevention. Active compounds are incorporated into films in other to prolong the keeping qualities of the products. Edible packaging is an active form of packaging which has the characteristics of interacting with the food product thereby exhibiting the ability to release scavenging compounds, removal of water vapor and harmful gasses [82]. Among the active packaging methods are oxygen scavenger, ethylene scavenger, antimicrobial packaging, carbon dioxide scavenger, antioxidant packaging etc [76]. Benbettaïeb et al. [83] observed that the films control the active compounds diffusivity at the food surface thereby reducing the quantity of preservatives in required in the food. Natural biopolymer films are majorly used due to their unique properties such as biodegradability, compostability, edibility as observed by Benbettaïeb et al. [83]. Suresh et al. [84] classified the types of biopolymers recently used in food packaging into three namely; the polysaccharides (starch, cellulose, pectin, algenate, chintin, gellan, carrageenan etc), protein (animal; whey and casein protein and plant; corn zein, soy and wheat protein) sources and the biobased polyester (polylactic and poly-ecaprolactone). Antioxidant packaging are useful in food industry for the major roles it plays during storage of food. It prevents lipid oxidation in food, hence, improves the keeping qualities of the products.

9. Production of films for antioxidant packaging on food product

Antioxidant packaging films had been developed over the years by researchers. According to Lai [85], active antioxidant food packaging films for food quality and shelf life extension have been developed and effective against oxidative activities. The major methods for production of edible films for antioxidant packaging include casting, injection, heat pressing and extrusion and blow-molding processes [86–89]. Vasile [90] explained that during casting the polymers are dissolved in a suitable solvent with simultaneous incorporation of the active compound. The mixture is poured on an inert surface for evaporation to take place and the resulting film emerges. In extrusion method, the polymeric materials and active compounds are blended together and passed through a single or double screw hot extruder to form a film [90]. Marcos et al. [91] developed biodegradable films consisting of α-tocopherol and olive leaf extract through blown film extrusion method. Belan et al. [92] used carboxymethylcellulose (CMC) and gelatin in *Antidesma bunius* (L.) Spreng, (*bignay*) phenolic extract to develop edible films using casting method for food application. Positive result was achieved on the antioxidant activity of the packaging film. López-de-Dicastillo et al. [93], used green tea extracts, ferulic acid, ascorbic acid and quercetin into an ethylene vinyl alcohol copolymer matrix. Reduction in water permeability was as a result of the ferulic and ascorbic acids present in the film. The green tea extract made the film more effective against lipid oxidation. Likewise, whey protein film was added to Portuguese green tea extract to preserve Latin-style fresh cheese and lipid oxidation was delayed effectively in the product [94]. Most studies conducted using antioxidant compounds in packaging films revealed improvement in the keeping qualities of the food products [95–100].

10. Mechanism of antioxidant release from packaging film to food products

Active compounds migrate, apply their specific roles and functions in the packaging products. Vasile [90] classified the active agents into two which are the migratory and non migratory active compounds. The embedded active compounds are used to control the release while the non migratory agents are stable in the packaging matrix. Yemmireddy et al. [101] stated that non migratory compounds antioxidants may be subjected to covalent immobilization which may not alter the organoleptic properties of the packaged products. The release of active compound into the product is controlled by the packaging films affinity in it and in the matrix, the morphology, porosity and the medium [90]. The major roles of antioxidants in packaging films are to prevent degradation of food products such as lipid oxidation which causes loss of food quality especially in fatty foods. Antioxidants are incorporated in packaging materials and released through controlled diffusion mechanisms. The antioxidants react with the contents, environment and absorb the unwanted substances from the contents. Antioxidant materials prevent lipid oxidation as well as protein denaturation [102,103]. The major advantage of using antioxidant packaging is that the antioxidants in the packaging films are gradual released into the food headspace rather than adding directly to food products [103,104]. The mechanisms of action of antioxidant packaging according to Nwakaudu et al. [103] involve scavenging free nitrogen and oxygen species thereby reducing the oxygen concentration, molecular oxidation potential and lipid peroxides to form non-radical products and binding metal ions to hinder release of free radicals. The antioxidant in the film packaging would release electrons to the free radicals leading to completion of the outer shell ions of the radical species. The release feature of antioxidants in food is the principal role for achieving the desired shelf life of the product [105]. López-de-Dicastillo et al. [93] and Gómez-Estaca et al. [106] explained that the discharge of antioxidants from the packaging film are dependent on the type and compatibility of antioxidant materials used and the food stimulant. Antioxidants are released at a targeted rate into the packaging interface in order to slow down the kinetics reactions of oxidation. The free radicals formed are neutralized through the scavenging activities of the antioxidant [107,108]. However, Park et al. [97] examined the release kinetics of liquid and gas phases of the films developed and reported effective liberation of antioxidants from the film leading to inhibition of oxidation. Gemili et al. [96] conducted a research on the rate of release of L-ascorbic acid and L-tyrosine from cellulose acetate films with different morphological features. It was discovered that increase in cellulose acetate in the medium reduced the pore size, porosity and diffusion rates of antioxidants. Gómez-Estaca et al. [106] concluded that antioxidant releasing packaging materials provide efficient protection on food products. The antioxidant activities of polyphenol is dependent on the initiation conditions, microenvironment of the medium and structure of the molecules [107].

11. Application of antioxidant in food

Natural antioxidants are gradually replacing synthetic antioxidants because of their negative effect on the human health. Most of the antioxidants are used in fish, meat cheese, fried food industries due to their quick susceptibility to lipid oxidation and microbial deterioration [109]. The use of herbs and spices (dill, cumin, rosemary, cloves, turmeric, ginger, thyme, cinnamon, beetroot, ginger oregano, black pepper etc) and essential oils from plants with strong antioxidant properties as alternative to synthetic antioxidants in food industry had been reported [110–113]. The activities of essential oils were reported to depend on extraction methods, synergistic effects and their concentration [114]. Spices and herbs can be used as extracts, ground, whole, as emulsion and encapsulated products [115]. Spices and herbs had successfully been incorporated into youghurt, ghee, cheese, butter and ice cream to impart flavor, increase the antioxidant and antimicrobial properties [113,115–117] According to Paur et al. [118], herbs and spices are phytochemicals used for their aromatic properties with low nutritive value and functions in antioxidant defense and redox signaling. Table 1.2 shows different types of antioxidant materials incorporated in films for active packaging. Positive results had been reported on the food products preserved with antioxidant packaging methods. Navikaite-Snipaitiene et al. [95] and Gemili et al. [96] observed stability in the color of beef stored in essential oil antioxidant packages. Other natural antioxidant materials used for food preservation are moringa extract, soy protein, papaya, tocophenol, nanoparticles etc on different food products as shown in Table 1.1. Improvement on the food products were reported in terms of the keeping qualities of the food products.

Table 1.2 Antioxidant packaging materials.

Antioxidant material	Product	Author
Corn-Zein and Phenolic compounds	Ground beef patties	Park et al. [97]
Gelatin base antioxidant with *Caesalpinia decapetala* and *Caesalpinia spinosa* "Tara"	Ground beef patties	Gallego et al. [119]
Eugenol	Fresh beef	Navikaite-Snipaitiene et al. [95]
Poly (ε-Caprolactone) and Almond Skin extract	Fried Almond	García et al. [88]
Moringa, papaya, soy protein isolate,	Fruits, cheese,	Rodríguez et al. [98]; Lee et al. [99]; Tesfay [100]
Green tea extract with ascorbic acid, quercetin	Brine sardine	López-de-Dicastillo et al. [93]
Herbs and Spices	Varieties of food products	Paur et al. [118]; Karakol & Kapi [120]
Soy edible film and essential oil	Meat patties	Coskun et al. [114]
Essential oil	Meat	Pateiro et al. [121] Pateiro et al. [122]
Tocophenol	Corn flakes	Paradiso et al. [123]
Sea buckthorn extract, grape seed extract	Pork patties	Kumar et al. [124]
Phenolic compounds	Beef	Barbosa-Pereira et al. [125]
Nanoparticles	Food products	Shankar et al. [126]
Phenolic acid	Pork meat	Hernández-García et al. [127]
Butylated hydroxytoluene (BHT) and butylated hydroxyanisole (BHA)	Fatty foods and other food products	Fasihnia et al. [128], Peighambardoust et al. [129]

12. Conclusions

Antioxidant packaging and properties were discussed in this chapter. Antioxidants retard lipid oxidations thereby prolong the shelf life of the products. There are various antioxidants agents which are classified based on their origin and nature. The use of natural antioxidants in packaging films in food industry is encouraged due to the harmful effect of synthetic antioxidants. Natural antioxidants had been used such as essential oils, herb and spices, moringa extract, papaya etc. Incorporation of antioxidants in the packaging film protects the food product by removing the free radicals and chelating the metal ions produced. Different types of antioxidant agents had been used in active packaging and improvement on nutritive values, color and organoleptic properties of the products had been reported. Therefore, direct use of antioxidant to food had been limited since the discovery of antioxidant packaging which releases its contents gradually into the headspace of the packaging materials.

References

[1] A.K. Arise, A.M. Alashi, I.D. Nwachukwu, O.A. Ijabadeniyi, R.E. Aluko, E.O. Amonsou, Antioxidant activities of Bambara groundnut (Vigna subterranea) protein hydrolysates and their membrane ultrafiltration fractions, Food Funct. 7 (5) (2016) 2431–2437.

[2] C.F. Ajibola, J.B. Fashakin, T.N. Fagbemi, R.E. Aluko, Effect of peptide size on antioxidant properties of African yam bean seed (Sphenostylis stenocarpa) protein hydrolysate fractions, Int. J. Mol. Sci. 12 (10) (2011) 6685–6702.

[3] K.K. Kattappagari, C.R. Teja, R.K. Kommalapati, C. Poosarla, S.R. Gontu, B.V. Reddy, Role of antioxidants in facilitating the body functions: a review, J. Orof. Sci. 7 (2) (2015) 71.

[4] J.K. Willcox, S.L. Ash, G.L. Catignani, Antioxidants and prevention of chronic disease, Crit. Rev. Food Sci. Nutr. 44 (4) (2004) 275–295.

[5] D.H. Li Niu, Guangzhou University, Guangzhou, China, Chemical Analysis of Antioxidant Capacity. Book, 2020, p. 175.

[6] S.E. Duncan, H.H. Chang, Implications of light energy on food quality and packaging Selection,Editor(s): jeyakumar Henry, Adv. Food Nutr. Res. 67 (2012) 25–73, https://doi.org/10.1016/B978-0-12-394598-3.00002-2. Academic press.

[7] M. Varvara, G. Bozzo, G. Celano, C. Disanto, C.N. Pagliarone, G.V. Celano, The use of ascorbic acid as a food additive: technical-legal issues, Italian J. food safety 5 (1) (2016) 4313, https://doi.org/10.4081/ijfs.2016.4313.

[8] W.M. Cort, Antioxidant properties of ascorbic acid in foods. Ascorbic acid: chemistry, metabolism and uses, Adv. Chem. (1982) 533–550, https://doi.org/10.1021/ba-1982-0200.ch022. American Chemical Society.

[9] M. Taniguchi, N. Arai, K. Kohno, S. Ushio, S. Fukuda, Anti-oxidative and anti-aging activities of 2-O-α-glucopyranosyl-l-ascorbic acid on human dermal fibroblasts, Eur. J. Pharmacol. 674 (2012) 126–131.

[10] X. Yin, K. Chen, H. Cheng, X. Chen, S. Feng, Y. Song, L. Liang, Chemical stability of ascorbic acid integrated into commercial products: a review on bioactivity and delivery technology, Antioxidants 11 (2022) 153, https://doi.org/10.3390/antiox11010153.

[11] D. Prakash, C. Gupta, in: D. Prakash, G. Sharma (Eds.), Carotenoids: Chemistry and Health Benefits. Phytochemicals of Nutraceutical Importance, CAB International, 2014, pp. 181–191, 2014, https://www.researchgate.net/publication/285808454_Carotenoids_Chemistry_and_health_benefits. (Accessed 22 May 2022).

[12] A. Fratianni, L. Cinquanta, G. Panfili, Degradation of carotenoids in orange juice during microwave heating, LWT—Food Sci. Technol. 43 (2010) 867–871.

[13] K. Murakami, M. Honda, R. Takemura, T. Fukaya, H. Kanda, M. &Goto, Effect of thermal treatment and light irradiation on the stability of lycopene with high Z-isomers content, Food Chem. 250 (2018) 253–258.

[14] L. Bogacz-Radomska, J. Harasym, β-Carotene—properties and production methods, Food Qual. Safety 2 (2) (2018) 69–74, https://doi.org/10.1093/fqsafe/fyy004.

[15] J. Shankaranarayanan, K. Arunkanth, K.C. Dinesh, Beta carotene -therapeutic potential and strategies to enhance its bioavailability, Nutri. Food Sci. Int. J. 7 (4) (2018) 555716, https://doi.org/10.19080/NFSIJ.2018.07.555716.

[16] G.M. Lowe, D.L. Graham, A.J. Young, Lycopene: chemistry, metabolism, and bioavailability, in: Lycopene and Tomatoes in Human Nutrition and Health, CRC Press, 2018, pp. 1–20.

[17] M. Ghellam, L. Koca, Lycopene: chemistry, sources, bioavailability, and benefits for human health, 1st Int. Cong. Eng. Life Sci. 11–14 April (2019) 298–505.

[18] E. Aziz, R. Batool, W. Akhtar, S. Rehman, T. Shahzad, A. Malik, M.A. Shariati, A. Laishevtcev, S. Plygun, M. Heydari, A. Rauf, S. Ahmed Arif, Xanthophyll: health benefits and therapeutic insights, Life Sci. J240 (2020) 117104, https://doi.org/10.1016/j.lfs.2019.117104.

[19] L. Dufossé, Microbial pigments from bacteria, yeasts, fungi, and microalgae for the food and feed industries, in: A.M. Grumezescu, A.M. Holban (Eds.), In Handbook of Food Bioengineering, Natural and Artificial Flavoring Agents and Food Dyes, Academic Press, 2018, pp. 113–132, https://doi.org/10.1016/B978-0-12-811518-3.00004-1.

[20] R. Qadir, F. Anwar, Chapter 3—lutein and zeaxanthin, in: M. Mushtaq, F. Anwar (Eds.), A Centum of Valuable Plant Bioactives, Academic Press, 2021, pp. 59–76, https://doi.org/10.1016/B978-0-12-822923-1.00017-0.

[21] C.T. Ho, Phenolic Compounds in Food. An Overview. Phenolic Compounds in Food and Their Effects on Health I ACS Symposium Series, American Chemical Society, Washington, DC, 1992.

[22] T. Ozcan, A. Akpinar-Bayizit, L. Yilmaz-Ersan, B. Delikanli, Phenolics in human health I, Int. J. Chem. Eng. Appl. 5 (5) (2014) 393–396, https://doi.org/10.7763/IJCEA.

[23] R. Tsao, Chemistry and biochemistry of dietary polyphenols, Nutrients 2 (2010) 1231–1246, https://doi.org/10.3390/nu2121231.

[24] J.J. Guo, H.Y. Hsieh, C.H. Hu, Chain-breaking activity of carotenes in lipid peroxidation: a theoretical study, J. Phys. Chem. B 113 (2009) 15699–15708.

[25] Y. Du, H. Guo, H. Lou, Grape seed polyphenols protect cardiac cells from apoptosis via induction of endogenous antioxidant enzymes, J. Agric. Food Chem. 55 (2007) 1695–1701.

[26] C. Manach, A. Scalbert, C. Morand, C. Rémésy, L. Jimenez, Polyphenols: food sources and bioavailability, Am. J. Clin. Nutr. 79 (2004) 727–747.

[27] P.D. Pekiner, Vitamin E as an antioxidant, J. Fac. Pharm, Ankara 32 (4) (2003) 243–267.

[28] R. Eitenmiller, J. Lee, Vitamin E: Chemistry and Biochemistry. Vitamin E Food Chemistry, Composition and Analysis, Marcel Dekker INC., New York, Basel, 2004, pp. 1–30, 2004.

[29] R. Yamauchi, Vitamin E: mechanism of its antioxidant activity-A review, Food Sci. Technol. Int. Tokyo 3 (4) (1997) 301–309, 1997.

[30] V.K.M.T. Husøy, M. Andreassen, I.T.L. Lillegaard, G.H. Mathisen, J. Rohloff, J. Starrfelt, M.H. Carlsen, T.G. Devold, B. Granum, J.D. Rasinger, C. Svendsen, E. Bruzell, Risk assessment of butylated hydroxytoluene (BHT). Opinion of the panel on food additives, flavourings, processing aids, materials in contact with food, and cosmetics of the Norwegian scientific committee for food and environment, in: VKM Report, vol 15, Norwegian Scientific Committee for Food and Environment (VKM), Oslo, Norway, 2019, pp. 2535–4019, 2019.

[31] IARC. Butylated hydroxyanisole (BHA), In some naturally occurring and synthetic food components, furocoumarins and ultraviolet radiation, in: IARC Monographs on the Evaluation of Carcinogenic Risk of Chemicals to Humans, vol 40, International Agency for Research on Cancer, Lyon, France, 1986, pp. 123–159.

[32] EFSA, EFSA panel on food additives and nutrient sources added to food (ANS); scientific opinion on the reevaluation of butylated hydroxytoluene BHT (E 321) as a food additive, EFSA J. 10 (2588) (2012) 43, https://doi.org/10.2903/j.efsa.2012.2588. Available online: www.efsa.europa.eu/efsajournal.htm.

[33] SCCS (Scientific Committee on Consumer Safety), Scientific Opinion on Butylated Hydroxytoluene (BHT), Preliminary Version of 27 September 2021, Final Version of 2 December 2021, SCCS/1636/21.

[34] D.J. Charles, Sources of natural antioxidants and their activities, Antioxi. Properties Spices, Herbs Other Sources (2012) 65–138, https://doi.org/10.1007/978-1-4614-4310-0_4.

[35] M.A. Shah, S.J.D. Bosco, S.A. Mir, Plant extracts as natural antioxidants in meat and meat products, Meat Sci. 98 (2014) 21–33, https://doi.org/10.1016/j.meatsci.2014.03.020.

[36] P.A. Fernandes, C. Le Bourvellec, C.M. Renard, F.M. Nunes, R. Bastos, E. Coelho, D.F. Wessel, M.A. Coimbra, S.M. Cardoso, Revisiting the chemistry of apple pomace polyphenols, Food Chem. 294 (2019) 9–18, https://doi.org/10.1016/j.foodchem.2019.05.006.

[37] L.C. Corrêa-Filho, S.C. Lourenço, D.F. Duarte, M. Moldão-Martins, V.D. Alves, Microencapsulation of tomato (Solanum lycopersicum L.) pomace ethanolic extract by spray drying: optimization of process conditions, Appl. Sci. 9 (2019) 612, https://doi.org/10.3390/app9030612.

[38] K.V. Peter, Introduction to herbs and spices, in: K.V. Peter (Ed.), Handbook of Herbs and Spices, vol. 2, Woodhead Publishing Limited, Abington Hall, Abington Cambridge CB1 6AH, England, 2004, pp. 16—24, 2004.

[39] P.N. Ravindran, A.K. Johny, K.N. Babu, Spices in Our Daily Life. Satabdi Smaranika 2002 vol. 2, Arya Vaidya Sala, Kottakkal, 2002.

[40] M.R. Shylaja, K.V. Peter, The functional role of herbal spices, in: K.V. Peter (Ed.), Handbook of Herbs and Spices, vol. 2, Woodhead Publishing Limited, Abington Hall, Abington Cambridge CB1 6AH, England, 2004, pp. 25—35, 2004.

[41] S. Raghavan, Forms, functions and application of spices, in: Handbook of Spices, Seasonings, and Flavorings, CRC Press Taylor & Francis Group 6000 Broken Sound Parkway NW, Suite 300 Boca Raton, FL, 2007, pp. 30—61, 33487-2742.

[42] A. Gramza, J. Korczak, R. Amarowicz, Tea polyphenols—their antioxidant properties and biological activity—A Review, Pol. J. Food Nutr. Sci. P 14/55 (3) (2005) 219—235.

[43] S.P.J.N. Senanayake, Green tea extract: chemistry, antioxidant properties and food applications—A review, J. Funct.Foods 5 (4) (2013) 1529—1541, https://doi.org/10.1016/j.jff.2013.08.011.

[44] S.C. Forester, J.D. Lambert, The role of antioxidant versus pro-oxidant effects of green tea polyphenols in cancer prevention, Mol. Nutr. Food Res. 55 (6) (2011) 844—854, https://doi.org/10.1002/mnfr.201000641.

[45] M.S. Brewer, Natural antioxidants: sources, compounds, mechanisms of action, and potential applications, Compr. Rev. Food Sci. Food Saf. 10 (4) (2011) 221—247.

[46] G. Rechkemmer, Funktionelle lebensmittel-zukunft de Ernahrung oder marketing-strategie, Forschungereport Sonderheft 1 (2001) 12—15.

[47] E.M. Yahia, The contribution of fruit and vegetable consumption to human health, in: L.A. de la Rosa Emilio Alvarez-Parrilla Gustavo, A. González-Aguilar (Eds.), Fruit and Vegetable Phytochemicals Chemistry, Nutritional Value and Stability, Blackwell Publishing, 2010, pp. 1—44.

[48] E. Sikora, E. Cieslik, K. Topolska, The sources of natural antioxidants, Acta Sci. Pol., Technol. Aliment. 7 (1) (2008) 5—17.

[49] G.G. Duthie, S.J. Duthie, J.A.M. Kyle, Plant polyphenols in cancer and heart disease: implications as nutritional antioxidants, Nutr. Res. Rev. 13 (1) (2000) 79—106.

[50] R.H. Liu, Potential synergy of phytochemicals in cancer prevention: mechanism of action, J. Nutr. 134 (2004) 3479S—3485S.

[51] E. Pastrana-Bonilla, C.C. Akoh, S. Sellappan, G. Krewer, Phenolic Content and antioxidant capacity of Muscadine grapes, J. Agric. Food Chem. 51 (2003) 5497—5503.

[52] D. Kammerer, A. Claus, R. Carle, A. Scheiber, Polyphenol screening of pomace from red and white grape varieties (*Vitis vinifera L.*) by HPLC-DAD-MS/MS, J. Agric. Food Chem. 52 (2004) 4360—4436.

[53] S.C. Lourenço, M. Moldão-Martins, V.D. Alves, Antioxidants of natural plant origins: from sources to food industry applications, Molecules 24 (22) (2019) 4132, https://doi.org/10.3390/molecules24224132.

[54] M.N. Irakli, V.F. Samanidou, I.N. Papadoyannis, Development and validation of an HPLC method for the simultaneous determination of tocopherols, tocotrienols and carotenoids in cereals after solid-phase extraction, J. Separ. Sci. 34 (12) (2011) 1375—1382.

[55] J. Klepacka, E. Gujska, J. Michalak, Phenolic compounds as cultivar- and variety-distinguishing factors in some plant products, Plant Foods Hum. Nutr. 66 (1) (2011) 64—69.

[56] A. Angioloni, C. Collar, Nutritional and functional added value of oat, Kamut, spelt, rye and buckwheat versus common wheat in breadmaking, J. Sci. Food Agric. 91 (7) (2011) 1283—1292.

[57] G.A. Annor, Z. Ma, J.I. Boye, Crops—legumes, in: S. Clark, S. Jung, B. Lamsal (Eds.), Food Processing: Principles and Applications, second ed., John Wiley & Sons, Ltd, 2014.

[58] B. Singh, J.P. Singh, A. Kaur, A. Kaur, N. Singh, Antioxidant profile of legume seeds, in: P. Guleria, V. Kumar, E. Lichtfouse (Eds.), Sustainable Agriculture Reviews 45. Sustainable Agriculture Reviews, vol 45, Springer, Cham, 2020, https://doi.org/10.1007/978-3-030-53017-4_4.

[59] E. Ryan, K. Galvin, T.P. O'Connor, A.R. Maguire, N.M. O'Brien, Fatty acid profile, tocopherol, squalene and phytosterol content of Brazil, pecan, pine, pistachio and cashew nuts, Int. J. Food Sci. Nutr. 57 (3–4) (2006) 219–228.

[60] E. Pongrácz, The environmental impacts of packaging, Environm. Consci. Mat. Chem. Proces. (2007) 237–278, https://doi.org/10.1002/9780470168219.ch9.

[61] K. Marsh, B. Bugusu, Food packaging—roles, Materials, and environmental issues, J. Food Sci. 72 (3) (2007) 39–55.

[62] G.L. Robertson, Introduction to food packaging. Food Packaging: Principles and Practice, CRC Press, 2013, pp. 1–8.

[63] A. Yaris, A.C. Sezgin, Food packaging: glass and plastic, in: Researches on Science and Art in 21st Century, Turkey, 2017, pp. 731–740.

[64] H. Sharma, Food Packaging Technology, AgriMoon.com, 2020, pp. 1–138.

[65] EPA, Environmental Protection Agency (U.S, Plastics.Washington,D.C.: EPA, 2006. Available from: http://www.epa.gov/epaoswer/non-hw/muncpl/plastic.htm.

[66] F. Debeaufort, Metal packaging, in: Packaing Materials and Processing for Food, Pharmaceuticals and Cosmetics, 2021, pp. 75–104, https://doi.org/10.1002/9781119825081.ch4.

[67] G.K. Deshwal, N.R. Panjagari, Review on metal packaging: materials, forms, food applications, safety and recyclability, J. Food Sci. Technol. 57 (7) (July 2020) 2377–2392, https://doi.org/10.1007/s13197-019-04172-z. Epub 2019 Nov 12. PMID: 32549588; PMCID: PMC7270472.

[68] M. Ahmed, J. Pickova, T. Ahmad, M. Liaquat, A. Farid, M. Jahangir, Oxidation of lipids in foods, Sarhad J. Agric. 32 (3) (2016) 230–238, https://doi.org/10.17582/journal.sja/2016.32.3.230.238.

[69] M. Repetto, J. Semprine, A. Boveris, Lipid peroxidation: chemical mechanism, biological implications and analytical determination, in: Lipid Peroxidation. IntechOpen, 2012, https://doi.org/10.5772/45943.

[70] K.A. Salman, S. Ashraf, Reactive oxygen species: a link between chronic inflammation and cancer, Asia Pac. J. Mol. Biol. Biotechnol. 21 (2) (2013) 42–49.

[71] K. Brieger, S. Schiavone, F.J. Miller Jr., K.H. Krause, Reactive oxygen species:from health to disease, Swiss Med. Wkly. 142 (2012) 13659.

[72] A. Ozcan, M. Ogun, Biochemistry of Reactive Oxygen and Nitrogen Species, Basic Principles and Clinical Significance of Oxidative Stress, 2015, https://doi.org/10.5772/61193.

[73] N. Wadhwa, B.B. Mathew, S.K. Jatawa, A. Tiwari, Lipid peroxidation: mechanism, models and significance, Int. J. Curr. Sci. 3 (2012) 11–17.

[74] H. Yin, L. Xu, N.A. Porter, Free radical lipid peroxidation: mechanisms and analysis, Chem. Rev. 111 (10) (October 12, 2011) 5944–5972, https://doi.org/10.1021/cr200084z. Epub 2011 Aug 23. PMID: 21861450.

[75] A. Ayala, M.F. Muñoz, S. Argüelles, Lipid peroxidation: production, metabolism, and signaling mechanisms of malondialdehyde and 4-hydroxy-2-nonenal, Oxid. Med. Cell. Longev. 2014 (2014) 360438, https://doi.org/10.1155/2014/360438.

[76] P.R. Salgado, L.D. Giorgio, Y.S. Musso, A.N. Mauri, Recent developments in smart food packaging focused on biobased and biodegradable polymers, Front. Sustain. Food Syst. 5 (2021).

[77] K.B. Biji, C.N. Ravishankar, C.O. Mohan, T.K. Srinivasa Gopal, Smart packaging systems for food applications: a review, J. Food Sci. Technol. 52 (10) (2015) 6125–6135, https://doi.org/10.1007/s13197-015-1766-7.

[78] P. Madhusudan, N. Chellukuri, N. Shivakumar, Smart packaging of food for the 21st century—a review with futuristic trends, their feasibility and economics, 2018, Mater. Today Proc. 5 (10) (2018) 21018–21022, https://doi.org/10.1016/j.matpr.2018.06.494. Part 1.

[79] D. Schaefer, W.M. Cheung, Smart packaging: opportunities and challenges. 51st CIRP conference on manufacturing systems, Procedia CIRP 72 (2018) (2018) 1022–1027, https://doi.org/10.1016/j.procir.2018.03.240.

[80] R.M. Han, J.P. Zhang, L.H. Skibsted, Reaction dynamics of flavonoids and carotenoids as antioxidants, Molecules 17 (2) (2012) 2140–2160.

[81] R. Dobrucka, R. Przekop, New perspectives in active and intelligent food packaging, J. Food Process. Preserv. 43 (2019) e14194, https://doi.org/10.1111/jfpp.14194.

[82] A. Trajkovska Petkoska, i D. Danilosk, N.M. D'Cunha, N. Naumovski, A.T. Broach, Edible packaging: sustainable solutions and novel trends in food packaging, Food Res. Int. 140 (2021) 109981, https://doi.org/10.1016/j.foodres.2020.109981.

[83] N. Benbettaïeb, T. Karbowiak, F. Debeaufort, Bioactive edible films for food applications: influence of the bioactive compounds on film structure and properties, Crit. Rev. Food Sci. Nutr. 59 (7) (2019) 1137–1153, https://doi.org/10.1080/10408398.2017.1393384.

[84] S. Suresh, C. Pushparaj, R. Subramani, Recent development in preparation of food packaging films using biopolymers, Food Res. 5 (6) (2021) 12–22.

[85] W.F. Lai, Design of polymeric films for antioxidant active food packaging, Int. J. Mol. Sci. 23 (2022) 12, https://doi.org/10.3390/ijms23010012.

[86] R. Campardelli, E. Drago, P. Perego, Biomaterials for food packaging: innovations from natural sources, Chem. Eng. Trans. 87 (2021) 571–576, https://doi.org/10.3303/CET2187096.

[87] A.C. Mellinas, A. Valdés, M. Ramos, N. Burgos, M.C. Garrigós, A. Jiménez, Active edible films: Current state and future trends, J. Appl. Polym. Sci. 133 (2016) 42631.

[88] A.V. García, N.J. Serrano, A.B. Sanahuja, M.C. Garrigós, Novel antioxidant packaging films based on poly (-Caprolactone) and almond skin extract: development and effect on the oxidative stability of fried almonds, Antioxidants 9 (2020) 629, https://doi.org/10.3390/antiox9070629.

[89] I. Lukic, J. Vulic, J. Ivanovic, Antioxidant activity of PLA/PCL films loaded with thymol and/or carvacrol using scCO2 for active food packaging, Food Packag. Shelf Life 26 (2020) 100578.

[90] C. Vasile, Polymeric nanocomposites and nanocoatings for food packaging: a review, Materials 2018 11 (2018) 1–49, https://doi.org/10.3390/ma11101834, 1834.

[91] B. Marcos, C. Sárraga, M. Castellari, F. Kappen, G. Schennink, J. Arnau, Development of biodegradable films with antioxidant properties based on polyesters containing α-tocopherol and olive leaf extract for food packaging applications, Food Packag. Shelf Life 1 (2) (2014) 140–150, https://doi.org/10.1016/j.fpsl.2014.04.002.

[92] D.L. Belan, L.E. Mopera, F.P. Flores, Development and characterisation of active antioxidant packaging films, Int. Food Res. J. 26 (2) (2019) 411–420.

[93] C. López-de-Dicastillo, J. Gómez-Estaca, R. Catalá, R. Rafael Gavara, P. Hernández-Muñoz, Active antioxidant packaging films: development and effect on lipid stability of brined sardines, Food Chem. 131 (2012) 1376–1384.

[94] J. Robalo, M. Lopes, O. Cardoso, A. Sanches Silva, F. Ramos, Efficacy of whey protein film incorporated with Portuguese green tea (*camellia sinensis L.*) extract for the preservation of Latin-style fresh cheese, Foods 11 (8) (2022) 1158, https://doi.org/10.3390/foods11081158.

[95] V. Navikaite-Snipaitiene, L. Lvanauskas, V. Jakstas, N. Rüegg, R. Rutkaite, E. Wolfram, S. Yildirim, Development of antioxidant food packaging materials containing eugenol for extending display life of fresh beef, Meat Sci. 145 (2018) 9–15, https://doi.org/10.1016/j.meatsci.2018.05.015.

[96] S. Gemili, A. Yemenicioglu, S.A. Altınkaya, Development of antioxidant food packaging materials with controlled release properties, J. Food Eng. 96 (2010) 325–332, 2010.

[97] H.Y. Park, S.J. Kim, K.M. Kim, Y.S. You, S.Y. Kim, J. Han, Development of antioxidant packaging material by applying corn-zein to LLDPE film in combination with phenolic, Compounds 77 (10) (2012) E273–E279, https://doi.org/10.1111/j.1750-3841.2012.02906.x, 2012.

[98] G.M. Rodríguez, J.C. Sibaja, P.J.P. Espitia, C.G. Otoni, Antioxidant active packaging based on papaya edible films incorporated with Moringa oleifera and ascorbic acid for food preservation, Food Hydrocol. 103 (2020) 105630.

[99] K.Y. Lee, H.J. Yang, K.B. Song, Application of a puffer fish skin gelatin film containing Moringa oleifera Lam. leaf extract to the packaging of Gouda cheese, J. Food Sci. Technol. 53 (2016) 3876–3883.

[100] S. Tesfay, The efficacy of combined application of edible coating and moringa extract in enhancing fruit quality in avocado (Persea americana Mill), in: South African Avocado Growers' Association Yearbook, South African Avocado Growers' Association, Tzaneen, South Africa, 2016, pp. 51–58.

[101] V.K. Yemmireddy, G.D. Farrell, Y.C. Hung, Development of titanium dioxide (TiO2) nanocoatings on food contact surfaces and method to evaluate their durability and photocatalytic bactericidal property, J. Food Sci. 80 (2015) N1903–N1911 ([CrossRef]).

[102] W. Torres-Arreola, H. Soto-Valdez, E. Peralta, J. Cardenas-Lopez, J. Ezquerra-Brauer, Effect of a low-density polyethylene film containing butylated hydroxytoluene on lipid oxidation and protein quality of sierra fish (Scomberomorus sierra) muscle during frozen storage, J. Agric. Food Chem. 55 (2007) 6140–6146.

[103] A.A. Nwakaudu, M.S. Nwakaudu, C.I. Owuamanam, N.C. Iheaturu, The use of natural antioxidant active polymer packaging films for food preservation, AppliedSignals Reports 2 (4) (2015) 38–50, 2015.

[104] A. Sanches-Silva, D. Costa, T.G. Albuquerque, G.G. Buonocore, F. Ramos, M.C. Castilho, H.S. Costa, Trends in the use of natural antioxidants in active food packaging: a review, Food Addit. Contam. 31 (3) (2014) 374–395, https://doi.org/10.1080/19440049.2013.879215.

[105] L. Kuai, F. Liu, B.S. Chiou, J. Avena-Bustillos, T.H. McHugh, F. Zhong, Controlled release of antioxidants from active food packaging: a review, Food Hydrocol. 120 (2021) 106992, 2021.

[106] J. Gómez-Estaca, C. López-de-Dicastillo, P. Hernández-Muñoz, R. Catalá, R. Gavara, Advances in antioxidant active food packaging, Trends Food Sci. Technol. 35 (1) (2014) 42–51, https://doi.org/10.1016/j.tifs.2013.10.008.

[107] J.M. Lü, P.H. Lin, Q. Yao, C. Chen, Chemical and molecular mechanisms of antioxidants: experimental approaches and model systems, J. Cell Mol. Med. 14 (4) (2010) 840–860, https://doi.org/10.1111/j.1582-4934.2009.00897.x.

[108] A. Balasubramanian, Antioxidant Packaging. Rutgers, The State University of, New Jersey, 2009.

[109] R. Domínguez, F.J. Barba, B. Gómez, P. Putnik, D.B. Kovačević, M. Pateiro, E.M. Santos, J.M. Lorenzo, Active packaging films with natural antioxidants to be used in meat industry: a review, Food Res. Int. 113 (2018) 93–101, https://doi.org/10.1016/j.foodres.2018.06.073.

[110] A. García, C. Mellinas, M. Ramos, N. Burgos, A. Jimenez, M. Garrigós, Use of herbs, spices and their bioactive compounds in active food packaging, RSC Adv. 5 (2015) 2021, https://doi.org/10.1039/C4RA17286H.

[111] Y. Flores, P. Pelegrín, J. Carlos, M. Ramos, A. Jimenez, M. Garrigós, Use of herbs and their bioactive compounds in active food packaging, in: Aromatic Herbs in Foods, 2021, https://doi.org/10.1016/B978-0-12-822716-9.00009-3.

[112] S.M. El-Sayed, A.M. Youssef, Potential application of herbs and spices and their effects in functional dairy products, Heliyon 5 (6) (2019) e01989, https://doi.org/10.1016/j.heliyon.2019.e01989.

[113] P. Srivastava, S.G.M. Prasad, N.A. Mohd, M. Prasad, Analysis of antioxidant activity of herbal yoghurt prepared from different milk, Pharma Innov. J. 4 (2015) 18–20.

[114] B.,K. Coskun, E. Çalikoğlu, Z.K. Emiroğlu, K. Candoğan, Antioxidant active packaging with soy edible films and oregano or thyme essential oils for oxidative stability of ground beef patties, J. Food Qual. 37 (3) (2014) 203–212, https://doi.org/10.1111/jfq.12089, 2014.

[115] M.E. Embuscado, Spices and herbs: natural sources of antioxidants—a mini review, J. Funct.Foods 18 (Part B) (2015) 811–819, https://doi.org/10.1016/j.jff.2015.03.005.

[116] L.D. Najgebauer, T. Grega, M. Sady, The quality and storage stability of butter made from sour cream with addition of dried sage and rosemary, Biotechnol. Anim. Husb. 25 (2009) 753–761.

[117] E. Akan, O. Yerlikaya, A. Akpinar, C. Karagozlu, O. Kinik, H.R. Uysal, The effect of various herbs and packaging material on antioxidant activity and colour parameters of whey (Lor) cheese, Int. J. Dairy Technol. 74 (3) (2021) 554–563, https://doi.org/10.1111/1471-0307.12778.

[118] I. Paur, M.H. Carlsen, B.L. Halvorsen, R. Blomhoff, Antioxidants in herbs and spices: roles in oxidative stress and redox signaling, in: I.F.F. Benzie, S. Wachtel-Galor (Eds.), Herbal Medicine: Biomolecular and Clinical Aspects, second ed., CRC Press/Taylor & Francis, Boca Raton (FL), 2011 (Chapter 2). Available from: https://www.ncbi.nlm.nih.gov/books/NBK92763/.

[119] M.G. Gallego, M.H. Gordon, F. Segovia, M.P.A. Pablos, Caesalpinia decapetala and tara as a coating for ground beef patties, Antioxidants 5 (10) (2016) 1–15, 2016.

[120] P. Karakol, E. Kapi, Use of Selected Antioxidant-Rich Spices and Herbs in Foods. Antioxidants—Benefits, Sources, Mechanisms of Action, IntechOpen, 2021, https://doi.org/10.5772/intechopen.96136.

[121] M. Pateiro, F.J. Barba, R. Domínguez, A.S. Sant'Ana, M.A. Khaneghah, M. Gavahian, B. Gómez, J.M. Lorenzo, Essential oils as natural additives to prevent oxidation reactions in meat and meat products: a review, Food Res. Int. 113 (2018) 156–166, https://doi.org/10.1016/j.foodres.2018.07.014.

[122] M. Pateiro, R. Domínguez, R. Bermúdez, P.E.S. Munekata, W. Zhang, M. Gagaoua, J.M. Lorenzo, Antioxidant active packaging systems to extend the shelf life of sliced cooked ham, Current Res. Food Sci. 1 (2019) 24–30, https://doi.org/10.1016/j.crfs.2019.10.002.

[123] V.M. Paradiso, F. Caponio, C. Summo, T. Gomes, Influence of some packaging materials and of natural tocopherols on the sensory properties of breakfast cereals, Food Sci. Technol. Int. 20 (2014) 161–170.

[124] V. Kumar, M.K. Chatli, R.V. Wagh, N. Mehta, P. Kumar, Effect of the combination of natural antioxidants and packaging methods on quality of pork patties during storage, J. Food Sci. Technol. 52 (2015) 6230–6241, https://doi.org/10.1007/s13197-015-1734-2.

[125] L. Barbosa-Pereira, G.P. Aurrekoetxea, I. Angulo, P. Paseiro-Losada, J.M. Cruz, Development of new active packaging films coated with natural phenolic compounds to improve the oxidative stability of beef, Meat Sci. 97 (2) (2014) 249–254, https://doi.org/10.1016/j.meatsci.2014.02.006.

[126] S. Shankar, L.F. Wang, J.W. Rhim, Effect of melanin nanoparticles on the mechanical, water vapor barrier, and antioxidant properties of gelatin-based films for food packaging application, Food Packag. Shelf Life 21 (2019) 100363.

[127] E. Hernández-García, M. Vargas, A. Chiralt, Starch-polyester bilayer films with phenolic acids for pork meat preservation, Food Chem. 385 (2022) 132650, https://doi.org/10.1016/j.foodchem.2022.132650.

[128] S.H. Fasihnia, S.H. Peighambardoust, S.J. Peighambardoust, A. Oromiehie, M. Soltanzadeh, D. Peressini, Migration analysis, antioxidant, and mechanical characterization of polypropylene-based active food packaging films loaded with BHA, BHT, and TBHQ, J. Food Sci. 85 (8) (2020) 2317–2328, https://doi.org/10.1111/1750-3841.15337.

[129] S.H. Peighambardoust, S.H. Fasihnia, S.J. Peighambardoust, M. Pateiro, R. Domínguez, J.M. Lorenzo, Active polypropylene-based films incorporating combined antioxidants and antimicrobials: preparation and characterization, Foods 10 (4) (2021) 722, https://doi.org/10.3390/foods10040722.

CHAPTER 2

Challenges and perspectives in application of smart packaging

Nadia Akram, Khawaja Taimoor Rashid, Tanzeel Munawar, Muhammad Usman and Fozia Anjum

Department of Chemistry, Government College University Faisalabad, Faisalabad, Punjab, Pakistan

1. Introduction

Packaging nowadays will have to do a lot more than just protect. Emerging technologies that are both active and intelligent ensure longer shelf lives, better marketing and brand reliability, and a quality experience. Performance standards that use gas monitoring, temperature control, and distinctive classification are added to technological innovations like oxidants and antimicrobial to ensure the products are safe and healthy. The rise of digital printing promises better quality assurance and batch processing and the chance to connect the support the decision-making and the customer through internet apps and media platforms. Shareholders of brands and Packaging are having a hard time [1–3].

It is to make sure that almost every product is safe and good. By keeping up with new technology to make sure that everyday products stay fresh, we develop smart packaging solutions that can help your packaging line and help you figure out which ingredients are suitable for the project. Our premium strategy for advanced products also involves creating custom-made products for your unique needs with the most up-to-date materials. Take a look at how Victory Packaging can display why packaging can be ecologically friendly, biodegradable, and cost-effective.

1.1 Smart packaging new technologies

Technology is changing the food industry in many ways, including smart packaging. When people bought something in the past, they didn't have as much information as now. Food is kept fresh with smart packaging that provides customers with information [2,4–6]. Here are some technologies that are new to smart packaging:
- Monitor the freshness of food.
- Improved shelf life.
- Keep an eye on quality and recalls with trackers.

New technologies will continue to expand the smart packaging and active packaging industries, and we will continue to utilize the most beneficial ones for you. Smart packaging will provide consumers with safer and more convenient options. A product's packaging protects it from damage due to external exposure and Use. Additionally, product

packaging can communicate with customers. As an interface, it has a variety of sizes and shapes that make it easy to use. Product packaging protects, communicates, simplifies, and keeps things together. These are some examples of packaging used in food products:
- To prevent spills or breakages and to prevent contamination.
- People will be able to reheat food inside, for example, in a microwave.
- Transport and handling must be kept safe.

As customers' expectations rise, products become more complex, and national and international efforts are made to reduce the carbon footprints of manufactured goods, traditional packaging may no longer be suitable. Innovative and functional packaging is needed to meet new consumer needs [7,8].

As the food supply is globalized and food safety regulations are becoming more rigorous, companies can avoid lawsuits if they use fewer preservatives, comply with stricter regulations and provide tracking. Additionally, it protects food against bioterrorism. In the literature, terms like "active packaging," "intelligent packaging," and "smart packaging" have emerged over the last two decades. Food bottles, drinks bottles, pharmaceutical bottles, cosmetic bottles, and many others [9–11].

According to Kerry et al. [2], "active packaging" refers to the Use of additives in packaging systems to extend shelf-life and maintain quality. Otles and Yalcin [12] define intelligent Packaging as that senses, detects, tracks, records, and communicates. Through the use of information provided in this site, it is easier to extend shelf lives, improve quality, increase safety, and avoid problems. Smart packaging is described by Otles and Yalcin [12] and Vanderroost et al. [13], and others as being capable of "intelligent and active monitoring."

2. Classification of packaging systems

It's not just the hardware that makes packaging base technologies different. They also vary in how much and what kind of data they can make. Sensors on smart Packagings, such as chemical or biological sensors, ensure that the food produced and consumed by consumers is of high quality and safe [13–15]. Classification of the packaging system is given in Fig. 2.1.

As described by Vanderroost and coauthors [13], is a complete packaging solution that uses intelligence to monitor its surroundings and takes appropriate actions. The sensors used in smart packaging can monitor many factors that influence food quality and safety, from freshness to pathogens to leaks, carbon dioxide to oxygen, pH levels to temperature, and time. A wide range of products can be packaged with smart packaging solutions, including foods, beverages, pharmaceuticals, and household items [2,16,17]. In turn, monitoring, communicating, or changing conditions becomes more challenging.

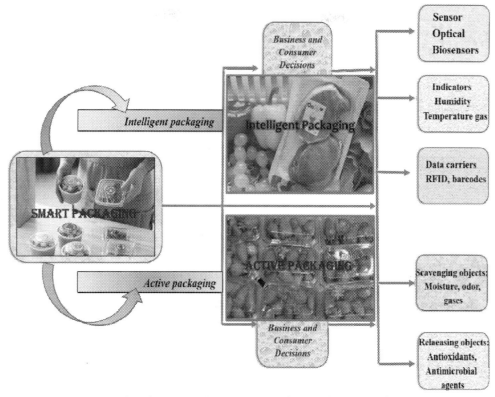

Fig. 2.1 Classification packaging system along with their application.

A product's lifecycle can be tracked via smart packaging. The environment inside or outside the package can be analyzed and controlled to provide manufacturers, retailers, and customers with relevant information about the product [14,18,19].

2.1 Active packaging

Traditional packaging can be replaced by active packaging. The packaging is intended to respond to changing consumer trends. Active packaging for food involves materials that release or absorb substances from or into the food or its environment to stay fresh for longer periods. Packaging perishable goods in active packaging have several benefits, including reducing the number of active chemicals, preventing cross-contamination of food from the packaging, and reducing industrial processes that create bacteria. The Use of oxygen and ethylene scavengers, flavors, odors, and antimicrobials in active packaging systems is quite common [5,20—22]. Active packaging is used, where secondary components are deliberately added to the packing materials or the headspace of the packet [23,24]. Active packaging systems are shown in Table 2.1.

Table 2.1 Active packaging systems.

Sr. No.	Active components	Applications
1	Absorber/emitter for humidity	The ripening of cheese; fresh meat purging pads
2	O_2 absorbers/emitters	Baking and meat products should be protected from oxidation by MAP
3	CO_2 absorber/emitters	Antimicrobial products
4	The absorbents of volatile odors	Products that improve fresh fruit quality
5	A selection of ethylene-based absorbers and adsorbers	Slow down ripening
6	Antimicrobials	By controlling microbes

Inactive food packaging materials have properties such as antimicrobial or antioxidative. Almost all food packaging materials have these properties. Adding active compounds into common packaging materials is one way to achieve this. You can apply coatings to the surface with the same functionality as they are added to the packaging material. In contrast, the second option maintains the bulk properties of the packaging materials [19,25].

"Smart" Packaging, also known as active packaging, seems to sense changes inside or outside the package and responds accordingly. A new type of packaging called "active packaging" was developed to prevent food from drying out. Desiccant stuff was placed inside dry food packets for the first time. This method involves the preservation of desiccants in moisture-permeable sachets, which means that they absorb moisture from the food and the sachet headspace. They are also capable of absorbing vapor that enters the package. Several packaging components are not always needed to complete a package, including sachets, pouches, patches, coupons, labels, and so forth [14,26–28].

In the food packaging industry, removing oxygen from the inside of the packaging is the most prevalent active packaging technique. Oxygen scavengers will reduce oxidative damage and are added to food or other packets to take out oxygen. Oxygen scavengers usually come in flexible, gas-permeable sachets with reduced iron particles. To remove air from packing packages, either a vacuum is created, or inert gases are sucked into them. The Use of oxygen scavenging in cars emerged in the last two decades of the 20th century [10,29–32].

The materials were arranged in a way so that they could be sold. Many different types of results were achieved when this was done. Several different types of beer and juice bottles were introduced in 2000. Unlike other food processing and packaging, active packaging makes food and drinks safer. During the manufacturing and packaging process, oxygen should be removed from the product so that the oxygen scavenging system can work efficiently. The materials and closures that keep oxygen out can be seen in the interior package and the construction of the package. Good oxygen-control performance depends on a good oxygen-scavenging process [33–35]. Some important applications of smart packaging materials are given in Fig. 2.2.

In addition to oxygen, several other factors can affect the quality of your food. Different factors play a role in food breakdown, including changes in moisture, oxidative reactions, microbial growth, and enzyme activity. It has been noted that similar efforts have been made to reduce oxygen inactive Packaging and are currently being investigated for other forms of active packaging. There are often carbon dioxide and ethylene scavengers used for bulk C.A. and M.A. storage that needs to be moved around for various reasons [10,33,36–40].

It has been found that modified atmosphere packaging (MAP) has not been very successful for emitters. In the bulk transportation category of fresh fruits, ethylene scavengers are available that are effective in active packaging. It is very important to maintain the right smell inside food products and prevent outside odors from entering. The Use of odor removers on this type of food is increasing, which are contained in the packaging of the food itself [41,42].

The smells are made when plastic is blocked by adding oxygen-blocking materials, such as tocopherols (vitamin E) or other antioxidants. Non-volatile tocopherols protect the body from oxygen. The best antioxidants are butylated hydroxyanisole and butylated hydroxytoluene (BHA/BHT), which can absorb into food and act as an antioxidant.

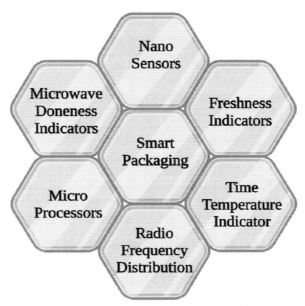

Fig. 2.2 Some important applications of smart packaging materials.

Scavengers or oxygen interceptors (chemical or physical) interact with oxygen, making different compounds. Oxygen absorbers can remove oxygen in any way. Antioxidants stop oxidation reactions caused by free radicals and peroxides. Inorganic catalysts are also prevented from speeding up undesirable oxidative reactions by sequestering agents [37,43].

2.2 The smart packaging

Food safety, shelf life extension, and overall quality are all emphasized by intelligent packaging. Smart packaging systems keep food fresh, extend shelf life, and give information about a product's quality during transportation and storage [8,14,15,19].

Smart packaging systems for food contain many hardware components and sensors, such as timers, thermometers, and gas detectors [14].

It's possible to use indicators and sensors to tell people what's going on. A change in pH level or temperature indicates that things have changed in a product's environment. The technology is often used to measure oxygen levels and freshness [15,44].

3. The applications and market opportunities

3.1 Applications

According to research by Pereira et al. [45], foodborne illnesses cause more than 9000 deaths in the United States every year. Food poisoning affects 60 people per 100,000 in Spain every year. Smart package technology can be used to detect pathogens in food through biosensors. Moisture absorbers, antimicrobial Packaging, CO_2 emitters, oxygen scavengers, antioxidants are also possible.

Smart packaging technology could be used for various purposes, ranging from food safety and drug use tracking to postal deliveries. It is believed that these opportunities present themselves to customers as a good thing from their point of view. An overview of applications is given in Fig. 2.3. Because people are always connected to the internet, new business opportunities have emerged with apps for tracking purchased goods [4,46,47]. Paquet et al. [28] said smart packaging could better detect supply chain problems, reduce costs, and increase performance.

3.2 Worldwide market prospect

There was $31.4 billion worth of packaging products sold worldwide in 2011. That number now increased to $33.3 billion in 2012, the most recently available statistics. Based on current forecasts, it is expected that this number will reach $44.3 billion in the coming years. By 2023, the electronic smart packaging market is predicted to reach $1.45 billion. In the United States, this type of packaging consumption will rise by 7.4% annually by the end of this decade [17,48–50].

Fig. 2.3 Opportunities and challenges of smart packaging research.

With an equivalent of U.S. $2360 million, Japan is the second-largest market for our products. Australia ranked fifth with US 1690 million, Germany ranked fifth with U.S. $1400 million, and the UK with the U.S. $1270 million, and the UK with the U.S. $1270 million [48,51–53].

4. The challenges and prospects of research

U.N. figures suggest that humans lose or waste over 1.3 billion metric tons of food every year. As a result, people do not do the proper thing when harvesting, transporting, storing, and storing the goods they produce [54]. The food industry spends a lot of money on throwing away food, so perishable packaging goods properly is important so that they can be stored and transported for a long time, thereby providing them with a longer shelf life [55,56]. Gray areas show where smart packaging offers opportunity from an industrial point of view, but making smart packaging that works well is difficult. Fig. 2.3 illustrates some of the challenges and opportunities in smart packaging research.

4.1 Challenges

Researchers and businesses are interested in antimicrobial packaging due to the potential to improve quality and safety [57,58]. Bio-preservatives, natural antimicrobial agents, and biodegradable solutions should be future research topics in active microbial packaging. It has already been demonstrated that using bioactive functional components in packaging can improve food quality and safety, resulting in new packaging innovations [59–61].

However, active materials' mechanical and barrier properties need to be improved. As a result, the food will also be safer and perform better on the shelf for longer [62]. Realini and Macros say biodegradable packaging materials will be used as a carrier polymer, and active compounds made from natural resources will be more common [17].

Thin-film electronics for packaging require intensive research and development. According to Vanderroost et al. [13], thin-film electronics can be used in printed and flexible sensor systems to keep track of conditions that are about to go bad. However, the question remains on how to improve the performance of the sensors. Businesses face a big problem because most smart packaging waste cannot be recycled.

Despite long-standing government policies, some packaging waste can be difficult to recycle [63]. Making smart packaging requires staying on top of the research on recycling and reusing packaging waste [64]. Perhaps biodegradable sensors and communication features can be created.

At some points in the supply chain, the food supply chain can lead to food waste. We have talked about the importance of food packaging when it enters the supply chain, as we have said. It ensures that the food will be safe, fresh, and of the highest quality. However, there is still much research to improve packaging [64].

Getting grain spilled out of sacks or attracting rodents can lead to infestations bigger than they already are if a product is not properly sealed. Some foods coming to the end of their life cycle may require different processing to give them a longer shelf life. It can cause disease outbreaks and be a very expensive source of disease, so it is not recommended to feed livestock [65].

Kuswandi et al. [14] say that food producer have difficulty updating packaging to include information. It has been suggested that more research is needed into this subject so that smarter packages can be developed:

Consumers want to know where products came from, the ingredients they contain, and how they were transported to the grocery store or their homes. Thin-film devices will be able to show food safety information directly to consumers through smart labels and stickers. In a smart packaging system, doctors and other medical providers get access to important patient data, which should ensure the safety and health of patients. They may also help stop people from misusing and stealing [66].

4.2 Opportunities

Shortly, nanotechnology will play a huge role in packaging, with the safety issue in mind. Majid et al. [67] suggest that packaging materials can be improved. Sensors in the packaging will control active agents through these materials. Its simplest form is an idea to connect objects to the cyber-physical world via a global network infrastructure. Devices with sensors and actuators can be tracked and controlled with it. RFID-tagged packaging from the factory to the customer can, among other things, be tracked easily [68].

In addition, the UPS website says that approximately 1% of all their online orders are lost or damaged. It is estimated that UPS sends about 4.6 billion packages every year. The postal service loses or damages approximately 4.6 million packages per year. Because of this, UPS parcels could become much less likely to get lost if smart packaging is used. By 2025, the Internet of Things will be more than just smartphones and computers. This includes items like food packages, furniture, machinery for manufacturing plants, and even whole factories [48,69,70].

People who can do things in real-time and have CPS. Technology that can keep an eye on, manage, and control the condition of goods in real-time is one of the main areas for more progress in packaging, and it needs to get better. Food safety, consumer health, and waste reduction will be profoundly affected.

One of the most important things that can be done is to ensure a good network infrastructure and the information and communication technologies that support that infrastructure. This covers the packaging as well as the entire supply chain. This creates cyber-physical distribution and production networks across several companies, both vertically (within a single company) and horizontally (between several companies).

4.2.1 Cybersecurity

Despite new technologies and things, cyber security remains a problem. There is a great deal of concern about security and privacy issues with current internet technologies. The application of smart packaging may never reach its full potential as one of the most exciting applications in the industry until these problems are solved. Multiple, fast, persistent, and sophisticated attacks distinguish this type of attack from others. This makes preventive security difficult to perform. We need automatic detection and response techniques for cyberattacks as part of defense-in-depth strategies. Smart packaging is only a part of cyber-physical systems, which is true for all cyber-physical systems. It is possible to run a business in new and innovative ways. Smart packaging offers many new opportunities that new business models will exploit. In an industry where people are looking for a good time, the industry has moved from products to product-service systems [11,51,68].

Businesses and value chains that have been around for a long time will face increasing pressure soon. The field has dealt with big data and new technology in the past. Observing the disruptive technologies that have occurred in the past, it is clear that the development of technologies is closely related to new business models. New business models and the latest technology are the keys to success (see Fig. 2.3). The smart packaging sector must also get on board to reach this goal. The manufacturing industry is a big fan of big data analytics, so it's no surprise that models that use data are highly sought after right now. As a whole, smart packaging includes active and intelligent elements [46]. Research into active and intelligent packaging will continue to excite basic and applied scientists for years to come.

5. Conclusion

During this chapter, we covered smart packaging opportunities as well as challenges. Several research areas emerge from it as well. A sensitive packaging uses sensors and materials to provide information about its quality, safety, shelf life, and usability. To improve current packaging technology, smart packaging must address a range of issues.

It would add value to the entire food packaging supply chain if sensors could be used with smart and traditional materials.

Nanotechnology, thin-film electronics, and smart materials will be used as packaging. They need to be easy to print and mass produces, inexpensive with the food's value, easy for people to use, environmentally friendly, and safe.

Food safety issues, like detecting microbial growth and oxidization and identifying open packages, should focus on new and more advanced smart packaging for food products. In addition to making a product more environmentally friendly, it should extend its shelf-life, make tracking easier, and enhance convenience.

There might be a way to recycle food waste and packaging by installing sensors built into the packages. Sensors can store information about the material of a package, when to ingest food, how much oxygen is present, etc. Those who manufacture or sell food or recycle packaging can access this information through the Industrial Internet of Things.

Incorporating smart packaging into an industrial internet of things network enables horizontal and vertical integration as part of the manufacturing sector. Furthermore, they face problems ensuring their internal Information Technology (IT) and Operational Technology (OT) are integrated efficiently and cost-effectively. Dispersed and independent keen industrial networks will be the norm in the upcoming, and IT and OT. convergence can be a step in the right direction. We must also find new ways to deal with cybersecurity and ensure data and intellectual property are safe and secure through all phases of a company's lifecycle.

References

[1] K.L. Yam, P.T. Takhistov, J. Miltz, Intelligent Packaging: concepts and applications, J. Food Sci. 70 (1) (2005) R1–R10.
[2] J. Kerry, M. O'Grady, S. Hogan, Past, current and potential utilization of active and intelligent packaging systems for meat and muscle-based products: a review, Meat Sci. 74 (1) (2006) 113–130.
[3] A.M. Youssef, S.M. El-Sayed, Bionanocomposites materials for food packaging applications: concepts and future outlook, Carbohydr. Polym. 193 (2018) 19–27.
[4] A.L. Brody, et al., Innovative food packaging solutions, J. Food Sci. 73 (8) (2008) 107–116.

[5] P. Prasad, A. Kochhar, Active Packaging in the food industry: a review, J. Environ. Sci., Toxicol. Food Technol. 8 (5) (2014) 1−7.
[6] S.F. SeyedReihani, H. Ahari, Study on Nano-Packaging Coatings and its Role on Quality Control of Food Products, 2020.
[7] A.L. Brody, et al., Scientific Status Summary: Innovative Food Packaging Solutions, 2008.
[8] C. Medina-Jaramillo, et al., Active and smart biodegradable packaging based on starch and natural extracts, Carbohydr. Polym. 176 (2017) 187−194.
[9] S. Hogan, J. Kerry, Smart Packaging of Meat and Poultry Products. Smart Packaging Technologies for Fast-Moving Consumer Goods, 2008, pp. 33−54.
[10] A. Dey, S. Neogi, Oxygen scavengers for food packaging applications: a review, Trend. Food Sci. Technol. 90 (2019) 26−34.
[11] S. Mihindukulasuriya, L.-T. Lim, Nanotechnology development in food packaging: a review, Trend. Food Sci. Technol. 40 (2) (2014) 149−167.
[12] S. Otles, B. Yalcin, Intelligent food packaging, LogForum 4 (4) (2008) 3.
[13] M. Vanderroost, et al., Intelligent food packaging: the next generation, Trend. Food Sci. Technol. 39 (1) (2014) 47−62.
[14] B. Kuswandi, et al., Smart Packaging: sensors for monitoring food quality and safety, Sens. Instrument. Food Qual. Safe. 5 (3) (2011) 137−146.
[15] K. Biji, et al., Smart packaging systems for food applications: a review, J. Food Sci. Technol. 52 (10) (2015) 6125−6135.
[16] In Meat biotechnology, in: M.N. O'Grady, J.P. Kerry (Eds.), Smart Packaging Technologies and Their Application in Conventional Meat Packaging Systems, Springer, 2008, pp. 425−451.
[17] C.E. Realini, B. Marcos, Active and intelligent packaging systems for a modern society, Meat Sci. 98 (3) (2014) 404−419.
[18] M. Ghaani, et al., An overview of the intelligent packaging technologies in the food sector, Trend. Food Sci. Technol. 51 (2016) 1−11.
[19] R. Dobrucka, R. Cierpiszewski, Active and intelligent packaging food-Research and development-A Review, Polish J. Food Nutrit. Sci. 64 (1) (2014).
[20] A. İçöz, Smart, Active and Sustainable Food Packaging, Center for Quality, 2017.
[21] E. Almenar, et al., Optimization of an active package for wild strawberries based on the release of 2-nonanone, LWT-Food Sci. Technol. 42 (2) (2009) 587−593.
[22] K.A. O'Callaghan, J.P. Kerry, Consumer attitudes towards applying smart packaging technologies to cheese products, Food Packag. Shelf Life 9 (2016) 1−9.
[23] S. Sarkar, K. Aparna, Food packaging and storage, Res. Trend. Home Sci. Exten. AkiNik Pub 3 (2020) 27−51.
[24] S.J. Lee, A.M. Rahman, Intelligent Packaging for food products, in: Innovations in Food Packaging, Elsevier, 2014, pp. 171−209.
[25] R. Dobrucka, The future of active and intelligent packaging industry, Log Forum 9 (2) (2013).
[26] G. Fuertes, I. Soto, R. Carrasco, M. Vargas, J. Sabattin, C. Lagos, Intelligent packaging systems: sensors and nanosensors to monitor food quality and safety, J. Sens. (2016) 2016.
[27] A. Pacquit, June frisby, danny diamond, king tong lau, alan farrell, brid quilty, and dermot diamond, developing smart packaging for monitoring fish spoilage, Food Chem. 2 (2007) 466−470.
[28] A. Pacquit, et al., Development of smart packaging to monitor fish spoilage, Food Chem. 102 (2) (2007) 466−470.
[29] I.S. Arvanitoyannis, A.C. Stratakos, Application of modified atmosphere packaging and active/smart technologies to red meat and poultry: a review, Food Bioproc. Technol. 5 (5) (2012) 1423−1446.
[30] S. Rossaint, S. Klausmann, J. Kreyenschmidt, Effect of high-oxygen and oxygen-free modified atmosphere packaging on the spoilage process of poultry breast fillets, Poult. Sci. 94 (1) (2015) 96−103.
[31] J.S. Lee, et al., Ascorbic acid-based oxygen scavenger in active food packaging system for raw meatloaf, J. Food Sci. 83 (3) (2018) 682−688.
[32] Z. Foltynowicz, et al., Nanoscale, zero-valent iron particles for application as an oxygen scavenger in food packaging, Food Packag. Shelf Life 11 (2017) 74−83.

[33] R.S. Cruz, G.P. Camilloto, A.C. dos Santos Pires, Oxygen scavengers: an approach on food preservation, Struct. Funct. Food Eng. 2 (2012) 21–42.
[34] S. Sängerlaub, et al., Palladium-based oxygen scavenger for food packaging: choosing optimal hydrogen partial pressure, Food Packag. Shelf Life 28 (2021) 100666.
[35] K.K. Gaikwad, Y.S. Lee, Effect of storage conditions on the absorption kinetics of non-metallic oxygen scavenger suitable for most food packaging, J. Food Measure. Characterizat. 11 (3) (2017) 965–971.
[36] D.S. Lee, Carbon dioxide absorbers for food packaging applications, Trend. Food Sci. Technol. 57 (2016) 146–155.
[37] K.K. Gaikwad, S. Singh, Y.S. Lee, Oxygen scavenging films in food packaging, Environment. Chem. Lett. 16 (2) (2018) 523–538.
[38] K.K. Gaikwad, S. Singh, Y.S. Negi, Ethylene scavengers for active packaging of fresh food produce, Environm. Chem. Lett. 18 (2) (2020) 269–284.
[39] H. Wei, et al., Ethylene scavengers for preserving fruits and vegetables: a review, Food Chem. 337 (2021) 127750.
[40] I.J. Church, A.L. Parsons, Modified atmosphere packaging technology: a review, J. Sci. Food Agricult. 67 (2) (1995) 143–152.
[41] J. Farber, Microbiological aspects of modified-atmosphere packaging technology-a review, J. Food Protect. 54 (1) (1991) 58–70.
[42] A.A. Kader, et al., Modified atmosphere packaging of fruits and vegetables, Critic. Rev. Food Sci. Nutrit. 28 (1) (1989) 1–30.
[43] I.U. Unalan, et al., Nanocomposite films and coatings using inorganic nano building blocks (NBB): current applications and future opportunities in the food packaging sector, RSC Advan. 4 (56) (2014) 29393–29428.
[44] L. Zheng, et al., Novel trends and applications of natural pH-responsive indicator film in food packaging for improved quality monitoring, Food Cont. (2021) 108769.
[45] D. Pereira de Abreu, J.M. Cruz, P. Paseiro Losada, Active and intelligent packaging for the food industry, Food Rev. Inter. 28 (2) (2012) 146–187.
[46] H. Cheng, et al., Recent advances in intelligent food packaging materials: principles, preparation, and applications, Food Chem. 375 (2022) 131738.
[47] S.Y. Lee, et al., Current topics in active and intelligent food packaging for the preservation of fresh foods, J. Sci. Food Agricult. 95 (14) (2015) 2799–2810.
[48] D. Schaefer, W.M. Cheung, Smart packaging: opportunities and challenges, Proc. CIRP 72 (2018) 1022–1027.
[49] H. Akram, Silver Nanoparticles Based Smart Packaging Materials for Detection of Meat Spoilage, 2021.
[50] Laflamme, M. et al., RockTenn (NYSE: RKT) Is One of North America's Leading Paperboard, Containerboard, and Consumer and Corrugated Packaging Manufacturers.
[51] A. Mirza Alizadeh, et al., Trends and applications of intelligent packaging in dairy products: a review, Crit. Rev. Food Sci. Nutrit. 62 (2) (2021) 383–397.
[52] L.M. Hoffman, B.S. Heisler, Airbnb, Short-Term Rentals, and the Future of Housing, Routledge, 2020.
[53] A.Q. Roya, M. Elham, Intelligent food packaging: concepts and innovations, Int. J. Chem. Tech. Res. 9 (2016) 669–676.
[54] L. Xue, et al., Missing food, missing data? A critical review of global food losses and food waste data, Environ. Sci. Technol. 51 (12) (2017) 6618–6633.
[55] P. Müller, M. Schmid, Intelligent Packaging in the food sector: a brief overview, Foods 8 (1) (2019) 16.
[56] P. Pessu, et al., The concepts and problems of postharvest food losses in perishable crops, Af. J. Food Sci. 5 (11) (2011) 603–613.
[57] P. Suppakul, et al., Active packaging technologies emphasize antimicrobial packaging and its applications, J. Food Sci. 68 (2) (2003) 408–420.
[58] J.H. Han, Antimicrobial food packaging, Novel Food Packag. Tech. 8 (2003) 50–70.

[59] A. Valdés, et al., Use of herbs, spices and their bioactive compounds in active food packaging, RSC Advanc. 5 (50) (2015) 40324–40335.
[60] A.G. Scannell, et al., Development of bioactive food packaging materials using immobilized bacteriocins Lacticin 3147 and Nisaplin, Int. J. Food Microbiol. 60 (2–3) (2000) 241–249.
[61] A. Lopez-Rubio, Bioactive food packaging strategies, Multifunct. Nanoreinfor. Polym. Food Packag. (2011) 460–482.
[62] M. Ozdemir, J.D. Floros, Active food packaging technologies, Crit. Rev. Food Sci. Nutrit. 44 (3) (2004) 185–193.
[63] M. Beccarello, G. Di, Foggia, Economic analysis of E.U. strengthened packaging waste recycling targets, J. Advanc. Res. L. & Econ. 7 (2016) 1930.
[64] A. Regattieri, et al., Innovative solutions for reusing packaging waste materials in humanitarian logistics, Sustainability 10 (5) (2018) 1587.
[65] N. Jabeen, I. Majid, G.A. Nayak, Bioplastics and food packaging: a review, Cog. Food Agricult. 1 (1) (2015) 1117749.
[66] S. Kumar, S.K. Gupta, Applications of biodegradable pharmaceutical packaging materials: a review, Middle-East J. of Sci. Res. 12 (5) (2012) 699–706.
[67] I. Majid, et al., Novel food packaging technologies: innovations and future prospective, J. Saudi Soc. Agricult. Sci. 17 (4) (2018) 454–462.
[68] L. Thames, D. Schaefer, Cybersecurity for Industry 4.0, Springer, 2017.
[69] L. Atzori, A. Iera, G. Morabito, The internet of things: a survey, Comput. Netw. 54 (15) (2010) 2787–2805.
[70] D. Schaefer, J. Walker, J. Flynn, A data-driven business model framework for value capture in Industry 4.0, Advanc. Manufact. Technol. XXXI (2017) 245–250. IOS Press.

CHAPTER 3

Innovations in smart packaging technologies for monitoring of food quality and safety

Biplab Roy[a], Deepanka Saikia[b], Prakash Kumar Nayak[b], Suresh Chandra Biswas[c], Tarun Kanti Bandyopadhyay[a], Biswanath Bhunia[d] and Pinku Chandra Nath[d]

[a]Department of Chemical Engineering, National Institute of Technology Agartala, Jirania, Tripura, India; [b]Department of Food Engineering and Technology, Central Institute of Technology, Kokrajhar, Assam, India; [c]Department of Home Science, Khowai, Tripura, India; [d]Department of Bio Engineering, National Institute of Technology Agartala, Jirania, Tripura, India

Abbreviations

KMnO₄ Potassium Permanganate
LDPE Low-Density Polyethylene
PET Polyethylene Terephthalate
RFID Radiofrequency Identification
SGS Soluble Gas Stabilization
TiO₂ Titanium Dioxide
TTIs Time Temperature Indicators

1. Introduction

To protect a product from degradation resulting from exposure and utilization in a surrounding environment, packaging serves as the principal protective barrier for that product. Additionally, product packaging is a powerful marketing tool for interacting with customers. It is obtainable in a diverse assortment of shapes and dimensions and provides users with both simplicities of use and convenience as a user interface [1]. The following are some prominent purposes of food packaging, for example:

- To Prevent leaking, breakage, and contamination of the product.
- To convey critical nutritional and other information about the food product.
- To make it easier for consumers to reheat the meal in the microwave.
- To facilitate transportation and handling, the provision of containment is recommended.

As a result of constantly rising customer demands, increased product intricacy, and, most notably, regional and global policies aimed at promoting a regenerative economic system and decreasing the CO_2 emissions of processed goods, traditionally used packaging seems to be no longer suitable [2]. Improved packaging featuring greater efficiency is indeed required to suit a wide range of new customer demands. An example will be to offer foods that have been processed under regulations and to package them in a way that allows for cardle-to-grave tracking, which can protect the company from lawsuits. Smart packaging is also a way to open up new markets in the global context, meet stricter food safety rules from both countries, and even protect against food bioterrorism. Active, smart, and intelligent packaging is all terms that have appeared in the literature during the past few decades. Maximum researchers refer to packaging systems that are used to keep food, beverages, medications, beauty products, and a variety of other items safe [3].

- It is defined by Kerry et al. [4] as "incorporating specific additives into packaging systems to enhance product value and longevity."
- To put it another way, intelligent packaging, according to Ref. [5], seems to be a packaging technology that can sense, identify, trace, track, and communicate to enhance shelf lifespan, purity, safety, give knowledge, and notify potential hazards.
- Other researchers [1,6], defined the smart technology (packaging) concept as "something that has the potential of combining intelligent and active packaging."

To give precise interpretations and categorizations for both intelligent and active packaging, together with related production technological advances up to commercial implementations, this research aims to provide a summary of significant advancements in packaged food products.

2. Overview of smart packaging technology

There are various distinct definitions of smart packaging technology that may be found in the literature. Packaging that is created by adding new functions to passive packaging [7] as well as a material that not only improves basic functions but also responds to external stimuli, can be classified as smart packaging [8]. Smart packaging refers to an active or intelligent strategy that contacts between package and food [9]. When used properly, active packaging can limit the progression of illness and destructive microorganisms, restrict the transfer of pollutants, and increase the longevity of a manufactured good while maintaining the protection and value of the product [10]. Interactive features can be found in the package created. Customers and suppliers can connect with the product itself through smart packaging. At least one of the electronic, mechanical, chemical, electrical, or internet technologies is used to accomplish this goal. These systems are geared toward enhancing the packaging's functionality in order to meet rising consumer expectations, tightening regulatory requirements, and a growing concern for product safety. In general, smart technology (packaging) can be classified into 2 types: active and intelligent packaging (Fig. 3.1).

2.1 Active packaging

The first alternative to standard packaging is active packaging. It's a new approach to food packaging that was developed in response to shifting consumer preferences and market

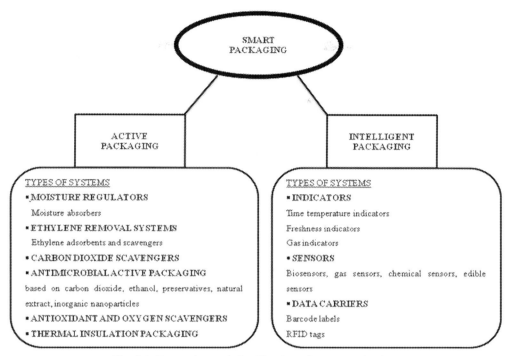

Fig. 3.1 Properties and classification of smart packaging.

conditions. Active packaging technology enables conserved foods and their surroundings to sustain their purity and shelf life for extended durations of time [11]. It's better to use active packaging with perishable commodities because it reduces the proportion of active ingredients, reduces particle activities and migrations from films to foodstuffs, and avoids needless industrial operations that may introduce microorganisms into the products [12]. Thus, active packaging's principal objective is to protect food from chemical and microbial contamination while preserving its visual and physiological features. Table 3.1 demonstrates examples of active food packaging that is accessible.

2.1.1 Moisture absorbers

Excess moisture in food packaging promotes microbiological growth and foggy film production. Water may accumulate inside the package if it has inadequate water mist permeability. Food packing can develop excess water because of fresh produce inhalation, temperature variations, or tissue liquid pouring from freshly sliced meat. Bacteria and fungus thrive in moist environments, reducing product shelf life and quality. Moisture scavengers physically absorb and hold water molecules. Drying agents absorb moisture from the air and reduce the relative humidity in the environment [13]. Water scavengers

Table 3.1 Food-active packaging that is commercially accessible.

Name (commercial)	Applications	Principle	Forms and materials
ATOX (www.artibal.com)	Cereal based products	Antioxidant	Film coating containing oregano essential oils
Green Box (www.greenbox.it)	Perishable type products	Phase transform materials	Vegetable oil based
Pure Temp (www.puretemp.com)	Frozen type food, frozen storage	Phase transform materials	Coconut oil, palm oil, and soybean oil based
Active Film™ (www.csptechnologies.com)	Vegetables and fruits	Moisture absorber	LDPE film
BIOPAC (www.biopac.com.au)	Fresh foodstuffs	Ethylene scavenger	Sachet in porous material mix by potassium permanganate
SANDRY® (www.hengsan.com)	Fruit, coffee, and fermented products	CO_2 absorber	Sachets
FreshPax® (www.multisorb.com)	Pre-cooked and processed food products	CO_2 emitter	Packets and films recognize with edible materials
ATCO® (www.emcotechnologies.co.uk)	Fresh products	CO_2 absorber	Film bags
Tenderpac® (www.sealpacinternational.com)	Meat based products	Moisture absorber	PET tray
Biomaster® (www.biomasterprotected.com)	Frozen and chilled products	Antimicrobial	Cool bags
Food-touch® (www.microbeguard.com)	All food items	Antimicrobial	Different form of paper products
Celox™ (www.grace.com)	Beverage type products	O_2 scavenger	Closure coatings and cans sealants
McAirlaid's CO_2Pad (www.mcairlaids.net)	Meat, fish, and fruit products	CO_2 emitter	Cellulose based pads

such as silica gel (Fig. 3.2), molecular sieves, natural clay, calcium oxide, calcium chloride, and modified starch are the most effective technique to manage excess water accumulation in food packaging that has a strong barrier against water vapor. Silica gel seems to be the most extensively used moisture scavenger because of its non-toxicity and corrosiveness in nature, making it the most environmentally friendly option. Silica gel physically

Fig. 3.2 Use silica gel to remove moisture from the air.

absorbs moisture, but calcium chloride chemically absorbs moisture, as seen in the examples above. In general, moisture scavengers are the most often utilized chemicals in food packaging applications, according to the FDA [14].

2.1.2 Oxygen scavengers

High quantities of O_2 in food packaging can promote microbial growth, odor production, color change, and nutritional loss, all of which can significantly impair the shelf-lifespan of the products they contain. As a result, it is critical to regulate the oxygen content of food products to prevent food spoiling and degradation. In addition to providing an alternative to vacuum or gas-flushing packaging, oxygen scavenging devices contribute to enhancing product shelf life and quality. As a result of lower packaging costs and more profitability, these products are also more expensive.

Typical oxygen absorption systems remove oxygen by oxidizing iron powder or using enzymes. Over the first instance, Fe is converted to the iron oxide by oxidation in a sachet. To maximize efficiency, the sachet's substance is extremely permeable to O_2 as well as water vapor. This technique, created by Mitsubishi Gas Chemical Company, is widely used and is the first in food packaging. Amounts of absorbent are dictated by the initial oxygen content in packaging, dissolved O_2 concentration in foodstuffs, packaging material permeability, size, shape, and weight. High, medium, or low-moisture, and lipid-containing meals can all benefit from iron-based oxygen absorption systems [9]. When using an enzyme-based oxygen scavenging technique, an enzyme interacts with substrates to eliminate O_2 (Fig. 3.3). Prices of enzymes used to remove oxygen are higher than iron-based methods. In addition to temperature and pH, enzymatic oxidation processes are very vulnerable to variations in water activity, solvents, and solvent concentrations [4].

Fig. 3.3 Food packaging enzymatic oxygen cleaning systems.

Sachets that absorb oxygen are not recommended for use with liquid foods since the immediate interaction of liquids with sachet may cause its contents to flow out. In addition, the sachets may be mistakenly consumed with the food or swallowed by children who are not paying attention. As a result, the FDA requires that oxygen scavenging sachets commercialized in the US bear the warning "do not consume" on the package (Fig. 3.4).

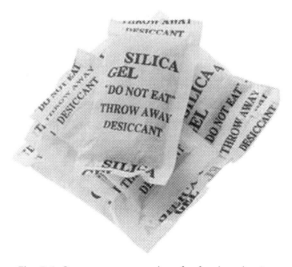

Fig. 3.4 Oxygen scavenger bag for food packaging.

2.1.3 Ethylene scavengers

Potassium permanganate (4%–6%) [15] based on inert matrix including silica gel or alumina is the most extensively utilized ethylene-scavenger compound [16]. Due to its toxicity [17], this chemical is frequently implanted with nanomaterials to boost its scavenging abilities and it has been packaged in porous sachets to prevent it from being inhaled [18]. Additionally, scavengers made of KMnO$_4$ are also available in a variety of forms including coatings, tube filters, and quilts, to name a few examples [13]. The oxidation of ethylene gas produces ethylene glycol, carbon dioxide, and water [19]. Photocatalytic oxidation of ethylene can be accomplished by using Ag, TiO$_2$, ZnO, Cu, and Pd nanomaterials that convert oxidized ethylene to CO$_2$ and H$_2$O [18]. It was created by Siripatrawan and Kaewklin [20], to alleviate the issues associated with a spontaneous agglomeration of TiO$_2$ nanomaterials and to provide an effective packaging system that has ethylene scavenging and antimicrobial characteristics in nature [13,18].

The amount of ethylene emitted by fruits and vegetables has a direct impact on the development and ripening processes [21,22]. The effect of ethylene in plant growth varies depending on the type of fruit, ripeness stage, and ethylene exposure [23]. Beyond increasing the ripening process, ethylene can cause over-ripening and even rot in fruits, lower the shelf life of produce and result in financial losses to the grower. A growing demand exists for long-term food preservation and the reduction of food rotting for a variety of reasons, including health and economic concerns [24]. To prevent fruits and vegetables from becoming overripe after harvest, it is vital to utilize anti-ethylene scavengers to halt the maturation of the fruits and vegetables (Fig. 3.5) [25].

2.1.4 CO$_2$ emitters and absorbers

CO$_2$ can be injected into the package atmosphere of some items, such as raw meat, chicken, seafood, butter, and other commodities [26], to reduce the growth of microbes in the food items. To reduce the rate of respiration of fresh food, CO$_2$ can be utilized in comparison with other techniques [27]. To extend the longevity of certain foods, high CO$_2$ levels (10%–80%) should be used [28]. A residual vacuum is created when O$_2$ is removed from the atmosphere, which can induce packaging collapse in flexible materials [29,30]. Due to CO$_2$ solubility at lower temperatures, whenever a package gets cleansed with a blend of oxygen and CO$_2$, a partial vacuum is created [31]. Soluble gas stabilization (SGS) is the process of dissolving enough CO$_2$ into a product for one to 2 h before the retail sale [32]. A muscular meal can be served in a conventional MAP tray with a pierced false bottom and a sachet Package collapse is prevented by the release of CO$_2$ from the exudates of the food [33]. Fig. 3.6 illustrates a carbon dioxide emitter in food packaging.

To remove extra CO$_2$ from packages, carbon dioxide scavengers are used. As a general rule, CO$_2$ scavengers have been used in newly ground coffees because they produce significant volumes of CO$_2$ while tightly wrapped into packs soon after roasting which

Fig. 3.5 Ethylene gas absorber for food packaging.

causes the package to rupture [34]. The MGC Company manufactures CO_2 scavenging sachets. CO_2 scavengers are used to replace the "aging" method that happens after the coffee is roasted. This prevents the loss of volatiles that are good for coffee [35].

2.1.5 Antioxidant releaser

Antioxidants are also being studied for their capacity to boost the stability of oxidation-sensitive foods. After bacteria development, oxidative breakdown causes food deterioration [36]. Oxidative processes are responsible for lowering the nutritive value through the breakdown of important fatty acids, proteins, and lipid-soluble vitamins, as well as for producing off-flavors and odors, as well as for color change owing to pigments deterioration in the food product [37,38]. There has been research conducted on the addition of antioxidant compounds in packaging as well as on the natural antioxidants that are now used in active food packaging, among other things. Also employed as carriers of natural

Fig. 3.6 Carbon dioxide emitter in food packaging.

antioxidants for lipid foods include edible and active coatings (such as chitosan, gelatin, galactomannans, and alginate) and films (such as alginate, chitosan derivatives, and cellulose derivatives) [39]. Including antioxidants in the packaging material outweighs the benefits of including them directly in food formulations in terms of overall value. Because of this, the majority of antioxidant systems are produced in the form of sachets, pads, or tags, or they can be found integrated into monolayer or multilayer materials used in packaging (Fig. 3.7) [16].

2.1.6 Antimicrobial packaging systems

A growing number of people are becoming interested in the usage of packaging materials that include antimicrobial compounds as a result of growing public concern about health. Consumers are looking for minimally processed, preservative-free packaged foods that have a longer shelf life than the competition [40]. Antimicrobials can be used to improve packaging materials, edible films, and coatings to create a protective barrier that will inhibit and delay the growth of microorganisms. Packaging materials act as carriers for antimicrobials that are delivered into the food in a controlled manner to enhance the shelf lifespan of food while also improving the quality and safety of food [41]. This way, the packaging can serve as a final line of defense against the development of food-borne microorganisms [42]. Furthermore, the majorities of natural antibacterial compounds are biodegradable and disintegrate quite quickly. Antimicrobial packaging can be created in a variety of ways. According to their interaction with the antimicrobial material utilized, as well as with the packaging and the food matrix, antimicrobial packaging systems

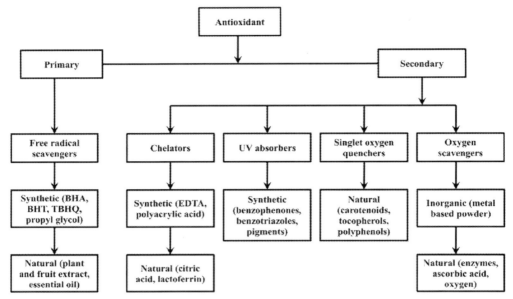

Fig. 3.7 Mechanism of antioxidant activity.

can either migrate into food or not migrate into it in general [43]. Antimicrobial packaging has already been found to promote global food safety and quality by inhibiting or limiting bacterial development in food [6,44].

2.1.7 Flavor/odor releasers and absorbers

Selective scavenging can be used to remove volatile organic compounds (VOCs) that build inside a package as a result of food degradation, such as aldehydes, amines, and sulfur dioxide [45]. During the transportation of mixed cargoes, strong odors are kept from spreading by flavor scavengers. Biji et al. [28], have created odor-proof packaging for transporting durian fruit. The package contains an odor-impermeable material (PET or polyethylene) of adequate width, an opening for inhaling gas passage, and an odor-absorbing sachet of Ni and charcoal. Fish muscle protein degradation produces volatile amines, which can be eliminated from the body by adding acidic chemicals such as citric acid into the polymer matrix [46]. A compound of Fe and organic acids is employed in the ANICO bags (Japan) to oxidize amines [47]. Sajilata et al. [48], investigated the flavor scalping of polyethylene. To inhibit the absorption of various non-food odors, the use of high-barrier packing materials can be beneficial [49].

2.1.8 Other active packaging

In the coming years, ready-to-eat meals wrapped in self-heating packaging might account for a major portion of the total volume of active packaging used. According to

Regulation (EC) No. 450/2009, the term "self-heating packaging" refers to packages that can heat their materials without the aid of additional heating systems or energy. Packages with self-venting capabilities regulate their internal pressure or steam, allowing steam to escape when the proper pressure and temperature levels are reached. By using shielding, field modulation, and susceptors, microwaveable active packaging is intended to improve the healing response of food while it is being prepared in the microwave [50]. As a result of the Al or SS steel being coated on substances including paperboard, the surface heats up and browns and crisps in a consistent manner [4,51,52]. Available commercial microwave susceptors include Sira Crisp™ (Sirane Ltd.) and SmartPouch® (VacPacInc.) [28]. Associated with each of the active microwave packs is a steam valve, which enables for rapid evacuation of steam mostly throughout microwave cooking. When used in a microwave oven, Flexis™ Steam Valve provides steam which can sometimes be successfully applied mostly to pliable food packaging sheets covered by a lid, enabling for steaming or heating of convenience meals. First, the food is protected, but as the cooking progresses, a hermetic seal is formed that allows it to be ingested without the need for additional ventilation. It maintains a progressive temperature equilibrium during the cooking process to keep food quality consistent throughout [28].

2.2 Intelligent packaging

The idea of preservatives moving inside food and the interaction feature of the package to help people make decisions are both relevant to intelligent packing [53]. "Intelligent materials and articles" refers to the material and articles which are capable of monitoring food or the atmosphere in which it is stored, according to EC/450/2009. Information about the food or its environment can be provided through intelligent packaging solutions (temperature, pH). Sensory packaging is an extension of the traditional package's communication function and can identify, sense, and capture the modification in the product's environment [54,55]. In contrast to active components, intelligent components don't want to put their components into the food. The intelligent packaging also can help improve HACCP and QACCP systems [56], which are used to detect unhealthy foods, diagnose possible health consequences, and set up a strategy to minimize or eradicate their incidence. It also helps in finding out which processes have a more impact on the quality of the food and can be utilized to rapidly progress the value of finished items [1]. Sensors, indicators and radiofrequency identification (RFID) devices are the three primary types of intelligent systems [1,4]. Table 3.2 illustrated food-intelligent packaging that is commercially accessible.

2.2.1 Indicators

Indicators tell the customer what they need to know. Whether or not a substance is present or not can be the subject of this information. It can also be about how two or more substances interact with each other. Usually, this kind of information is shown by things

Table 3.2 Food-intelligent packaging that is commercially accessible.

Applications	Trade name	Company
TTIs	CheckPoint®	Vitsab
	Colour-Therm	Colour-Therm
	Cook-Chex	Pymah Corp.
	Timestrip®	Timestrip Plc
	Fresh-Check®	Temptime Corp.
Integrity indicators	Best-by™	FreshPoint Lab
	O$_2$ Sense™	FreshPoint Lab
	Timestrip	Timestrip Ltd.
Freshness indicators	Fresh Tag	COX Technologies
	RipeSense	RipSense™
RFID tags	Temptrip	Temptrip LLC
	Intelligent Box	Mondi Pic
	CS8304	Convergence Systems Ltd.

like different color densities or the spread of the dye across the shape of the indicator. Indicators have a unique feature: the type of information that is important is either qualitative or semi-quantitative. There are a lot of different types of indicators, but they can indeed be broken down into three groups: indicators of time-temperature, indicators of freshness, and indicators of gas. These indicators make products better and more valuable [57,58].

2.2.1.1 Time temperature indicators (TTIs)

Using TTIs, food manufacturers can monitor and convey the quality of their products to consumers because of their simplicity, cheap cost, affordability, and efficiency. As part of the TTI implementation process, kinetic studies of food quality and response loss ratios are necessary [59]. Various classes of TTIs trading have been developed and tested based on enzymes, polymers, and biological mechanisms [60]. To assure the safety and quality of the food which requires a specific temperature, it's indeed important to observe time and temperature variables from preparation to the ultimate customer [61]. TTI can sometimes be applied to shipping containers or single containers in the form of tiny stickers; if the food is exposed to a temperature that differs from the required temperatures, an irreversible chemical alteration will be displayed (Fig. 3.8). The chiller is a very significant control point throughout transportation and storage [62].

2.2.1.2 Freshness indicators

These indicators make it possible to keep an idea about the quality of food products during storage and transportation. Freshness depletion may happen as a consequence of being exposed to dangerous conditions or surpassing the shelf lifespan of commodities. Freshness indicators provide immediate feedback on the freshness of goods based on

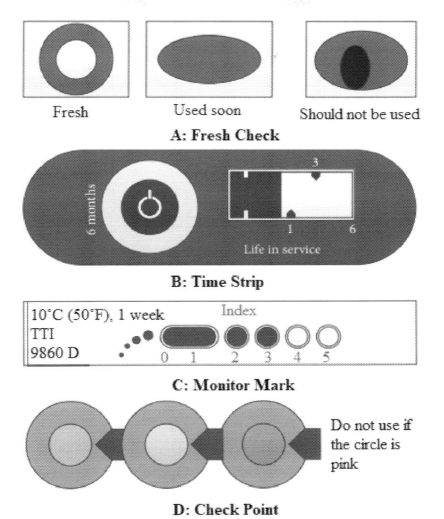

Fig. 3.8 Schematic diagram of the TTIs.

microbial growth or chemical changes. It accomplishes this by detecting volatile amines, which are created when food spoils, via conduct oximetry, pH change, or other comparable ways. Freshness indicators work on the concept that the organic acid, carbon dioxide, and volatile nitrogen compounds produced by microbial growth in the structure of spoiled food alter the chemical structure of the indicator's dye. Typically, the dye undergoes a color shift as a result of the reaction between the dye and its degradation metabolites. This enables the creation of visible, easily identifiable freshness measurement devices. Freshness indications include hydrogen sulfide, ethanol, diacetyl, and carbon dioxide.

2.2.1.3 Integrity indicators

The cleanliness of packaging is ensured across the manufacturing and distribution processes by using a leak indicator on the package (Fig. 3.9). According to Kerry et al. [4], MAP foodstuffs with low commencing oxygen had poor visible oxygen indicators. Oxygen indicators that use redox dyes change color when there is more or less oxygen in the air. This approach has the disadvantage of requiring extremely sensitive equipment, and the remaining oxygen inside the packaging can be exploited to detect problems. The natural bacteria that exist in the foods may well be able to consume some of the oxygen which escapes out from the foodstuff [63]. Ageless Eye® (a division of MGC Company) is a type of oxygen indicator capsule that turns color to reflect whether or not there is enough oxygen in the bloodstream. It turns pink when oxygen is deficient (0.01%). Whenever the O_2 content is higher than 0.6%, the tablet will become blue. The incidence of O_2 can be detected in less than 5 min, whereas the transition from blue to pink could take up to 180 min or higher [64]. Reverse and non-reversible O_2 indicator labels have been released by the packaging company EMCO Packaging (UK) [28].

2.2.1.4 Barcodes and RFID tags

In addition to facilitating information flow throughout the food supply chain, these devices that do automatic identification also improve efficiency in terms of food value and safety. However, rather than providing details on the food's effectiveness, the goal of these labels, which are also known as data carriers, is to automate the process, ensure traceability, and secure the product from theft or counterfeiting. It is customary for them to be placed on an outside packaging (box, parcel, pallet, etc.) that is separate from the product's packaging. In terms of labeling, barcodes and RFID labels seem to be the most popular options. Because of their low cost and convenience of use, barcodes are commonly utilized in retail and wholesale establishments for inventory control and stock tracking. When creating a barcode, it is necessary to organize parallel spaces and

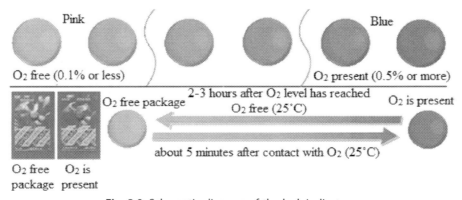

Fig. 3.9 Schematic diagram of the leak indicators.

bars to represent 12-digit data. RFID tags are an example of a data carrier tag that is more advanced. The utmost significant component in an RFID system is the tag, which is comprised of a microchip and tiny antenna. The other two components are the reader, which sends out radio waves and gets replies from tags in accordance, and the middleware, which communicates with the RFID through a local network or web server. Fig. 3.10 depicts an example of an RFID tag.

2.2.2 Sensors
The term "sensor" refers to devices that recognize, navigate, or measure energy or substance by responding to the identification or assessment of a physio-chemical attribute whereby the device responds [65]. Continuous signal output is provided through sensors. Receptors and transducers are the most common functional components of most sensors.

2.2.2.1 Biosensors
Biological analytes are used in biosensors, which use sensors to detect changes in the product and then turn those changes into an electrical signal using a transducer. Sensors that use biological analytes to monitor the quality of beef and chicken fillet, as well as those that use nylon-6 nanofiber membranes operationalized by glucose oxidase to detect glucose in various beverages, are examples of biosensors that may be used to monitor food quality [66,67]. Fig. 3.11 is a schematic representation of biosensors.

2.2.2.2 Gas sensors
Compounds that can be detected by gas sensors include carbon dioxide, nitrous oxide, and other gases. Indicator paint and α-naphtholphthalein, for example, are employed to detect correct carbon dioxide levels with a carbon dioxide sensor. These sensors are

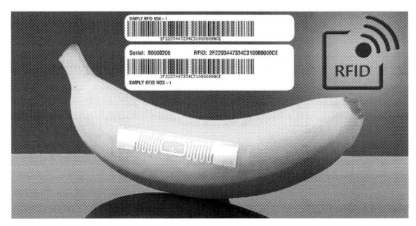

Fig. 3.10 Tags of RFID.

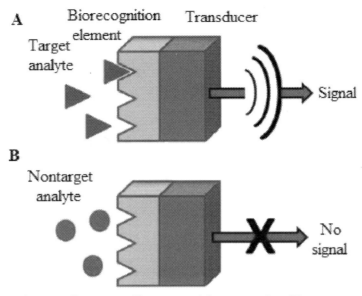

Fig. 3.11 Schematic illustration of biosensors, (A) target analyte (B) non-target analyte.

also known for their toxicity, which is produced by the removal of dyes from the sensor's surface [68,69]. There are many several types of gas sensors. The photochemical sensor, sol-gel sensor, colorimetric sensor, photoluminescence sensor, and colorimetric sensor are just a few examples of what is commonly found.

2.2.2.3 Chemical sensors

Using surface adsorption, a chemical sensor or receptor can be used to determine whether a certain chemical or gas is present and active as well as the composition and concentration of the substance. It is being monitored whether specific compounds are present, and the presence of these chemicals is being turned into signals by a transducer. It is determined if transducers are passive/active based on the amount of exterior power required for quantity [1]. Due to the high specific surface area and outstanding electrical and mechanical properties of carbon nanomaterials, they are commonly used in chemical sensors [4]. In addition to detecting viruses, chemical pollutants, spoilage, and product manipulation, Nanotechnology-based sensors can be implemented to trace the substances or commodities throughout the manufacturing process [70–72]. The usage of optical sensors, which don't really need electrical energy and can even be checked out over a distance employing visible, infrared, or ultraviolet light, is a recent advancement in sensor technology. Silicon-based optical transducers of this type are made up of optical circuits that are incorporated into silicon semiconductor material [73].

2.2.2.4 Edible sensors
Smart food packaging may benefit greatly from edible sensors that detect deterioration in food because they use only renewable and biodegradable ingredients that have no adverse or detrimental effects on humans. As an illustration, a sensor based on a pectin matrix and a colorimetric indication of red cabbage extract has been designed. As a polysaccharide produced from apples and citrus fruit, pectin is frequently utilized in the food sector to enhance the emulsification of foods. Anthocyanins, particularly cyanidin glycoside derivatives, are abundant in the red cabbage extract. Anthocyanins are well-known pigments that can detect amines by changing color when exposed to pH variations. Gelatin and gellan gum has been used by other researchers to create an edible film that changes color from orange-red to yellow when it is exposed to the gas. Milk deterioration and fish spoiling can be detected in real-time using this edible sensor, which detects gas produced by anaerobic bacteria and changes in the color of the film due to protein degradation by microorganisms, causing a color variation.

2.2.3 Legal aspects of intelligent packing
Food contact materials are prohibited from transferring elements to food in such amounts to damage human health under the terms of Regulation (EC) No 1935/2004, as stated in Article 3. Substances that cause an undesirable alteration in the composition of a substance, as well as substances that lead to an impairment in organoleptic properties of the product Regulation No. 450/2009 of the European Commission explains that the specific substances or groups/combinations of substances that start making the active or smart constituent must be protected and must comply with all applicable necessities of regulation nos. 135/2004 and 450/2009 of the European Commission. As specified in Articles 4(d) and 11 of the European Union's No 450/2009 Regulation, active and smart substances must be marked as non-edible to prevent inadvertent eating. Additional considerations include ensuring that the packaging does not deceive the end user.

3. Conclusion
In recent years, many smart packaging technologies have emerged, which are currently being implemented into packaging technologies to fulfill the food distribution chain. By implementing effective food packaging technology, the food manufacturing industry can increase food superiority and protection while also providing consumers with information about their food. Researchers are hoping to find ways to improve the current system by conducting research into smart packaging technologies. Future smart packaging has the potential to provide consumers with convenience and benefits.

References

[1] M. Vanderroost, P. Ragaert, F. Devlieghere, B. De Meulenaer, Intelligent food packaging: the next generation, Trend. Food Sci. Technol. 39 (1) (2014) 47−62.

[2] W.M. Cheung, J.T. Leong, P. Vichare, Incorporating lean thinking and life cycle assessment to reduce environmental impacts of plastic injection moulded products, J. Clean. Product. 167 (2017) 759−775.

[3] K.L. Yam, P.T. Takhistov, J. Miltz, Intelligent packaging: concepts and applications, J. Food Sci. 70 (1) (2005) R1−R10.

[4] J. Kerry, M. O'grady, S. Hogan, Past, current and potential utilisation of active and intelligent packaging systems for meat and muscle-based products: a review, Meat Sci. 74 (1) (2006) 113−130.

[5] S. Otles, B. Yalcin, Intelligent food packaging, LogForum 4 (4) (2008) 3.

[6] J. Jung, G.M. Raghavendra, D. Kim, J. Seo, One-step synthesis of starch-silver nanoparticle solution and its application to antibacterial paper coating, Int. J. Biol. Macromole. 107 (2018) 2285−2290.

[7] J. Brockgreitens, A. Abbas, Responsive food packaging: recent progress and technological prospects, Comprehens. Rev. Food Sci. Food Safe. 15 (1) (2016) 3−15.

[8] C.N. Priyanka, D.N. Parag, Intelligent and active packaging, Int. J. Eng. Manage. Sci. 4 (4) (2013) 417−418.

[9] New approaches in smart packaging technologies, in: A. Ozcan (Ed.), Proceedings of the 10th International Symposium on Graphic Engineering and Design, GRID, 2020.

[10] M. Ozdemir, J.D. Floros, Active food packaging technologies, Crit. Rev. Food Sci. Nutrit. 44 (3) (2004) 185−193.

[11] S. Haghighi-Manesh, M.H. Azizi, Active packaging systems with emphasis on its applications in dairy products, J. Food Proc. Eng. 40 (5) (2017) e12542.

[12] S. Yildirim, B. Röcker, M.K. Pettersen, J. Nilsen-Nygaard, Z. Ayhan, R. Rutkaite, et al., Active packaging applications for food, Comprehen. Rev. Food Sci. Food Safe. 17 (1) (2018) 165−199.

[13] K.K. Gaikwad, S. Singh, A. Ajji, Moisture absorbers for food packaging applications, Environm. Chem. Lett. 17 (2) (2019) 609−628.

[14] T. Anukiruthika, P. Sethupathy, A. Wilson, K. Kashampur, J.A. Moses, C. Anandharamakrishnan, Multilayer packaging: advances in preparation techniques and emerging food applications, Comprehens. Rev. Food Sci. Food Safe. 19 (3) (2020) 1156−1186.

[15] K.K. Gaikwad, S. Singh, Y.S. Negi, Ethylene scavengers for active packaging of fresh food produce, Environm. Chem. Lett. 18 (2) (2020) 269−284.

[16] C. Vilela, M. Kurek, Z. Hayouka, B. Röcker, S. Yildirim, M.D.C. Antunes, et al., A concise guide to active agents for active food packaging, Trend. Food Sci. Technol. 80 (2018) 212−222.

[17] K.K. Gaikwad, Y.S. Lee, Current scenario of gas scavenging systems used in active packaging-A review, Korean J. Packag. Sci. Technol. 23 (2) (2017) 109−117.

[18] K. Sadeghi, Y. Lee, J. Seo, Ethylene scavenging systems in packaging of fresh produce: a review, Food Rev. Int. 37 (2) (2021) 155−176.

[19] W. Utto, Mathematical Modelling of Active Packaging Systems for Horticultural Products: A Thesis Presented in Partial Fulfilment of the Requirements for the Degree of Doctor of Philosophy in Packaging Technology at Massey University, Massey University, New Zealand, 2008.

[20] U. Siripatrawan, P. Kaewklin, Fabrication and characterization of chitosan-titanium dioxide nanocomposite film as ethylene scavenging and antimicrobial active food packaging, Food Hydrocoll. 84 (2018) 125−134.

[21] M. Dan, M. Huang, F. Liao, R. Qin, X. Liang, E. Zhang, et al., Identification of ethylene responsive miRNAs and their targets from newly harvested banana fruits using high-throughput sequencing, J. Agricult. Food Chem. 66 (40) (2018) 10628−10639.

[22] M. Sun, X. Yang, Y. Zhang, S. Wang, M.W. Wong, R. Ni, et al., Rapid and visual detection and quantitation of ethylene released from ripening fruits: the new use of Grubbs catalyst, J. Agricult. Food Chem. 67 (1) (2018) 507−513.

[23] N. Pathak, O.J. Caleb, M. Geyer, W.B. Herppich, C. Rauh, P.V. Mahajan, Photocatalytic and photochemical oxidation of ethylene: potential for storage of fresh produce—a review, Food Bioproc. Technol. 10 (6) (2017) 982−1001.

[24] H. Wei, F. Seidi, T. Zhang, Y. Jin, H. Xiao, Ethylene scavengers for the preservation of fruits and vegetables: a review, Food Chem. 337 (2021) 127750.
[25] N. Pathak, Photocatalysis and Vacuum Ultraviolet Light Photolysis as Ethylene Removal Techniques for Potential Application in Fruit Storage, 2019.
[26] L. Vermeiren, F. Devlieghere, M. van Beest, N. de Kruijf, J. Debevere, Developments in the active packaging of foods, Trend. Food Sci. Technol. 10 (3) (1999) 77–86.
[27] C.A. Phillips, Modified atmosphere packaging and its effects on the microbiological quality and safety of produce, Int. J. Food Sci. Technol. 31 (6) (1996) 463–479.
[28] K. Biji, C. Ravishankar, C. Mohan, T. Srinivasa Gopal, Smart packaging systems for food applications: a review, J. Food Sci. Technol. 52 (10) (2015) 6125–6135.
[29] C. Mohan, Recent Advances in Packaging of Fishery Products, 2021.
[30] C. Mohan, C. Ravishankar, T.S. Gopal, Active Packaging of Fishery Products: A Review, 2010.
[31] M. Sivertsvik, J.T. Rosnes, W.K. Jeksrud, Solubility and absorption rate of carbon dioxide into non-respiring foods. Part 2: raw fish fillets, J. Food Eng. 63 (4) (2004) 451–458.
[32] B.T. Rotabakk, S. Birkeland, W.K. Jeksrud, M. Sivertsvik, Effect of modified atmosphere packaging and soluble gas stabilization on the shelf life of skinless chicken breast fillets, J. Food Sci. 71 (2) (2006) S124–S131.
[33] J.W. Han, L. Ruiz-Garcia, J.P. Qian, X.T. Yang, Food packaging: a comprehensive review and future trends, Comprehens. Rev. Food Sci. Food Safe. 17 (4) (2018) 860–877.
[34] B.P. Day, Active Packaging of Food. Smart Packaging Technologies for Fast Moving Consumer Goods vol. 1, 2008.
[35] V. Eyiz, I. Tontul, Bioactive components structure and their applications in active packaging for shelf stability enhancement, in: Active Packaging for Various Food Applications, CRC Press, 2021, pp. 23–36.
[36] J. Gómez-Estaca, C. López-de-Dicastillo, P. Hernández-Muñoz, R. Catalá, R. Gavara, Advances in antioxidant active food packaging, Trend. Food Sci. Technol. 35 (1) (2014) 42–51.
[37] L.J. Bastarrachea, D.E. Wong, M.J. Roman, Z. Lin, J.M. Goddard, Act. Packag. Coat. Coat. 5 (4) (2015) 771–791.
[38] A. Sanches-Silva, D. Costa, T.G. Albuquerque, G.G. Buonocore, F. Ramos, M.C. Castilho, et al., Trends in the use of natural antioxidants in active food packaging: a review, Food Add. Contam.: Part A. 31 (3) (2014) 374–395.
[39] S. Ganiari, E. Choulitoudi, V. Oreopoulou, Edible and active films and coatings as carriers of natural antioxidants for lipid food, Trend. Food Sci. Technol. 68 (2017) 70–82.
[40] R. Irkin, O.K. Esmer, Novel food packaging systems with natural antimicrobial agents, J. Food Sci. Technol. 52 (10) (2015) 6095–6111.
[41] M. Corrales, A. Fernández, J.H. Han, Antimicrobial Packaging Systems. Innovations in Food Packaging, Elsevier, 2014, pp. 133–170.
[42] A. Guarda, J.F. Rubilar, J. Miltz, M.J. Galotto, The antimicrobial activity of microencapsulated thymol and carvacrol, Int. J. Food Microbiol. 146 (2) (2011) 144–150.
[43] V. Muriel-Galet, G. López-Carballo, R. Gavara, P. Hernández-Muñoz, Antimicrobial food packaging film based on the release of LAE from EVOH, Int. J. Food Microbiol. 157 (2) (2012) 239–244.
[44] E.A. Kandirmaz, A. Ozcan, Antibacterial effect of Ag nanoparticles into the paper coatings, Nordic Pulp & Paper Res. J. 34 (4) (2019) 507–515.
[45] A. Bhardwaj, T. Alam, N. Talwar, Recent advances in active packaging of agri-food products: a review, J. Postharvest Technol. 7 (1) (2019) 33–62.
[46] B. Kuswandi, Active and Intelligent Packaging, Safety, and Quality Controls. Fresh-Cut Fruits and Vegetables, 2020, pp. 243–294.
[47] B.P. Day, Active packaging, in: Food Packaging Technology, Blackwell Publishing Ltd p, Oxford, UK, 2003, pp. 282–302.
[48] M. Sajilata, K. Savitha, R. Singhal, V. Kanetkar, Scalping of flavors in packaged foods, Comprehen. Rev. Food Sci. Food Safe. 6 (1) (2007) 17–35.
[49] F. Soares, A.C.S. Pires, G.P. Camilloto, P. Santiago-Silva, P.J.P. Espitia, W.A. Silva, Recent patents on active packaging for food application. Recent patents on food, Nutrit. Agricult. 1 (2) (2009) 171–178.
[50] M. Regier, Microwavable Food Packaging. Innovations in Food Packaging, Elsevier, 2014, pp. 495–514.

[51] R. Ahvenainen, Novel Food Packaging Techniques, Elsevier, 2003.
[52] M.R. Perry, R.R. Lentz, Susceptors in microwave packaging, in: Development of Packaging and Products for Use in Microwave Ovens, Elsevier, 2020, pp. 261–291.
[53] M.N. O'Grady, J.P. Kerry, Smart packaging technologies and their application in conventional meat packaging systems, in: Meat Biotechnology, Springer, 2008, pp. 425–451.
[54] D. Restuccia, U.G. Spizzirri, O.I. Parisi, G. Cirillo, M. Curcio, F. Iemma, et al., New EU regulation aspects and global market of active and intelligent packaging for food industry applications, Food Contr. 21 (11) (2010) 1425–1435.
[55] C.E. Realini, B. Marcos, Active and intelligent packaging systems for a modern society, Meat Sci. 98 (3) (2014) 404–419.
[56] J.K. Heising, M. Dekker, P.V. Bartels, M. Van Boekel, Monitoring the quality of perishable foods: opportunities for intelligent packaging, Crit. Rev. Food Sci. Nutrit. 54 (5) (2014) 645–654.
[57] M. Ghaani, C.A. Cozzolino, G. Castelli, S. Farris, An overview of the intelligent packaging technologies in the food sector, Trend. Food Sci. Technol. 51 (2016) 1–11.
[58] Innovation of oxygen indicator for packaging leak detector: a review, in: A. Sari, E. Warsiki, I. Kartika (Eds.), IOP Conference Series: Earth and Environmental Science, IOP Publishing, 2021.
[59] M. Giannakourou, P. Taoukis, Application of a TTI-based distribution management system for quality optimization of frozen vegetables at the consumer end, J. Food Sci. 68 (1) (2003) 201–209.
[60] A. Mirza Alizadeh, M. Masoomian, M. Shakooie, M. Zabihzadeh Khajavi, M. Farhoodi, Trends and applications of intelligent packaging in dairy products: a review, Crit. Rev. Food Sci. Nutrit. 62 (2) (2021) 383–397.
[61] E. De Boeck, L. Jacxsens, H. Goubert, M. Uyttendaele, Ensuring food safety in food donations: case study of the Belgian donation/acceptation chain, Food Res. Int. 100 (2017) 137–149.
[62] Cold chain management in meat storage, distribution and retail: a review, in: I. Nastasijević, B. Lakićević, Z. Petrović (Eds.), IOP Conference Series: Earth and Environmental Science, IOP Publishing, 2017.
[63] N. Turkmen, S. Ozturkoglu-Budak, Novel packaging technologies in dairy products: principles and recent advances, in: Technological Developments in Food Preservation, Processing, and Storage, IGI Global, 2020, pp. 65–85.
[64] U.D. Venkatesh, O.A.A. Alsamuraaiy, Adoption of Smart Packaging: Case Study Analysis from Retailer's Perspective, 2019.
[65] L. Wang, Z. Wu, C. Cao, Technologies and fabrication of intelligent packaging for perishable products, Appl. Sci. 9 (22) (2019) 4858.
[66] M. Scampicchio, A. Arecchi, N.S. Lawrence, S. Mannino, Nylon nanofibrous membrane for mediated glucose biosensing, Sens. Actuat. B: Chem. 145 (1) (2010) 394–397.
[67] Z. Fang, Y. Zhao, R.D. Warner, S.K. Johnson, Active and intelligent packaging in meat industry, Trend. Food Sci. Technol. 61 (2017) 60–71.
[68] V.A. Pereira Jr., I.N.Q. de Arruda, R. Stefani, Active chitosan/PVA films with anthocyanins from Brassica oleraceae (Red Cabbage) as Time–Temperature Indicators for application in intelligent food packaging, Food Hydrocoll. 43 (2015) 180–188.
[69] S. Kalpana, S. Priyadarshini, M.M. Leena, J. Moses, C. Anandharamakrishnan, Intelligent packaging: trends and applications in food systems, Trend. Food Sci. Technol. 93 (2019) 145–157.
[70] K. Nachay, Analyzing Nanotechnology, Food technology, 2007.
[71] A. Balan, R.-K. Kadeppagari, Biopolymers for nano-enabled packaging of foods, J. Handbook Polym. Ceram. Nanotechnol. (2021) 839–854.
[72] Y. Liu, S. Chakrabartty, E.C. Alocilja, Fundamental building blocks for molecular biowire based forward error-correcting biosensors, Nanotechnology 18 (42) (2007) 424017.
[73] N.A. Yebo, S.P. Sree, E. Levrau, C. Detavernier, Z. Hens, J.A. Martens, et al., Selective and reversible ammonia gas detection with nanoporous film functionalized silicon photonic micro-ring resonator, Optics Exp. 20 (11) (2012) 11855–11862.

CHAPTER 4

Food safety guidelines for food packaging

Kashish Gupta
Noida International University, Greater Noida, Uttar Pradesh

1. Introduction to food packaging

As the trend of eating outside has increased, so the increase in food borne illness due to the use of toxic ingredients or due to transport delays, spoilage of food stuff. WHO has reported that about 18 lakh people die every year due to food borne diseases specially diarrhea due to contaminated drinking water or infected food [1]. WHO clearly stated that Food borne illness are serious problem in developing and developed countries, raises cost of health care systems, threat to economy of national and international development. In Rome, International conference on nutrition (1992) was held in which new right was enacted, i.e., "Access to safe food is a right of each individual" [2]. Availability of safe, consumable food is the top most priority laid by government industry and consumers. Most prone individuals belong to population group of elderly people, sick, pregnant women, and infants, especially in hygiene deficient countries. Not only this is the limit new food related hazards such as food allergy, food toxin related diseases are discovered each year which are directly associated with the contamination added during processing, packaging or spoilage of food during transport [3]. Traditional methods of packaging are carried out with the objective of the protection, marketing [4,5] So, food packaging plays a substantial role in food safety and availability. The only limitation of food packaging is that sometimes it is more costly than the food itself. Food consumption is also dependent on the customer aesthetic satisfaction for food packaging., in term of color, shape, of the packaging materials. These days, innovative methods of packaging such as active packaging and intelligent packaging are used [6]. With the changing trends, demands and awareness of manufacturer and consumers, food safety is prime concern. Smart packaging is the call of the hour and includes both active and intelligent packaging [7,8].

2. Food packaging role in food safety

In order to improvise the food safety and prevention from occurrence of any disease, certain regulations are being made. Food loss and waste has been a serious issue, if proper packaging has not occurred. Need of the hour is the search of smart and suitable packaging. Smart packaging tools are responsible for the overall quality, safety and maintain

Fig. 4.1 Smart packaging and its types.

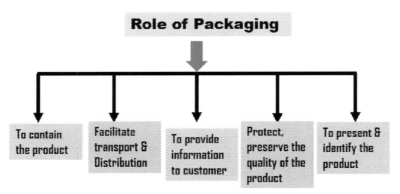

Fig. 4.2 Role of packaging.

sustainability of food during the supply chain. Figs. 4.1 and 4.2 clearly demonstrates the smart packaging types and its role respectively.

Furthermore, to strengthen the food safety regulation and curb the food risks, several international standards has been implemented. Food safety legislation worldwide International standard ISO 9000 is primarily concerned with food quality. Quality includes all the characteristics that make food acceptable to customers in terms of flavor, texture and appearance. Food contamination related illness or food borne illness has been a major concern since it puts huge pressure on the health industry and economy of the country. At various levels, food consumption related health risks are comprising such as inadequate hygiene during packaging and distribution, during the supply chain, which needs to be checked [9,10].

3. International legislation related to food safety

European countries have the best legislation related to food safety encompassing the US food legislation. Under EC food law, Food law Act 1990, makes sure that the food available is safe to eat. This act also mentions any additives harmful for human use is strictly prohibited. Regulation (EC) 178/20028 (General Food Law) clearly defined the responsibilities of the various players [11–16].

4. Food safety laws

Several food safety laws has been enacted for preventing the wide scale loss of food. Several measures taken under for curbing the deceptive practices. Law focusses on keeping stringent watch on practices which can misguide the consumer. Food safety guidelines takes into account food quality check in market for both human and animal usage. Food safety measures for import is also taken into consideration. HACCP principles need to be followed: Basically, seven principles are added under it, viz., [17,18].

(A) Identifying the hazards related to the product and its processing activities.
(B) To identify the critical points, whose regulation is essential and has critical limits better named as critical control points or CCP.
(C) Self-evaluation of CCP monitoring.
(D) Trouble shooting in case of any malfunctioning in any processing.
(E) Set of protocols and procedures are available to verify and validate the system used in processing for auditing purposes.
(F) Documentation of data records for procedures used, for better functioning, preventing malfunctioning and also traceability for quick and efficient problem solving, if any.

Europe has the simplest regulations for food but with more wholesaler and retailer responsibility, in terms of planning schemes for evaluating the suppliers for reference standards, named as, GLOBAL G. A. P (Good Agricultural practices). or TNC (Transnational corporations). Range of private standards are also available for ease in use for producers and promotes adopting the relevant approach for dual purposes., viz., food safety management (responsible approach) and other is standard approach as per the market requirements.

4.1 Food packaging legislation in India

Packaging and labeling guidelines came into force after August 5, 2011. Several key points are used under this regulation.

Best Before: It indicates the time period for which the food can be kept safely and is marketable and consumable.

Date of Manufacturing is the date when the food product is packaged.

Date of Packaging is the date when the food product is distributed in the market for selling purposes.

Infant: Child below 12 months of age.

Lot no/code age no/batch no: This is the specific code by which we can identify particular product is distributed.

Multipiece Package: packaging which comprises of two or more individually packaging or labeled pieces of same commodity which further tells whether the product is retail or whole sale.

Non-veg Food: food comprising of animal origin such as birds, eggs with the exclusion of any milk or any milk products.

Prepackaged and Packed foods: Special foods where contents are guaranteed to remove same until unless some tempering is done.

Principal display panel: Package which are displayed or presented and examined by customer.

Use by date or recommend last consumption date or expiry date which indicate the time period, during which food can be stored, and safe to consume.

Veg food: Food is vegetarian in origin.

Non-veg: Food source is animal or living source.

Deliberate attempts have been made to incorporate the substances which can release or absorb the substances which can release or absorb during the packaging production. Such materials are called as intelligent materials. List include, adhesives, ceramics, cork, rubbers, resins, textiles, During the manufacturing of these materials, it is predetermined the safety of the materials, in terms of its transfer their content to the food which is packed. Certain directives are also enlisted for that as given below.

(a) Purity standard for substances is maintained.
(b) Specific limits of substance migrated permission levels.
(c) Additional guidelines for tracing the material of packaging is also available.

5. Food contact legislation

For the first time, concept of food contact legislation was introduced in India [19]. In this regard, prevention of Food Adulteration Act. of 1954 was enacted and rules of 1955 [20]. Maximum emphasis these days are given to labeling in packaging.

6. Condition for sale and license

There are several conditions given for the sale and license of food materials in India.

(1) Each utensil used for manufacturing, preparation of any type of foodstuff for sale purposes should be maintained well in clean good hygiene conditions. Such containers are strictly restricted for such purposes.

(2) It is strictly prohibited to use the tempered containers for sale, manufacturing and storage.

(3) All container intended for food storage should be provide the tight-fitting lid or kept closed by perfect tight fitting.

(4) Utensil or container from metals are regarded as unfit for human consumption.

(5) Rusty and chipped containers are banned for use.

(6) Containers made from plastic should conform to some specifications whether for storing, packing purposes.

(7) Tin and plastic containers are strictly prohibited for its second use.

(8) Copper utensils are preferred for tinning of sugar confectionery products for human consumption.

(9) Lactic acid is only sold under Indian Standard Institution marks and not allowed to be sold free.

Certain IS specification for food stuffs are enlisted below:

IS:20 specifications are also mentioned regarding use of cast aluminum and AL based alloys for containers.

IS:20 specifies the use of wrought aluminum and based alloys as utensils.

IS:1046 specification mentions clauses for polyethylene use for foodstuff.

IS:10142 Specification mentions cluses for use of styrene polymers as contact materials.

IS:10151 gives specifications for the PVC food contact material.

IS:10910 mentions specifications for the polypropylene for foodstuff.

IS:11434 gives specifications for the ionomer resins for food stuff.

IS:11704 gives specifications for ethylene acrylic acid copolymer.

IS:12252 mentions specifications for polyalkylene terephathalates (PET);

IS:12247 Specifications for Nylon 6 Polymer;

IS:13601 Specifications for ethylene vinyl acetate (EVA);

IS:13576 Gives Specifications for ethylene methacrylic acid (EMAA).

For asafoetida packaging limit of weight packaging is 1 kg with label is only permitted under rule no 42.

Food grade titanium dioxide is only allowed under the Indian standard institution certification mark.

Bakery products used should be in conformation under Rule no 42 (T).

Edible common salt or any iodized salt/fortified salt containing anticaking agent are only sold in package and labeled condition under sub rule of Rule 42 (V, 10 A). Moreover, iron fortified common salt is only sold in the high-density polyethylene bag (HDPE) (14 mesh, density 100 kg/m3, unlaminated) packaging bearing the label as per specified in sub-rule (VV) of Rule 42.

All edible oils are only permitted for consumption which are under the label declaration in rule 42 (W). **Blended edible vegetable oils** are not allowed in loose form, only sealed packaging for maximum weight 5 kg is permitted. Bearing the AGMARK certification mark including the label declarations as provided under Rule 42 and Rule 44 are essential.

Dried glucose syrup with sulfur dioxide limit greater than optimum limit 40 ppm can't be sold open as per as the sub rule (X) of Rule 42.

Sale and purchase of insecticides are prohibited in the area close to food storage, packaging area under the Insecticide Act, 1968 m (46 of 1968).

Mineral oil of food grade is only sold under Indian Standards Institution certification mark.

Confectionary items more than 500 gm are only sold in packed condition and confectionery sold in pieces can be provided in glass and other suitable material container.

Sugar boiled confectionery such as chewing and bubble gums, lozenges are covered under sub rules 17 and 18.

Milk products such as condensed milk, skimmed milk is only sold under Indian Standards Institution certification mark.

Infant milk is denoted by label "MOTHERS MILK IS BEST FOR THE BABY".

Infant milk food and infant formula are not allowed for sell., manufacture, sell store except under bureau of Indian standards Institution certification mark.

Protein rich atta and Maida are only sold under labeled packaging including the ingredients of the packaging.

6.1 Packing and labeling of foods

Primary condition for a packaged product is to carry a label on it and each label should specify name, trade name, description of food and any ingredient. Moreover, the ingredients used in the product should have label and ingredients should be mentioned in descending order composition by weight or volume. In case the food contains any flavoring agent, it should be mentioned whether natural or artificial flavor. Similarly, if gelatine is included, it should be mentioned in label as animal origin.

Declaration is available for this by name of symbol and color code to denote the veg an d non veg status of food product. Brown color filled circle denotes the specifications for non veg food under the clause (16) of sub-rule (ZZZ) of Rule 42.

Symbol is prominently displayed on the package on a contrasting background.

In case certain flavoring agents are used, in accordance to tule 24 and Rule 64 BB, labels are mentioned. Such as, contains permitted natural colors.

Contains permitted synthetic flavoring color/flavor.

Contains permitted added flavor/color.

Contains permitted natural, synthetic, added flavoring color or flavor.

Products containing birds or fresh water or marine animals' egg, declaration indicates non veg food. Its denotation is given by circle containing single chord passing through its center on top left-hand side extending to the right diagonally (Table 4.1).

7. Packaging label requirements of oils

Edible oils are denoted by expressions such as Super refined, Extra Refined, Macro-Refined, Double-Refined, Ultra-Refined, Anti-cholesterol, Cholesterol Fighter, Soothing to Heart, Cholesterol Friendly, Saturated Fat Free.

Table 4.1 Food safety laws and directives.

S.No	Directive No	Purpose
1.	Directive 79/112/EEC	Labeling, presentation and advertising of foodstuff
2.	Directive 85/374/EEC	Principle of producers' liability for defective products to be made obligatory for primary agricultural products production
3.	Directive 89/109	Framework directive in food packaging
4.	Directive 92/59/EEC	Rapid alert system for immediate risks to health and safety of consumers
5.	Directive 92/59/EEC	General product safety
6.	Directive 93/43/EEC	On the hygiene of foodstuff
7.	Directive 97/258/EC	Novel foods
8.	Com (97) 176 final	General principal of food law in the European Union s(commission green paper)
9.	Com from the commission	Consumer health and food safety

Indian Standards for Direct Food Contact Materials: Bureau of Indian Standards BIS standards 9833-1981are applied to permitted pigments ad colorants for use with plastics, which is directly in contact with food packaged, also pharmaceuticals and also drinking water.

Specific and overall migration limits are also mentioned for the heat sellable films, non-sealable films, containers and closures for sealing as lid, in accordance to the BIS standards 9845:1998.

BIS Standard 10171:1986 mentions the guidelines on suitability of plastics for food.

BIS Standard 10146:1982 — Specifies the guidelines for the following plastic resins polyethylene, polypropylenes, ionomers, acid copolymers, nylon, polystyrene, polyester, EVA.

8. Packaging labeling regulations in different countries

Integrated approach for food safety is adopted by EU which aims at full assurance of high level of food safety, animal health, animal welfare under the Directive 90/496/EEC, the Health and Consumer Protection Directorate General Services (January 2004). In continuation of these act, WHO released the consultation document on nutrition labeling of food-stuffs which contains targeted considerations. Furthermore, the food Supplements Directive 2002/46/EC came in July 2002. Several Directives and relevant regulations executed from August 1, 2005. One of the provisions of this Directive also includes lists of nutrients and their sources that can be included in food supplements.

8.1 Active packaging and intelligent packaging

EFSA has described intelligent packaging as 'materials and articles that monitor the condition of packaged food or the environment surrounding food. Materials can sense the he environmental conditions of the packaging but show inertness toward the food. Basically, it helps in monitoring the product and acknowledging the updated info to the consumers. Information regarding the condition of Package, time of manufacturing and storage conditions are also delivered. It can be used as a part of the primary, secondary or tertiary depending on the type of packaging [21–25]. Commercial application of intelligent packaging requires the legal criteria to be fulfilled. Stricter laws in EU prevents easy introduction of any food packaging norms to be passed from outside even America. Certain rules on good manufacturing practices for materials are granted permission for interaction with food as per Regulation EC No 2023/2006. Another in line regulation No 450/2009 is concerned with the approval and mandatory requirements for any active and intelligent materials for food. Intelligent packaging is primarily divided into three sub types.

8.2 Carriers, sensors and indicators

Active packaging extends the shelf life of the food while maintaining its quality. Intelligent packaging acts on indicate, monitor the freshness the status of food in packaged products. These include all those methods, which are unique way to package the product, providing the quick, cheap and efficient methodology for monitoring the status of the environ of the packaged food during the transport and its distribution [26–28]. Smart packaging methods prevents waste of food and maintains the quality of food. Packaging demands directly depends on the consumer interest and packaging manufacturer user interface for both the product purchase and product utility. Consumer convenience plays important role in areas in terms of storage, quality and ease of disposal, in day-to-day life. Mainly there are six stages of consumer experience at level of any customer product, simplicity, convenience, safety and risk, fun and image, environmentally friendly. These parameters provides the consumer saving of time, save me and support me are the key points s using during the smart packaging, user convenience is the top most priority. Hectic lifestyle led to change in consumer demands in food packaging is the light weight, portability, ease to operate and reusable.

8.3 Legislation related to smart packaging

Article 3 of EC 1935/2004 verifies whether the food contact material should not transfer any component in the packaged food, especially dangerous to human health. Materials which can bring unusual/unacceptable changes the food quality in terms of organoleptic properties are strictly prohibited.

Another commission regulation No 450/2009 clearly states that the food contact material used as a part of active and intelligent packaging should be safe for humans and also in compliance with the framework regulation No 135/2004 and regulation no 450/2009.

Another in series, articles 4 (d) and 11 of EC no 450/2009 states the all materials under the smart packaging should be labeled for customer use and should be non-edible, so that no accidental consumption happens.

8.4 Food safety regulation in active packaging

Frame work regulation of food contact materials 1935/2004 and 450/2009, 2020 under EC.

8.5 Nanotechnology in food packaging and its regulation

Nanotechnology intends to extent the shelf life of any food product by sustaining release of antimicrobials, enzymes, flavors, nutraceuticals and antioxidants [26]. Major scope is in the agro-food industry [27]. Kuswandi et al., 2011, reported the leading food packaging based on smart packaging is able to sense the environment and self-repair holes and tears. Since biobased materials has certain limitations (poor barrier and mechanical properties) [28], so nanocomposites are used to compensate for its limitation and enhances the usability of edible and biodegradable films and also extend the shelf life [29–36] Nearly 500 nano-packaging products are in commercial use and also rising at insurmountable pace [28,37,38].

8.6 Food safety regulation in intelligent packaging

Frame work regulation on food contact materials 1935/2004 and 450/2009, 2020. States to control the state of packaged foods, and the surrounding environ.

8.7 Future trends and scope in smart packaging

Latest trend is the use of smart packaging, smart labels, addition of more natural additives, minimal processing of food, down gauging of food without compromising the functionality and also use of edible films. Sustainability approach in food packaging is also focused on. Nanomaterials in food and beverage industry. Although a new technology but increasing at insurmountable pace [34]. More than 400 companies around the world are focused on research toward use of the nanotechnology in food packaging [39]. Although in India and other countries, norms, food safety has been made, execute, still the legislation is at its nascent stage. Need of strong compliance to the food legislation is needed. Undoubtedly, all HACCP and BRC practices are adhered with the food packaging, supply and its distribution. More efforts should be made for batch traceability aspect for food contact legislation. Acceptability criteria for the Food laws are dependent on the man power skilled and awareness with in the government bodies (FDA). Limiting factor becomes the lack of awareness at the level of packaging experts and also the limited audits performed by them. Higher attention and focus are laid to the food labeling and its ingredient demonstration as compared with the food packaging and its labeling. More stringent rules and regulation are the need of the hour for the food contact legislation is on top priority. Use of certain bio-based plastics has ruled the market because of its environmentally friendly nature and biocompatibility. Sustainable packaging that comprises of biodegradable materials for food packaging will be the popular global trend.

Future, better monitoring of contaminated package, sophisticated product tracking and efficient method of tamper evidence. Also, consideration of impact of technology used in packaging in the environment. Although among all packaging material, plastics have been the relatively cheap, light in weight.

References

[1] https://apps.who.int/iris/handle/10665/43592 https://apps.who.int/iris/bitstream/handle/10665/175942/WHA46_6_eng.pdf?sequence=1&isAllowed=y.

[2] T. Caon, S.M. Martelli, F.M. Fakhouri, New trends in the food industry: application of nanosensors in food packaging, in: Nanobiosensors, Academic Press, 2017.

[3] A.L. Brody, B. Bugusu, J.H. Han, C.K. Sand, T.H. McHugh, Innovative food packaging solutions, J. Food Sci. 73 (2008).

[4] C.N. Priyanka, D.N. Parag, Intelligent and active packaging, Int. J. Eng. Manage. Sci. 4 (4) (2013) 417−418. Biji et al., 2000).

[5] K.B. Biji, C.N. Ravishankar, C.O. Mohan, T.K. Srinivasa Gopal, Smart packaging systems for food applications: a review, J. Food Sci. Technol. 52 (10) (2015) 6125−6135, https://doi.org/10.1007/s13197-015-1766-7.

[6] J.M. Lagaron, R. Catalá, R. Gavara, Structural characteristics defining high barrier properties in polymeric materials, Mater. Sci. Technol. 20 (1) (2004) 1−7, https://doi.org/10.1179/026708304225010442.

[7] R. Ahvenainen, R. Boca, Novel Food Packaging Techniques, CRC Press, 2003, p. 1.

[8] S. Boqvist, K. Söderqvist, I. Vågsholm, Food safety challenges and one health within Europe, Acta. Vet. Scand. 60 (2018) 1, https://doi.org/10.1186/s13028-017-0355-.

[9] T. Maberry, A Look Back at 2018 Food Recalls. Food Safety Magazine, ENewsletter, February 19, 2019. Available online at: https://www.foodsafetymagazine.com/enewsletter/a-look-back-at-2018-food-recalls-outbreaks/ (accessed July 4, 2019).

[10] https://ec.europa.eu/food/index_en.

[11] https://libguides.lib.msu.edu/c.php?g=212831&p=1411064.

[12] M. Broberg, European Food Safety Regulation and the Developing Countries Regulatory Problems and Possibilities, 2009.

[13] https://leap.unep.org/countries/eu/national-legislation/regulation-ec-no-1782002-european-parliament-and-council-laying.

[14] https://eur-lex.europa.eu/LexUriServ/LexUriServ.do?uri=CONSLEG:2002R0178:20080325:en:PDF.

[15] https://www.fao.org/food-safety/food-control-systems/policy-and-legal-frameworks/codex-alimentarius/en/.

[16] https://www.fda.gov/food/international-interagency-coordination/international-cooperation-food-safety.

[17] https://www.ag.ndsu.edu/foodlaw/overview/history/milestones.

[18] https://fssai.gov.in/cms/food-safety-and-standards-act-2006.php#:~:text=It%20is%20an%20Act%20to,and%20wholesome%20food%20for%20human.

[19] https://www.corpseed.com/knowledge-centre/food-laws-and-regulations-in-india.

[20] M.A. Pascall, J.H. Han (Eds.), Packaging for Nonthermal Processing of Food, second ed., John Wiley & Sons Ltd, 2018. Published 2018 by John Wiley & Sons Ltd.

[21] 10. Unilever Reducing Food Loss and Waste, 2019. Available online at: https://www.unilever.com/sustainable-living/reducing-environmental-impact/waste-andpackaging/reducing-food-loss-and-waste/ (accessed March 1, 2019).

[22] World Health Organization. Available online at: https://apps.who.int/iris/bitstream/handle/10665/193736/9789241509763_eng.pdf?sequence=1 (accessed Oct 14, 2019).
[23] B. Lipinski, C. Hanson, J. Lomax, L. Kitinoja, R. Waite, T. Searchinger, Reducing Food Loss and Waste. Working Paper, Installment 2 of Creating a Sustainable Food Future, World Resources Institute, Washington, DC, 2013. Available online at: http://www.worldresourcesreport.org.
[24] S. Lebersorger, F. Schneider, Food loss rates at the food retail, influencing factors and reasons as a basis for waste prevention measures, Waste Manag. 34 (2014) 1911–1919, https://doi.org/10.1016/j.wasman.2014.06.013.
[25] A. Fanar Hamad, J.-H. Han, B.-C. Kim, A. Irfan. Rather, the Intertwine of Nanotechnology with the Food Industry.
[26] Saudi Journal of Biological Sciences 25 (Issue 1) (2018) 27–30, https://doi.org/10.1016/j.sjbs.2017.09.004. ISSN 1319-562X.
[27] https://www.sciencedirect.com/science/article/pii/S1319562X17302267).
[28] C.E. Realini, B. Marcos, Active and intelligent packaging systems for a modern society, Meat Sci. 98 (2014) 404–419.
[29] A. Lamba, V. Garg, Recent innovations in food packaging: a review, Int. J. Food Sci. Nutr. Int. 4 (2019) 123–129.
[30] D. Videira-Quintela, F. Guillén, O. Martin, G. Montalvo, Antibacterial LDPE films for food packaging application filled with metal-fumed silica dual-side fillers, Food Packag. Shelf Life 31 (2022) 100772.
[31] N. Dasgupta, S. Ranjan, M. Deepa, C. Ramalingam, R. Shanker, A. Kumar, Nanotechnology in agro-food: from field to plate, Food Res. Int. 69 381 (2015) 400, https://doi.org/10.1016/j.foodres.2015.01.005. ISSN 0963-9969.
[32] B. Kuswandi, Nanotechnology in Food Packaging, 2016, https://doi.org/10.1007/978-3-319-39303-2_6.
[33] S. Alfadul, A.A. Elneshwy, Use of nanotechnology in food processing, packaging and safety review, Afr. J. Food, Agric., Nutriti. Develop. 10 (6) (2010) 10, https://doi.org/10.4314/ajfand.v10i6.58068 (ISSN: 1684-5358).
[34] T. Singh, S. Shukla, K. Pradeep, W. Verinder, K. Bajpai Vivek, A. Rather Irfan, Application of Nanotechnology in Food Science: Perception and Overview Frontiers in Microbiology =8, 2017. https://www.frontiersin.org/article/10.3389/fmicb.2017.01501.
[35] S.D.F. Mihindukulasuriya, L.-T. Lim, Nanotechnology development in food packaging: a review, Trend. Food Sci. Technol. 40 (Issue 2) (2014) 149–167, https://doi.org/10.1016/j.tifs.2014.09.009. ISSN 0924-2244.
[36] S.K. Ameta, A.K. Rai, D. Hiran, R. Ameta, S.C. Ameta, Use of nanomaterials in food science, Biogen. Nano-Part. Their Use Agro-Ecosyst. (2020) 457–488, https://doi.org/10.1007/978-981-15-2985-6_24.
[37] S. Neethirajan, D.S. Jayas, Nanotechnology for the food and bioprocessing industries, Food Bioprocess. Technol. 4 (2011) 39–47.
[38] J.K. Momin, C. Jayakumar, J.B. Prajapati, Potential of nanotechnology in functional foods, Emir J. Food Agricult. 25 (1) (2012) 10.
[39] S. Neethirajan, D.S. Jayas, Nanotechnology for the food and bioprocessing industries, Food Bioproc. Technol. 4 (2011) 39–47, https://doi.org/10.1007/s11947-010-0328-2.

CHAPTER 5

Industrial barriers for the application of active and intelligent packaging

Partha Pratim Sarma[a], Kailash Barman[b] and Pranjal K. Baruah[a]
[a]Department of Applied Sciences, GUIST, Gauhati University, Guwahati, Assam, India; [b]Department of Food Engineering and Technology, Central Institute of Technology, Kokrajhar, Assam, India

1. Introduction

Packaging may be considered as a procedure that imparts an informative and protective covering to the product, protects the product during storage, material handling, and movement, and also contains pertinent information about the content of the package. Different packaging techniques are used in food packaging such as paper and carton packaging, film packaging, foam packaging, textile packaging, plastic boxes, containers, etc. However, different types of packing materials used in food packaging systems are glass, metals, rubbers, plastics, fibrous materials, foils, films and laminates, blister packs, textile, etc. [1]. The basic functions that packaging aims for are protection and promotion of the product. The more deftly the product is packaged the more efficient is its individuality and identity. However, the packaging itself can act as a medium for sales promotion. In most cases, packaging acts as a vehicle that carried out the brand of the product to the consumer [2]. That is why food packaging industries spent a huge amount of money on branding and packaging, as this can help to sell the product which looks more attractive to consumers. Trademarks are used in packaging to promote and sell the products.

Maintaining the quality of the food, extension in shelf life and enhancing safety are the key factors for food industries [1]. Consumer attention to health and environmental awareness, food safety, and the content of additives used in food products have increased significantly. Consumers are increasingly attracted to unprocessed, natural food products of high standards with long shelf life without the use of preservatives [3]. Shelf life is the time period for which the food product can be stored without affecting the eating quality, considering the organoleptic properties and safety concerns [4]. Packaging plays a vital role in storage, transportation, and ensuring food product quality and safety. The shelf life of food depends upon various intrinsic and extrinsic factors [5]. Intrinsic factors involve moisture content, nutrient content, pH, biological structure, respiratory rate, etc. While extrinsic factors involve relative humidity, storage temperature, surrounding gas composition, light, dust, pests, etc. [1,6]. The traditional food packaging system acts as a passive barrier to these extrinsic factors and helps in prolonging the shelf life of food [7,8]. It also helps to minimize food spoilage and waste and helps in reducing the conservation cost. The oxidation process, enzymatic activity, microbial activity, different types of deleterious chemical reactions, moisture content, temperature, etc. Are the key factors that disrupt the natural properties of food products. In addition, storage condition, and transportation can often affects the quality of food. Active and intelligent packaging technologies emerged as an excellent tool for solving these problems [1,3]. In the field of food packaging, active and intelligent packaging has taken a distinct place as an alternative to conventional packaging around the world but has lagged behind in its expected promotion, dissemination, and successful implementation on a commercial and industrial basis. However, the lack of knowledge and promotion about the useful features of smart packaging systems among consumers has hampered the application and adoption of this promising technology by consumers in daily life.

This chapter makes special emphasis on the numerous factors that impede the successful and widespread industrial application and market implementation of the active and intelligent packaging system. A brief overview is also made on the role of smart packaging in maintaining the quality standards of food products. However, along with the differences between active and intelligent packaging, the usefulness and various important components of active and intelligent packaging are also discussed. Furthermore, an overview of the traditional packaging systems and their advantages and disadvantages is also presented.

2. Need of packaging

The prime objective of food packaging is to serve as a carrier for food and protection from external contamination; inhibition and protection from chemical, mechanical and biological changes; microbial spoilage and to defend from manipulation and theft [3,4]. Thus the key features of food packaging can be summarized as protection, containment, convenience, economical, and communicative [9,10]. In most cases, food packaging contains all the essential information about the product such as nutritional content, product composition, volume of the substance, color, expiration date, manufacturing date, tracking, storage recommendations, recycling or disposal, mode of transport, etc which makes it to serve as an efficient media to communicate with the customer [11–13]. Thus we can conclude that the main objectives of packaging are barrier protection from the extrinsic factors, information transmission, marketing, security, convenience, agglomeration or containment which generally includes a grouping of small items together in a single package for easy and efficient handling. Generally, powders, granular and liquid materials required containment. Factors that can include convenience in storage, distribution, operation, sale, opening, closing, and handling can be added to the packaging [12].

3. Traditional packaging

In the ancient days, packaging was mainly used to hold and store edible items for a short course of time. During prehistoric times various natural materials like shells, leaves, animal skins, hollowed logs, gourds, etc. Were used to hold and package the foods. Earthenware was the first and enormous invention in the Neolithic age as synthetic packaging materials because of their durability, anti-worm-eaten and antiseptic properties. Packaging materials like leaves, bamboo, etc. Also influence the smell and flavor of the product. But nowadays things are changing drastically [14]. With the evolution of men's lifestyle, packaging techniques also evolved. The development of trade concepts led to the development of boxes, wooden crafts, clay pots, barrels, stem baskets, fabrics, ceramics, etc. For long time storage and transportation of liquids and dry products. Packaging materials may vary with the process of industrialization, geographical location, economy, urbanization, preferences, and users.

The rapid industrial revolution brings out tremendous changes in food packaging techniques. Globalization, market internationalization, quality control, consumer health concerns, growing interest in environmental safety, and hygiene of food products with prolonged shelf life are the key factors for the modernization and renovation of the packaging concepts [15,16]. Modern packaging techniques play a vital role in maintaining food quality for a large period. The natural and hand-made approaches in food packaging are replaced by numerous machine-made materials like metal cans, glass, cardboard boxes, aluminum foils, etc. But the arrival of polyethylene and plastics in the 19th century changes the whole scenario of the food packaging techniques and become the principal packaging material for storage and transportation of both liquid and solid products.

The key feature of the traditional packing material is its inertness. Traditional packing materials are usually made up of inert materials that act as a passive barrier specially designed to manipulate the deleterious effects of various extrinsic factors of the environment on the packaged product [17,18]. However, an ideal packaging material is expected to bear some special attributes like mechanical, thermal, and easy transportation features, and to obtain such features different types of substances like polymer are incorporated into the main matrix [1]. Traditional packing material does not communicate with the content packaged. The primary function of this packaging material is to serve as a protective and transportation tool [19].

Polyethylene, polypropylene, polyamide, ethylene, terephthalate, vinyl alcohol, polystyrene, and polyvinylchloride are the different types of synthetic polymers used drastically for the synthesis of food packing materials due to their useful functional attributes and physicochemical properties like barrier property, pliability, tensile strength, mechanical robustness, etc [20,21]. However, environmental pollution because of these materials is of huge concern. The extended use of these polymeric materials in packaged processed foods produces an enormous amount of waste materials. These can persist on the earth's top layer for an extended period of time which is threatening to ecological balance and invokes adverse environmental consequences because the degradability and recycling are a matter of great concern for these types of materials [21,22].

These Environmental concerns have imposed a ban on these materials globally and led to enhance the interest of the researchers in the development of improved packaging concepts with sustainable, more biodegradable, and environment friendly materials [22–24].

4. Advantages and disadvantages of different types of packing materials [12]

Slno	Materials	Advantages	Disadvantages
1	Glass	(1) Chemically inert in nature and inorganic material (2) Transparent and can be given any shape without harming the mechanical properties. (3) High thermal stability. (4) Possess admirable barrier properties to water vapor and gases. (5) Non-polluting and recyclable for many times.	(1) Manufacturing process of glass requires high temperature. (2) Heavy and breakable
2	Metals	(1) Possess admirable barrier properties toward moisture and gas. (2) Recyclable. (3) Metals have high thermal stability and can be sterilized.	(1) Possess an energy intensive manufacturing process. (2) Comparatively expansive.
3	Paper and cardboard	(1) Biodegradable and recyclable. (2) Light weight. (3) Possess excellent printing properties.	(1) Possess poor barrier properties for moisture, air and gases. (2) Low tensile strength.

		(4) Can be burned without effecting the environment.	
4	Synthetic plastic	(1) Manufacturing process is cost-effective. (2) Have good barrier properties for water vapor, air, and gases. (3) Elastic, lightweight, transparent, high tensile strength, aesthetic.	(1) Non biodegradable. (2) Causes serious environmental consequences.
5	Biodegradable plastics	(1) Biodegradable polymers are of two types. One of them undergoes enzymatic degradation by the action of microorganisms. Other type is made up of synthetic polymer incorporated with edible materials like starch. During degradation, only edible parts undergo decomposition to gi various products like carbon dioxide, water, and oxygen in aerobic conditions and give methane and water in anaerobic circumstances. (2) Non-polluting.	Relatively expensive.
6	Laminates	(1) These types of packing materials contain multiple thin layers of metal, papers, wood plastic	Difficult to recycle and reuse.

Continued

7	Active and intelligent packaging	films, paperboards, etc. That possess admirable barrier properties for water vapor, air, and gases. (2) Each thin layer possesses different properties hence it can protect food materials for a longer period of time. (3) High tensile strength. (1) These types of materials possess admirable barrier properties for both the intrinsic and extrinsic factors of the environment. (2) In many cases these materials can interact with the packaged materials and extend the shelf-life of the food materials. Generally, ethylene, moisture or oxygen scavengers, carbon dioxide emitters, antioxidants, and antimicrobial agents are used in active and intelligent packaging.	(1) Relatively expensive (2) Difficulty in the large scale production.
8	Nanocomposites	(1) Possess superior mechanical strength as well as barrier properties for gases, air, and moisture. (2) Can be reused.	Relatively expensive

5. Smart packaging

The extent of smart packaging is wide that includes both active and intelligent packaging which keeps an eye on the external and internal transformations that arise in a product and further reacts corresponding to it [19,25]. Besides the key features of packaging like protection, containment, convenience, and marketing, smart packaging is also characterized by emerging active functions, which grant packaging to perform a vital role during the storage and processing of foods [26]. A smart packaging system can sense, inform, react and reduce the undesirable internal and external changes that take place in the food products [27]. Thus smart packaging can be regarded as the complete package that brings all the solutions for food packaging which senses and observe the transformations in the product and its surrounding environment (intelligent packaging) as well as react to these changes (active packaging) [19]. Based on the functional features, smart packaging can be classified into two categories-*active packaging* and *intelligent packaging*.

The key features of smart packaging are discussed below.

1. Prolong the shelf life of the food product without disturbing the integrity and quality of the food matrix, thereby reducing food spoilage.
2. Enhance the quality like flavor, appearance, texture, and taste.
3. Readily react in response to any transformation in packaging product and environment.
4. Bring all the essential information about the product to the consumers easily.
5. Increased sustainability.
6. Assure genuineness of the product, Anti-counterfeiting and Improve compliance, authentication, traceability thereby increasing product safety.

6. Active packaging

An active packaging system includes the incorporation of different compounds in packaging which are able to absorb or release substances from/into the food matrix to avoid food spoilage. Active substances are capable of inhibiting microbial growth, controlling the respiration rate, delaying the ripening process and migration of gases, moisture, etc. Which enhances the freshness, safety, and quality of the food and prolongs the shelf-life of the food as well [13,28,29]. Active packaging not only provides admirable barrier properties but also act as a packaging system that executes various active functional properties to elevate or sustain the quality and safety of the food matrix. It acts as a leading substitute for the traditional packaging system [30]. Active packaging systems are usually classified as active emitter or releasing system which are capable of incorporating active compounds into the packaged food or its surrounding environment and active scavengers or absorbers which can absorb or remove undesirable components from the food matrix or its surrounding environment [31]. Functional characteristics of the active packaging system are mainly based on the physicochemical properties of the active compound and the polymer material utilized in packaging [13]. Active biopolymers can be regarded as precious in the field of the active packaging system. The benefit of using biopolymer in packaging techniques is that industrial waste can be transformed into value-added active elements thereby enhancing sustainability and reducing the waste [32]. Many of the waste products are enriched with proteins, lipids, polysaccharides, and other active compounds like antioxidants, antimicrobials, etc. Which make them useful source of the active ingredients [33]. Certain nanocomposites are also used in active packaging as a barrier to different gases like oxygen, carbon dioxide, and moisture. Active packaging is a specially designed packaging system which involves the incorporation of various active ingredients possessing the ability to interact with the food items or with the internal vicinity of the packaging in a positive way by emitting or absorbing different types of compounds [34]. Thus active packaging is a system where the package, product, and its surrounding environment collaborate in a beneficial way to achieve prolonged shelf-life and enhance product quality.

Based on the functions of the active compounds used in the packaging, active packaging may be categorized into various classes as shown below (Fig. 5.1).

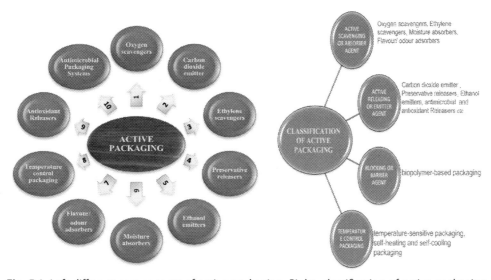

Fig. 5.1 Left-different components of active packaging. Right- classification of active packaging.

6.1 Oxygen scavenger

The existence of oxygen inside the packaged food accelerates the oxidation of vitamins, fats, etc which results in rapid deterioration of food, and can inhibit the growth of aerobic microorganisms which may lead to change in taste, odor, and color of food [28,35]. The prime role of oxygen scavenger is to reduce or remove the oxygen present inside the package by reacting with it. Various methods such as nitrogen gas filling, hot filling, vacuum packaging etc [36] are employed for the removal of oxygen level in packaged food but the results are not satisfactory. The most familiar oxygen scavenger system is the iron-based oxygen scavenging system. Oxygen scavengers are usually applied in case of cooked and dried meat products, bakery products, vegetables, fruit juices, nuts, cheese, fried snacks, tea, coffee, oils, etc. Which is useful in the inhibition of mole growth, maintaining vitamin C content, prohibit decolorization, browning, and rancidity [3]. The mechanism of action of the iron-based scavenger involves the complete oxidation of iron to ferric oxide trihydrate complex. Nano iron can show oxygen scavenging properties both in anhydrous or in the presence of moisture [37]. The iron powder however can be incorporated into the packaging material. Foltynowicz et al. evaluated the oxygen scavenging property of nanoscale zero valent iron particles at dry condition and at 100% relative humidity. For the analysis of oxygen absorption capacity iron nanoparticles were blended with a silicone matrix and reported that iron nanoparticles dispersed in silicon matrix show oxygen absorption rate at about ten times higher in comparison with commercially available oxygen scavenger based on iron blended in a polypropylene or polyethylene polymer matrix [38]. Palladium and platinum because of their low toxicity can also be used as oxygen scavengers as both the metal act as a catalyst in the conversion of oxygen and hydrogen into water [39]. Unsaturated hydrocarbon can also be used as an oxygen scavenger system. For example 1,4-polybutadiene in presence of the catalyst cobalt neodecanoate acts as an oxygen scavenger system [40]. Other non metal based scavenger systems involved the use of ascorbic acid, ascorbate salts or catechus, gallic acid, salicylic acid, unsaturated fatty acids, and enzymes like glucose oxidase, laccase, oxalate oxidase etc [41]. The scavenging mechanism of ascorbic acid involved the oxidation of ascorbic acid to dehydroascorbic acid [42]. Mahieu et al. synthesizes a multilayer film with the core layer consisting of thermoplastic starch and the skin layer consists of poly(ϵ-caprolactone) (PCL). 1.5% w/w iron powder and 15% w/w ascorbic acid were incorporated as active oxygen scavenging agents and reported that the multilayer film shows active oxygen scavenging properties however, the active film without PCL layers show faster oxygen scavenging rate as compared to the multilayered film [43].

6.2 Carbon dioxide emitter

A definite concentration of CO_2 in food packages is useful for the extension of shelf life thereby inhibiting the growth of microorganism. The higher the solubility of CO_2 on food higher will be its antimicrobial effect [44]. CO_2 is also used to stop the oxidation reaction taking place in food. Along with N_2, it is used in antioxidant food packaging [45]. However, excess amount of CO_2 can affect the quality and integrity of the food martrix. The reduction of the concentration of CO_2 produced because of fermentation, respiration, and roasting of food product in gaseous form can be accomplished by physical absorption, chemical absorption, and membrane separation [13]. Carbon dioxide scavengers are usually employed in fresh vegetables, fish, meat, cheese, coffee, baked goods, crisps, nuts, etc. Alkaline salt solutions are generally used as chemical absorbers. Among these Calcium Hydroxide $Ca(OH)_2$ is the most prominent which acts according to the following reaction [46].

$$Ca(OH)_2 + CO_2 \rightarrow CaCO_3 + H_2O$$

Sodium carbonate (Na_2CO_3) is another class of alkaline salt that is used as a CO_2 scavenger in presence of water which acts by the following reaction- [44].

$$Na_2CO_3 + CO_2 + H_2O \rightarrow 2NaHCO_3$$

Amino acid salt solutions can also serve as carbon dioxide scavengers. Among these sodium glycinate which is used to enhance the flavor of processed food is the most prominent one [47]. However, physical absorbers involved zeolite, activated carbon, and silica [48]. These carbon dioxide scavengers can be used by incorporating on the film used for packaging. The absorber materials in the form of pellets or beads keeping inside a bag are placed in a rigid package. CO_2 scavengers are widely used in the case of roasted coffee as it produces a large amount of CO_2. Hence if immediately sealed the roasted coffee sloe release of the CO_2 gas can create pressure inside the package which cause the bursting of packages [36].

6.3 Ethylene scavengers

Ethylene a colorless and odorless alkene is a multifunctional phytohormone that influences plant growth, seed germination, root growth and also increase the rate of respiration thereby accelerating the senescence and ripening process of food [49]. However, even at a lower concentration, it can cause yellowing of green vegetables and fruits which leads to a decrease in shelf life and affect the flavor and nutrients of the food product [13]. Thus to increase the shelf life and to maintain the organoleptic properties of such food products, removal or reduction of ethylene is very significant. Ethylene scavengers are usually employed for climacteric fruit, vegetables, and different types of horticultural products. Usually, ethylene absorbers, ethylene oxidizing agents, nanoparticles, etc. Are used as ethylene scavenger system [50]. Among which potassium permanganate is most prominent. Ethylene is very prone to oxidation reaction. The scavenging property of $KMnO_4$ is based upon oxidation reaction as represented below- [51].

$$KMnO_4 + C_2H_4 \rightarrow MnO_2 + KOH + CO_2 + H_2O$$

The reactivity of $KMnO_4$ is increased by immerging it with SiO_2 or Al_2O_3. Nanoparticles such as Ag, Cu, Pd, ZnO, TiO_2, etc, can also act as ethylene scavengers through oxidation of ethylene to carbon dioxide and water via photocatalytic reaction [52]. Spricigo et al. impregnated silica (SiO_2) and alumina (Al_2O_3) nanoparticles with potassium permanganate ($KMnO_4$) to indicate ethylene removal through a color change which is able to scavenging ethylene more efficiently for 1 h exposure time. Reduction of potassium permanganate acts as the indicator of ethylene removal [53].

6.4 Moisture absorber/scavengers

Moisture absorbers are the materials that are able to absorb water molecules physically, present in the environment surrounding the food inside the package. However in case of moisture scavenger absorption of water molecule takes place through a chemical reaction. Moisture absorption by silica and zeolite involved physical absorption however calcium chloride absorbs water by means of a chemical reaction- $CaO + H_2O \rightarrow Ca(OH)_2$ [54]. Excess moisture content inside the package may contribute to the growth of microorganisms which can lead to quality deterioration, and affect texture and nutrition value thereby reducing the shelf-life of food [55]. Active moisture scavenger materials may be organic, inorganic, and polymer-based materials. Organic based materials involve fructose, sorbitol, diethanolamine, triethanolamine, cellulose, and derivative of cellulose such as sodium carboxymethyl cellulose, ammonium carboxymethyl cellulose, etc. Inorganic based materials involved silica, zeolite, calcium chloride, sodium chloride, zinc chloride, magnesium chloride, activated alumina, etc. Polymer based materials involved the use of polyvinyl alcohol, polyacrylic acid, starch copolymers, etc [54]. Moisture scavengers are used to regulate the moisture content inside a package in the form of films, trays, pads, and sachets [54]. Absorption pads are most commonly used in the case of raw meat and fish products. Bovi et al. prepared Fruitpad containing different concentrations of fructose and applied for the packaging of fresh strawberries and was found to be effective in absorbing moisture from the package headspace containing strawberries. FruitPad with 30% fructose showed highest amount of moisture absorption of 0.94 g of water/g of pad at 20°C and 100% RH. Weight loss of packaged strawberry was less than 0.9% which was much below the acceptable limit of 6% for strawberry [56].

6.5 Ethanol emitters

Ethanol possesses extensive antimicrobial properties. It can inhibit the growth of yeast, bacteria and most prominently inhibit or slow down the mold growth process in food thereby prolonging the shelf life of the food matrix [13,57]. It can be used in food packaging as a germicidal agent by directly spraying upon food products before packaging. However, sachets and films that can slowly emit ethanol inside the package are also used [58]. Ethanol emitters are widely used in bakery products like cookies, breads, cakes, pizza, pastries, etc and in fish products as an antibacterial and anti-fungal agent. Ethanol emitting sachets firstly absorb moisture from the food matrix and then release the ethanol vapor [59]. Along with ethanol, sometimes flavoring agents like vanilla are also added in trace amount in sachets to hide the odor of alcohol [58].

6.6 Antimicrobial and antioxidant packaging

The use of antimicrobial and antioxidant packaging is most prominent and is growing significantly in the field of active packaging. Active antioxidant and antimicrobial compounds are used in food packaging to prohibit the oxidation reaction taking place in food items and to inhibit microbial growth that can lead to food spoilage respectively [3,60,61].

The antimicrobial and antioxidant compounds can be directly incorporated as active compounds into the packaging film which release slowly inside the package or as a coating on the surface of the food matrix. The other approaches involve the use of sachets or pads containing volatile antimicrobial compounds [31]. Organic acids, polysaccharides, alcohols, bacteriocins, essential oils, plant extracts, and nanoparticles are widely used as active ingredients in antimicrobial and antioxidant packaging [62,63]. Common organic acids that are used as antimicrobial agents involved benzoic acid, sorbic acid, propionic acid, acetic acid, ascorbic acid, and citric acid [64]. The use essential oils and plant extracts in antioxidant and antimicrobial active packaging become significant nowadays because of their non-toxic, edible, and non expansive properties [13,36]. The secondary metabolites obtained in different sections like leaves, roots, fruits, flowers, and barks of plants act as a rich source of alkaloids, terpenoids, flavonoids, anthocyanins, and phenolic compounds which can serve as a natural antioxidant and antimicrobial agents [65,66]. Hence plant extracts and essential oils comprising these compounds can be incorporated directly into the polymeric packaging film to maintain the freshness and to prolong the shelf life of food by inhibiting the growth of pathogenic and food spoiling microbes. Wrona et al. synthesized low-density polyethylene (LDPE) packaging film by incorporating two essential oils and seven vegetable oils as active compounds and investigated the antioxidant properties of the films. Essential oils are extracted from rose seeds and ginger root and vegetable oils are extracted from avocado, flaxseed, grape seed, milk thistle, pomegranate, starflower and walnut. Avocado oil, grape seed oil and milk thistle oil, incorporated films possess higher antioxidant activity. However, flaxseed oil incorporated film extended the shelf-life of fresh meat by 22% [67]. Nontoxic metal and metal oxide nanoparticles like silver, gold, copper, zinc oxide, titanium dioxide, and magnesium oxide are widely used as an active antimicrobial and antioxidant agents in polymeric packaging films [31].

7. Intelligent packaging

An intelligent packaging system generally involve the hardware elements like sensors, indicators, time-temperature indicators, radio frequency identification devices, etc which provide the ability to communicate with the consumer [68]. Intelligent packaging involves the use of components that consistently observe the quality and situation of the food items inside the package and their surrounding internal environments [28]. Intelligent packaging is capable of continuously monitoring the food matrix and its interior environments like pH, temperature, gas composition, moisture content, and microbial growth to furnish the information regarding the physical, chemical, biological, microbiological, and biochemical quality of the food [11]. It possesses attractive intelligent properties such as tracing, sensing, detecting, communicating, and recording of crucial information which give assurance of quality and safety of the product, and pack integrity and are employed in application like product traceability, product authenticity, antitheft, etc [69,70]. The characteristics of intelligent packaging are also useful to examine the efficacy, integrity, and durability of the packaging system [70].

Different classes of intelligent packaging systems are summarized below (Fig. 5.2).

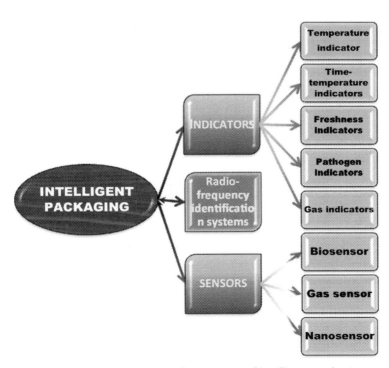

Fig. 5.2 Components of intelligent packaging.

7.1 Sensors

A sensor may be defined as a device employed to locate, detect and assess the energy of matter by producing a signal corresponding to a definite chemical and physical property that need to be detect against which it responds [71]. A sensor usually consists of four components – [72].

(1) Receptor which acts as the sensing portion for a sensor where all the analytical physicochemical information is obtained.
(2) Transduction element which acts as the measuring portion for a sensor and converts the physicochemical information to a fruitful analytical signal.
(3) Signal processing unit.
(4) Signal display unit.

An ideal sensor must possess characteristics such as high sensitivity, fast response, selectivity, extended lifetime, small size, and low cost. However, high cost and large size associated with the sensor make their use limited in the field of food packaging [73].

Sensors can be classified as chemical, biological, and gas sensors. In chemical sensors, a chemical compound acts as a receptor. Chemical sensors are used to detect the change in PH, production of different types of gases, and volatile compounds such as H_2, NH_3, CO_2, NO_2, CH_4, dimethylamine, trimethylamine, etc [74]. In biosensors, biological components like enzymes, antigens, nucleic acids, etc. Act as a receptor. Commercially biosensors are widely used in the field of clinical diagnostics. However, their uses are limited in the field of food packaging [72]. Gas sensors are employed to detect the presence of different types of gases which commonly involve CO_2, O_2, H_2S, and volatile amines [63]. These can produce inside the package due to some physicochemical and biological processes and the concentration may change with time. Thus the concentration of these gases may provide information about the quality of the food. Different types of luminescent compounds and pH-sensitive dyes are used for the development of gas sensors [13].

7.2 Indicators

Indicators are the tools employed in food packaging to signify the existence of a component, concentration of a certain component or a specific class of compound, rate of reaction between two or more components through an instant visual transformation like color change [75]. Indicators directly convey the information about the quality of the packaged food to the consumers. Different types of indicators applied on food packaging are time-temperature indicators, temperature indicators, freshness indicators, gas indicators, and microbial growth indicators [13]. **The working principle is generally based on chemical, microbial, mechanical and electrochemical changes which are expressed in the form of color change or mechanical deformation** [72]. Lactic acid, dimethyl sulfoxide, isopropyl palmitate, glycerol tributyrate, glucose 1-hydrate, carbamide, etc. Are used as active indicator components in time-temperature indicators in case of fruits, vegetables, perishable fruits, fish, and meat products [13].

7.3 Radio frequency identification (RFID)

Radio frequency identification (FRID) technology which can be attached to assets like containers, pellets, castles, etc is used to store, transmit, and carry accurate real-time data without any human intervention [71]. It can also bear complicated information about quality, integrity, nutritional value, relative humidity, temperature, and condition of the surrounding environment during transportation and storage [63]. It is most commonly used for tracking and recognition purposes. FRID is most commonly used in meat products. FRID has the ability to store data up to 1 MB [63]. It stores some simple information like identification numbers on the basis of which a consumer can fetch the information against the identification number from the database and can use it accordingly [76].

8. Differences between active and intelligent packaging

Active packaging	Intelligent packaging
(1) Active packaging primarily concentrates on the components with active functions that can improve the quality of the food.	(1) Intelligent packaging concentrates on communication.
(2) Active packaging can directly interact with the food component or with the environment surrounding the food to expand the shelf life [57].	(2) Intelligent packaging does not interact directly with the food matrix. It only monitors the quality of the food matrix.
(3) Active packaging components perform some action on the food matrix.	(3) Intelligent packaging components only sense and share the information of the food matrix [57,77].
(4) Active packaging is an emerging and innovative approach to expand the storage time, increase the shelf-life of food products and delayed ripening without harming the safety, quality, and integrity of the food product [3].	(4) Intelligent packaging is specifically advantageous for tracing the information and observing the quality of the product that it encloses. Also, it eases the access of data as well as information exchange through changing the conditions outside or inside the packaging [70].

9. Advantages of active and intelligent packaging materials

(1) The growing health concern among the consumers appeals for hygiene and fresh food with prolonged shelf-life of the food matrix. These are the driving factors of the smart packaging technology for food matrixes. The active and intelligent packaging technologies possess amazing properties to overcome the growing demands of consumers with wide applications [15,16].

(2) Active and intelligent packaging possess the advantages of increases shelf life of the food matrix without affecting the quality of the food, easy to use, reduced counterfeiting.

(3) Active materials that are used in active and intelligent packaging reduce the spoilage of the food matrix without disturbing the organoleptic properties and composition of the food matrix. This can mislead the customers.

(4) Active materials are the materials which are specially designed to prolong the shelf life or to preserve or enhance the packaged food quality. Active substances can be able to absorb or release components from or into the food matrix.

(5) However packaging materials can monitor the quality of the food matrix and also the surrounding environment of the food.

(6) The prime objective of maximum smart packaging technologies is to improve the quality and organoleptic characteristics of the food matrix by highly sustaining food safety for a long duration by scaling down unwanted food reactions. However, these were found to be the fundamental value that motivates the purchase intent of consumers. The prolongation of Shelf-life helps to sustain the healthfulness, freshness, and naturalness of the food product. This also provides special expansion in the consumption time period which diminishes the food spoilage [63].

(7) New technologies such as smart and active packaging based on biopolymers have been developed to curb these concerns. These packaging materials contained active compounds like antioxidants, antimicrobial, engineered nanomaterials, absorbers or emitters, etc which can expand the shelf life of the foodstuff by controlling oxidation, microbial growth, moisture content, and various reactions that affect the quality of the food [32].

While intelligent packaging monitors and acts as a detector for the quality and freshness of the foodstuff in real-time by emitting optical, chemical, colorimetric, and electrical signals.

10. Industrial barriers for the application of active and intelligent packaging

The actives and intelligent packaging technologies were launched in the mid-seventies in the markets of Japan. However, growing attention of the industries of the USA and Europe was started in the mid-nineties [62,72].

Global smart packaging market and industries are growing day by day because of the growing interest in sustainable packaging, convenience foods like ready-to-cook food, and tough food safety legislation. But still, the commercialization, industrial, and market implementation of active and intelligent packaging technologies are in their infancy, limited to a large extent by numerous factors especially including costs, acceptance, regulations, safety, etc. Growing health awareness among consumers, food wastage, manufacturers' concern for a longer shelf life of the food products, and supply chain inefficiencies are the other factors fueling the growth of the market [78]. Various factors that can interrupt the industrial application of active and intelligent packaging technologies may be categorized as economic, social, environmental, legislative, technological, and sustainability [79]. However, the knowledge among the consumers, high capital cost for the security issues, and development of new manufacturing techniques, instalments, sensors, and indicators that are congruent with current packaging standards is the restrictions and key challenges for the growth of smart packaging industries [80–82]. The following are considered key factors that can act as a barrier to the industrial application of active and intelligent packaging

(i) Sustainability (ii) Safety concerns (iii) Lake of knowledge about active and intelligent packaging (iv) Move from laboratory to industrial scale (v) Various prohibitive regulations for industries (vi) The gap between the industry and consumers (vii) Costs (viii) Need for new manufacturing techniques (ix) Acceptance.

10.1 Sustainability

There are three crucial elements for sustainability-economic sustainability, social sustainability, and environmental sustainability [83]. Sustainability is a prominent factor in the development of food packaging industries [84]. However, in most cases, nonbiodegradable plastic-based materials are used in the field of active and intelligent packaging which can cause environmental problems. Hence, it is a great matter of concern to developing a sustainable and environment-friendly solution [83]. This can be achieved by using fiber-based materials possessing or by incorporating active and intelligent properties. Till now, in the fiber-based industries accessible implication of smart packaging is limited. However, one drawback of fiber-based materials is that the barrier properties of these types of materials are less effective as compared to plastic-based materials [80]. However, it is significant to figure out the important factors that can impede the successful industrial and market implementation of active and intelligent food packaging.

Many components used in active and intelligent packaging technologies do not adequately suit the eco-friendly nature. For example, many FRID and the elements used in sensors are not perfectly sustainable by nature causing difficulties in the separation and recycling process [81]. Hence it is very important to study, understand and measure the impact of these materials on the environment. Therefore industries should use easily separable and recyclable electronics.

10.2 Safety concern

The active and intelligent packaging system was put in place around 30 years ago. However, research is still going on and diverse research publications are available till now. But the field is a bit newer to the industries and rather few market applications are available [85,86]. As an example nanocellulose can be considered. By incorporating nanocellulose into the fiber-based materials antimicrobial properties can be introduced without affecting the sustainable properties [87]. Nanotechnology can also play a vital role in active and intelligent packaging. But the issue of safety concern is associated with the engineered nanomaterials because of the lack of knowledge about their toxicological effects when consumers have come into contact [88,89].

There are generally two types of migration of active compounds to the food matrix may possible- [82].
(1) Direct migration of active and intelligent components into food.
(2) Migration of their reaction or degradation products into food.

Hence the study of toxicological effects of both the active compound and also their degradation product is important.

10.3 Lake of knowledge about active and intelligent packaging

The lack of knowledge and awareness about the usefulness, impacts, quality, and functions of active and intelligent packaging can also act as a barrier in the market and industrial applications of the active and intelligent packaging [80]. The extensively available research or scientific publications are not enough to overcome it. When introducing a new Active packaging material it may be useful and helpful to provide an accepted demonstrator to display its potentiality. However, discussion about the scope and opportunities, benefits, and impacts of active and intelligent packaging technologies among the consumers, different types of people associated with the food packaging industries as well as marketing and purchasing departments is very important to improve the knowledge and awareness [80]. If the usefulness, working procedure, advantages, disadvantages, and the hidden features of the active and intelligent packaging are not known to consumers then these technologies may be apprehensive and may not be interested to adopt it. However, the experts and the researchers must be more aware and should engage in the improvement of smart packaging.

10.4 Move from lab to industrial scale

The limitation of the research works on active and intelligent packaging material only on the laboratory scale is another factor that can act as a barrier to the industrial application of it. From the lab-scale results, the expectations for the efficacy of these active elements in the practical field become too high. In practice, the effectiveness is influenced by a number of parameters such as food product, material characteristics, and storage condition, chemical, mechanical and biological factors [86]. Hence, the successful transformation from the laboratory to the industry is quite complicated. However, a research project is devoted to a distinct application and is not conductible to a broader application.

10.5 Various prohibitive regulations for industries

During the synthesis and application of food materials, certain rules and regulations must be followed to maintain the quality and consumer health concerns. For example in European Union for the application of Active and intelligent packaging in food must follow European Framework Regulation (EC) No. 1935/2004 [82]. **Smart packaging involves the use of a wide range of components hence it is exposed to more strict rules and regulations as compared to traditional packaging because of safety concerns.** During the application of the active and intelligent packaging materials in food packaging, there is a possibility that their constituent materials may interact and transfer to the food product under certain conditions [82]. Which may cause- (i) Threat to consumer health (ii) Intolerable change in the food matrix (iii) Decrease in the quality of the food. Thus, to get rid of these negative impacts industries must follow certain hard and fast rules and regulations during the manufacturing process. However, it may not be possible for all the industries to maintain these Legislations. Hence many companies may miss the market opportunities [80].

However, food packaging industries must work through some hard and fast rules and regulations and must not misguide customers through the advertisement, identity, or presentation. Only authorized substances should be used in the production of active and intelligent materials. Substances should be added only after the safety has been examined by respective authorities [72].

Before using any active material for food packaging manufacturers have to examine the composition against proper legislation. However, proper tests for overall migration of the components in food, its impact on consumer health should also have to perform before market implementation [90]. The compounds which are deliberately released into the food matrix must obey the food additive regulations like Regulation 1333/2008/EC [82]. Food additives are substances added to food to maintain or improve its freshness, safety, texture, taste, or appearance that are presumed to become a constituent of the food [82].

10.6 The gap between the industry and consumers

The successful incorporation of active and intelligent features in the packaging material is not the extreme success for their industrial and market implementation. However, Consumers are more concerned with the advantages and applicability of smart packaging for their personal assets instead of industries and retailers. Lack of awareness and trust of consumers in technologies, i.e., the gap between the consumers and industries is another factor that can hinder the industrial application of active and intelligent packaging [80]. Different components like absorbers, emitters, antimicrobial and antioxidant components which are used to improve the quality of the food are unknown to the consumer and the lack of knowledge may be the reason for the decline in the use of active and intelligent packaging materials. The features of the smart packaging material are not so complicated hence there should not be any scope for misunderstanding among the consumers. However, the route of representation of the information may also impact consumer trust. For example, when emitters are used as active features in packaging materials, it may release substance to the food matrix which may be suspicious for the consumers when the advantages are unaware and unclear [80]. This can create doubt about safety and health concerns among the consumer and may go for alternatives. The active and intelligent material incorporated in the packaging technology sometimes may not be edible [57,82]. So it is necessary to adequately label the nonedible component for the easy identification of the consumer.

10.7 Costs

Cost is another barrier factor for the industrial application of the active and intelligent packaging [62]. **Compared to the traditional packaging materials smart packaging materials are not cheap.** The materials used for active and intelligent are quite expansive as a result cost of the end product become higher as compared to the traditional packing [81]. Since the active and intelligent packaging technologies are highly specialized and sophisticated, hence the cost associated with them is generally high, which hinders the commercialization and industrialization process [72]. For the broader application cost must be drastically reduced. As compared to the traditional packaging materials like fossil-based plastics, metals, glass, etc. The active and intelligent packaging materials are quite expansive [81]. The extensive costs of raw materials for the fabrication of flexible and environment-friendly packaging materials are likely to impede the market growth of smart packaging in near future.

10.8 Need for new manufacturing techniques

Since the smart packaging technologies are still in the initial phase of development, so there is a lack of long-term proven success which makes it challenging to establish a strong and winning trade and industrial framework [81]. However, it is also challenging for the overall assessment about the total expenditures of possession for active and intelligent packaging industries. Before the application of the active and intelligent packaging materials in a wide range, industries have to develop new manufacturing techniques like in the case of indicators and sensors which are congruent with current packaging standards [81].

10.9 Acceptance

Consumer adoption and acceptance play a key role in the commercial market implementation of new technology [78]. Due to lack of knowledge and awareness sometimes consumers do not recognize the active and intelligent materials as a beneficial one. Consumers often believe that food materials possessing shorter shelf life are fresher and the extension of the shelf life of food using active compounds seems not attractive to the consumers [72].

11. Conclusion

Active and intelligent packaging has emerged as a promising and environment-friendly resource in the field of food packaging. These technologies have the potential to be used as excellent material in the future to improve food quality, safety, vitality, increase in shelf life and prevent food wastage. It is already being successfully applied and various researches are going on in laboratories and on a small scale in different parts of the world which is of great significance in the field of food packaging. Smart packaging like oxygen scavengers, carbon dioxide scavengers, antimicrobial and antioxidant packaging, etc. Are being used on a widespread basis in this field. Sensors have emerged as the most widely used and advanced components in intelligent packaging by providing fast, accurate and authentic information about food products. The incorporation of nanoparticles into active and intelligent packaging has gained widespread popularity and shown promising performance. There is also growing research interest in the use of natural, environment friendly and biodegradable components in food packaging materials. The rapid use of plastics poses a serious threat to the ecological environment around the world. The development of biodegradable polymer-based active and edible packaging film incorporated with active components is also a promising one and can serve as an excellent alternative for plastics. However, as mentioned above various factors have hampered the progress in industrialization and its application on a commercial basis. There is still a gap between industrial applications, laboratory research activities, consumer awareness and food commodities. For the acceleration of industrial adoption of active and intelligent packaging an efficient collaboration among them is very crucial. In the health sector, food products and packaging, however, smart packaging has undoubtedly opened up huge opportunities for new innovations and developments. At the same time, new challenges are emerging daily in communicating the benefits of this technology to the consumer and its commercial application without halting the pace of development. Therefore, these factors must be removed through appropriate research and analysis as well as effective promotion and dissemination measures to enable the successful and widespread application of active and intelligent packaging worldwide. The successful industrial application of Active and Intelligent Packaging will usher in a new revolution in global food packaging.

References

[1] H. Shekarchizadeh, F.S. Nazeri, Active nanoenabled packaging for the beverage industry, Nanotechnol. Beverage Indus. (2020) 587–607, https://doi.org/10.1016/B978-0-12-819941-1.00020-1 (Chapter 20).

[2] E.B. Alfaro, D.V. Craveiro, K.O. Lima, H. Leão, G. Costa, D.R. Lopes, C. Prentice, Intelligent packaging with pH indicator potential, Food Eng. Rev. 11 (2019) 235–244, https://doi.org/10.1007/s12393-019-09198-9.

[3] S. Yildirim, B. Rocker, M.K. Pettersen, J.N. Nygaard, Z. Ayhan, R. Rutkaite, T. Radusin, P. Suminska, B. Marcos, V. Coma, Active packaging applications for food, Compr. Rev. Food Sci. Food Saf. 17 (1) (2017) 165–199, https://doi.org/10.1111/1541-4337.12322.

[4] K. Galic, M. Scetar, M. Kurek, The benefits of processing and packaging, Trend. Food Sci. Technol. 22 (2011) 127–137, https://doi.org/10.1016/j.tifs.2010.04.001.

[5] Functional polymers in food science: from technology to biology, in: G. Cirillo, U.G. Spizzirri, F. Iemma (Eds.), Food Packaging, vol. 1, Wiley, 2015, ISBN 978-1-118-59489-6.

[6] B.P.F. Day, Active packaging of food, in: J. Kerry, P. Butler (Eds.), Smart Packaging Technologies for Fast Moving Consumer Goods, John Wiley & Sons, England, 2008, pp. 1–18.

[7] D.A. Pereira de Abreu, J.M. Cruz, P.P. Losada, Active and intelligent packaging for the food industry, Food Rev. Intl. 28 (2) (2012) 146–187, https://doi.org/10.1080/87559129.2011.595022.

[8] S. Yildirim, Active Packaging for Food Biopreservation. Protective Cultures, Antimicrobial Metabolites and Bacteriophages for Food and Beverage Biopreservation, 2011, pp. 460–489, https://doi.org/10.1533/9780857090522.3.460.

[9] D. Schaefer, W.M. Cheung, Smart packaging: opportunities and challenges, Procedia CIRP 72 (2018) 1022–1027, https://doi.org/10.1016/j.procir.2018.03.240.

[10] G.L. Robertson, Food Packaging: Principles and Practice, CRC Press, Boca Raton, FL, 2003.

[11] R. Priyadarshi, P. Ezati, J.W. Rhim, Recent advances in intelligent food packaging applications using natural food colorants, ACS Food Sci. Technol. 1 (2021) 124–138, https://doi.org/10.1021/acsfoodscitech.0c00039.

[12] G. Ghoshal, Recent trends in active, smart, and intelligent packaging for food products, Food Pack. Preser. (2018) 343–374, https://doi.org/10.1016/B978-0-12-811516-9.00010-5 (chapter 10).

[13] B. Kuswandi, Jumina, Active and intelligent packaging, safety, and quality controls, Fresh-Cut Fruits Vegetab. (2020) 243–294, https://doi.org/10.1016/B978-0-12-816184-5.00012-4 (Chapter 12).

[14] M. Mustafa, S. Nagalingam, J. Tye, A.S. Hardy, J. Dolah, Looking back to the past: revival of traditional food packaging, in: 2nd Regional Conference on Local Knowledge (KEARIFAN TEMPATAN), 15–16 October, Jerejak Island Rainforest Resort, Penang, 2012.

[15] J. Končar, A. Grubor, R. Marić, G. Vukmirović, N. Đokić, Possibilities to improve the image of food and organic products on the AP Vojvodina market by introducing a regional quality label, Food and Feed Res. 46 (1) (2019) 111–124, https://doi.org/10.5937/FFR1901111K.

[16] T.J. Gerpott, Relative fixed Internet connection speed experiences as antecedents of customer satisfaction and loyalty: an empirical analysis of consumers in Germany, Manage. Market.-Challeng. Knowled. Soci. 13 (4) (2018) 1150–1173.

[17] A. Grubor, J. Končar, R. Marić, Challenges of introducing intelligent packaging to the retail market of AP Vojvodina, Strateg. Manage. 25 (2) (2020) 018–026, https://doi.org/10.5937/StraMan2002018G.

[18] A.L. Brody, B. Bugusu, J.H. Han, C.K. Sand, T.H. McHugh, Innovative food packaging solutions, J Food Sci 73 (2008) R107–R116, https://doi.org/10.1111/j.1750-3841.2008.00933.x.

[19] M. Vanderroost, P. Ragaert, F. Devlieghere, B. De Meulenaer, Intelligent food packaging: the next generation, Trend. Food Sci. Technol. 39 (1) (2014) 47–62, https://doi.org/10.1016/j.tifs.2014.06.009.

[20] M.A. Sani, M.A. Lalabadi, M. Tavassoli, K. Mohammadi, D.J. McClements, Recent advances in the development of smart and active biodegradable packaging materials, Nanomaterials 11 (5) (2021) 1331, https://doi.org/10.3390/nano11051331.

[21] K.J. Groh, T. Backhaus, B. Carney-Almroth, B. Geueke, P.A. Inostroza, A. Lennquist, H.A. Leslie, M. Maffini, D. Slunge, L. Trasande, A.M. Warhursth, J. Munckea, Overview of known plastic packaging-associated chemicals and their hazards, Sci. Total Environ. 651 (2019) 3253–3268, https://doi.org/10.1016/j.scitotenv.2018.10.015.

[22] M.I. Din, T. Ghaffar, J. Najeeb, Z. Hussain, R. Khalid, H. Zahid, Potential perspectives of biodegradable plastics for food packaging application-review of properties and recent developments, Food Addit. Contam, Part A 37 (4) (2020) 665–680, https://doi.org/10.1080/19440049.2020.1718219.

[23] A. Iordanskii, Bio-based and biodegradable plastics: from passive barrier to active packaging behavior, Polymers 12 (7) (2020) 1537, https://doi.org/10.3390/polym12071537.

[24] J.-W. Rhim, H.-M. Park, C.-S. Ha, Bio-nanocomposites for food packaging applications, Prog Polym Sci 38 (2013) 1629–1652, https://doi.org/10.1016/j.progpolymsci.2013.05.008.

[25] R.G. Patil, V. Birhade, V. Jadhav, Innovative food packaging, Int. J. Sci. Res. 6 (4) (2017) 1287–1291.

[26] R. Ahvenainen, Novel Food Packaging Techniques, CRC Press, Boca Raton, FL, USA, 2003.

[27] S.Y. Lee, S.J. Lee, D.S. Choi, S.J. Hur, Current topics in active and intelligent food packaging for preservation of fresh foods, J. Sci. Food Agric. 95 (14) (2015) 2799–2810, https://doi.org/10.1002/jsfa.7218.

[28] K.B. Biji, C.N. Ravishankar, C.O. Mohan, T.K. Srinivasa Gopal, Smart packaging systems for food applications: a review, J. Food Sci. Technol. 52 (10) (2015) 6125–6135, https://doi.org/10.1007/s13197-015-1766-7.

[29] K.A.M.O. Callaghan, J.P. Kerry, Consumer attitudes towards the application of smart packaging technologies to cheese products, Food Packag. Shelf Life 9 (2016) 1–9, https://doi.org/10.1016/j.fpsl.2016.05.001.

[30] S. Chen, S. Brahma, J. Mackay, C. Cao, B. Aliakbarian, The role of smart packaging system in food supply chain, J. Food Sci. 85 (3) (2020) 517–525, https://doi.org/10.1111/1750-3841.15046.

[31] S. Yildirim, B. Rocker, Active packaging, Nanomater. Food Packag. (2018) 173–202, https://doi.org/10.1016/B978-0-323-51271-8.00007-3 (Chapter 7).

[32] M. Asgher, S.A. Qamar, M. Bilal, H.M.N. Iqbal, Bio-based active food packaging materials: sustainable alternative to conventional petrochemical-based packaging materials, Food Res. Int. 137 (2020) 109625, https://doi.org/10.1016/j.foodres.2020.109625.

[33] N. Bhargava, V.S. Sharanagat, R.S. Mor, K. Kumar, Active and intelligent biodegradable packaging films using food and food waste-derived bioactive compounds: a review, Trend. Food Sci. Technol. 105 (2020) 385–401, https://doi.org/10.1016/j.tifs.2020.09.015.

[34] European Comission, EU Guidance to the Commission Regulation (EC) No. 450/2009 of 29 May 2009 on Active and Intelligent Materials and Articles Intended to Come into the Contact with Food (Version 1.0), 2009.

[35] Z. Kordjazi, A. Ajji, Development of TiO2 catalyzed HTPB based oxygen scavenging films for food packaging applications, Food Contr. 121 (2021) 107639, https://doi.org/10.1016/j.foodcont.2020.107639.

[36] T. Janjarasskul, P. Suppakul, Active and intelligent packaging: the indication of quality and safety, Critic. Rev. Food Sci. Nutrit. 58 (5) (2018) 808–831, https://doi.org/10.1080/10408398.2016.1225278.

[37] H. Mu, H. Gao, H. Chen, F. Tao, X. Fang, L. Ge, A nanosised oxygen scavenger: preparation and antioxidant application to roasted sunflower seeds and walnuts, Food Chem. 136 (1) (2013) 245–250, https://doi.org/10.1016/j.foodchem.2012.07.121.

[38] Z. Foltynowicza, A. Bardenshteinb, S. Sängerlaubc, H. Antvorskovb, W. Kozaka, Nanoscale, zero valent iron particles for application as oxygen scavenger in food packaging, Food Packag. Shelf Life 11 (2017) 74–83, https://doi.org/10.1016/j.fpsl.2017.01.003.

[39] J. Yu, R.Y.F. Liu, B. Poon, S. Nazarenko, T. Koloski, T. Vargo, A. Hiltner, E. Baer, Polymers with palladium nanoparticles as active membrane materials, J. Appl. Polym. Sci. 92 (2) (2004) 749–756, https://doi.org/10.1002/app.20013.

[40] H. Li, K.K. Tung, D.R. Paul, B.D. Freeman, M.E. Stewart, J.C. Jenkins, Characterization of oxygen scavenging films based on 1, 4-polybutadiene, Industr. Eng. Chem. Research 51 (21) (2012) 7138–7145, https://doi.org/10.1021/ie201905j.

[41] B.J. Ahn, K.K. Gaikwad, Y.S. Lee, Characterization and properties of LDPE film with gallic-acid-based oxygen scavenging system useful as a functional packaging material, J. Appl. Polym. Sci. 133 (43) (2016), https://doi.org/10.1002/app.44138.

[42] A.A. Dey, S. Neogi, Oxygen scavengers for food packaging applications: a review, Trend. Food Sci. & Technol 90 (2019) 26–34, https://doi.org/10.1016/j.tifs.2019.05.013.

[43] A. Mahieu, C. Terrie, N. Leblanc, Role of ascorbic acid and iron in mechanical and oxygen absorption properties of starch and polycaprolactone multilayer film, Packaging Res 2 (2017) 1–11, https://doi.org/10.1515/pacres-2017-0001.

[44] D.S. Lee, Carbon dioxide absorbers for food packaging applications, Trend Food Sci. Technol. 57 (A) (2016) 146–155, https://doi.org/10.1016/j.tifs.2016.09.014.

[45] P. Singh, A.A. Wani, A.A. Karim, H.-C. Langowski, The use of carbon dioxide in the processing and packaging of milk and dairy products: a review, Int. J. Dairy Technol. 65 (2011) 161–176, https://doi.org/10.1111/j.1471-0307.2011.00744.x.

[46] R. Rodriguez-Aguilera, J.C. Oliveira, Review of design engineering methods and applications of active and modified atmosphere packaging systems, Food Eng. Rev. 1 (2009) 66–83, https://doi.org/10.1007/s12393-009-9001-9.

[47] H.J. Wang, D.S. An, J.-W. Rhim, D.S. Lee, A multi-functional biofilm used as an active insert in modified atmosphere packaging for fresh produce, Packag. Technol. Sci. 28 (12) (2015) 999–1010, https://doi.org/10.1002/pts.2179.

[48] J.W. Han, L. Ruiz-Garcia, J.P. Qian, X.T. Yang, Food packaging: a comprehensive review and future trends, Compr. Rev. Food Sci. Food Saf. 17 (2018) 860–877, https://doi.org/10.1111/1541-4337.12343.

[49] Z. Zhu, Y. Zhang, Y. Shang, X. Zhang, Y. Wen, Preparation of PAN@ TiO2 nanofibers for fruit packaging materials with efficient photocatalytic degradation of ethylene, Materials 12 (6) (2019) 896, https://doi.org/10.3390/ma12060896.

[50] S. Chopra, S. Dhumal, P. Abeli, R. Beaudry, E. Almenar, Metal-organic frameworks have utility in adsorption and release of ethylene and 1-methylcyclopropene in fresh produce packaging, Postharvest Biol. Technol. 130 (2017) 48–55, https://doi.org/10.1016/j.postharvbio.2017.04.001.

[51] K. Sadeghi, Y. Lee, J. Seo, Ethylene scavenging systems in packaging of fresh produce: a review, Food Rev. Int. 37 (2) (2021) 155–176, https://doi.org/10.1080/87559129.2019.1695836.

[52] P.S. Casey, S. Boskovic, K. Lawrence, T. Turney, Controlling the photoactivity of nanoparticles, NSTI-nanotech 3 (2004) 370–374.

[53] P.C. Spricigo, M.M. Foschini, C. Ribeiro, D.S. Corrêa, M.D. Ferreira, Nanoscaled platforms based on SiO2 and Al2O3 impregnated with potassium permanganate use color changes to indicate ethylene removal, Food Bioproc. Technol. 10 (2017) 1622–1630, https://doi.org/10.1007/s11947-017-1929-9.

[54] K.K. Gaikwad, S. Singh, A. Ajji, Moisture absorbers for food packaging applications, Environ. Chem. Lett. 17 (2019) 609–628, https://doi.org/10.1007/s10311-018-0810-z.

[55] S.B. Murmu, H.N. Mishra, Selection of the best active modified atmosphere packaging with ethylene and moisture scavengers to maintain quality of guava during low-temperature storage, Food Chem. 253 (2018) 55–62, https://doi.org/10.1016/j.foodchem.2018.01.134.

[56] G.G. Bovi, O.J. Caleb, E. Klaus, F. Tintchev, C. Rauh, P.V. Mahajan, Moisture absorption kinetics of FruitPad for packaging of fresh strawberry, J. Food Eng. 223 (2018) 248–254, https://doi.org/10.1016/j.jfoodeng.2017.10.012.

[57] E. Drago, R. Campardelli, M. Pettinato, P. Perego, Innovations in smart packaging concepts for food: an extensive review, Foods 9 (11) (2020) 1628, https://doi.org/10.3390/foods9111628.

[58] B.P.F. Day, L.L. Potter, Active Packaging. Food and Beverage Packaging Technology, second ed., 2011, pp. 251–262, https://doi.org/10.1002/9781444392180.

[59] K.V.P. Kumar, J. Suneetha W, B.A. Kumari, Active packaging systems in food packaging for enhanced shelf life, J. Pharmacogn. Phytochem. 7 (6) (2018) 2044–2046.

[60] L. Motelica, D. Ficai, A. Ficai, O.C. Oprea, D.A. Kaya, E. Andronescu, Biodegradable antimicrobial food packaging: trends and perspectives, Foods 9 (10) (2020) 1438, https://doi.org/10.3390/foods9101438.

[61] T. Radusin, S. Torres-Giner, A. Stupar, I. Ristic, A. Miletic, A. Novakovic, J.M. Lagaron, Preparation, characterization and antimicrobial properties of electrospun polylactide films containing Allium ursinum L.extract, Food Packag. Shelf Life 21 (2019) 100357, https://doi.org/10.1016/j.fpsl.2019.100357.

[62] D. Dainelli, N. Gontard, D. Spyropoulos, E. Zondervan-van den Beuken, P. Tobback, Active and intelligent food packaging: legal aspects and safety concerns, Trend. Food Sci. Technol. 19 (Suppl. 1) (2008) S103–S112, https://doi.org/10.1016/j.tifs.2008.09.011.

[63] M.S. Firouz, K.M. Alden, M. Omid, A critical review on intelligent and active packaging in the food industry: research and development, Food Res. Int. 141 (2021) 110113, https://doi.org/10.1016/j.foodres.2021.110113.

[64] M. Parish, L. Beuchat, T. Suslow, L. Harris, E. Garrett, J.N. Farber, F.F. Busta, Methods to reduce/eliminate pathogens from fresh and fresh-cut produce, Compr. Rev. Food Sci. Food Saf. 2 (1) (2003) 161–173, https://doi.org/10.1111/j.1541-4337.2003.tb00033.x.

[65] N. Anupama, G. Madhumitha, Green synthesis and catalytic application of silver nanoparticles using Carissa carandas fruits, Inorg. Nano-Metal Chem. 47 (1) (2017) 116–120, https://doi.org/10.1080/15533174.2016.1149731.

[66] S. Farhadi, B. Ajerloo, A. Mohammadi, Low-cost and eco-friendly phyto-synthesis of silver nanoparticles by using grapes fruit extract and study of antibacterial and catalytic effects, Int. J. Nano Dimens. 8 (1) (2017) 49.

[67] M. Wrona, F. Silva, J. Salafranca, C. Nerín, M.J. Alfonso, M.A. Caballero, Design of new natural antioxidant active packaging: screening flowsheet from pure essential oils and vegetable oils to ex vivo testing in meat samples, Food Contr. 120 (2021) 107536, https://doi.org/10.1016/j.foodcont.2020.107536.

[68] J.P. Kerry, M.N. O'grady, S.A. Hogan, Past, current and potential utilization of active and intelligent packaging systems for meat and muscle-based products: a review, Meat Sci. 74 (1) (2006) 113–130, https://doi.org/10.1016/j.meatsci.2006.04.024.

[69] C.E. Realini, B. Marcos, Active and intelligent packaging systems for a modern society, Meat Sci. 98 (3) (2014) 404–419, https://doi.org/10.1016/j.meatsci.2014.06.031.

[70] R. Dobrucka, The future of active and intelligent packaging industry, LogForum 9 (2) (2013) 103–110. URL: http://www.logforum.net/vol9/issue2/no4.

[71] I. Ahmed, H. Lin, L. Zou, Z. Li, A.L. Brody, I.M. Qazi, L. Lv, T.R. Pavase, M.U. Khan, S. Khan, L. Sun, An overview of smart packaging technologies for monitoring safety and quality of meat and meat products, Packag. Technol. Sci. 31 (7) (2018) 449–471, https://doi.org/10.1002/pts.2380.

[72] M. Ghaani, C.A. Cozzolino, G. Castelli, S. Farris, An overview of the intelligent packaging technologies in the food sector, Trend. Food Sci. Technol. 51 (2016) 1–11, https://doi.org/10.1016/j.tifs.2016.02.008.

[73] G. Hanrahan, D.G. Patil, J. Wang, Electrochemical sensors for environmental monitoring: design, development and applications, J. Environ. Monit. 6 (2004) 657–664, https://doi.org/10.1039/B403975K.

[74] K. Lee, S. Baek, D. Kim, J. Seo, A freshness indicator for monitoring chicken-breast spoilage using a Tyvek® sheet and RGB color analysis, Food Packag. Shelf Life 19 (2019) 40–46, https://doi.org/10.1016/j.fpsl.2018.11.016.

[75] S. Hogan, J. Kerry, Smart packaging of meat and poultry products, in: J. Kerry (Ed.), Smart Packaging Technologies for Fast Moving Consumer Goods, 2008, pp. 33–59.

[76] V. Todorvoic, M. Neag, M. Lazarevic, On the usage of RFID tags for tracking and monitoring shipped perishable goods, Proc. Eng. 69 (2014) 1345–1349, https://doi.org/10.1016/j.proeng.2014.03.127.

[77] K.L. Yam, P.T. Takhistov, J. Miltz, Intelligent packaging: concepts and applications, J. Food Sci. 70 (1) (2005) R1–R10, https://doi.org/10.1111/j.1365-2621.2005.tb09052.x.

[78] S.J. Lee, Rahman, A.T.M.M. Rahman, Intelligent Packaging for Food Products. Innovations in Food Packaging, second ed., 2014, pp. 171–209, https://doi.org/10.1016/B978-0-12-394601-0.00008-4 (Chapter 8).

[79] ActInPak Network. General Roadmap Facing Challenges for Market Implementation of Active and Intelligent Packaging. Available online: http://www.actinpak.eu/roadmaps/(accessed on 10 September 2020).

[80] S. Tiekstra, A.D. Parada, H. Koivula, J. Lahti, M. Buntinx, Holistic approach to a successful market implementation of active and intelligent food packaging, Foods 10 (2) (2021) 465, https://doi.org/10.3390/foods10020465.

[81] Greehy, et al. GLOPACK, Position Paper on Barriers of Market Introduction of Active and Intelligent Packaging, D1.3 Report on the Mapping and Analysis of Stakeholders' Preferences, Acceptances and Expactations, 2020.

[82] D. Restuccia, U.G. Spizzirri, O.I. Parisi, G. Cirillo, M. Curcio, F. Iemma, F. Puoci, G. Vinci, N. Picci, New EU regulation aspects and global market of active and intelligent packaging for food industry applications, Food Contr. 21 (11) (2010) 1425–1435, https://doi.org/10.1016/j.foodcont.2010.04.028.

[83] Z. Boz, V. Korhonen, C.K. Sand, Consumer considerations for the implementation of sustainable packaging: a review, Sustainability 12 (6) (2020) 2192, https://doi.org/10.3390/su12062192.

[84] F. Licciardello, Packaging, blessing in disguise. Review on its diverse contribution to food sustainability, Trend. Food Sci. Technol. 65 (2017) 32–39, https://doi.org/10.1016/j.tifs.2017.05.003.

[85] M. Rooney, History of active packaging, in: Intelligent and Active Packaging for Fruits and Vegetables, CRC Press, Abingdon, UK, 2007, pp. 11–30.

[86] Y. Selçuk, B. Röcker, M.K. Pettersen, J. Nilsen-Nygaard, Z. Ayhan, R. Rutkaite, T. Radusin, P. Suminska, B. Marcos, V. Coma, Active packaging applications for food, Compr. Rev. Food Sci. Food Saf. 17 (1) (2018) 165–199, https://doi.org/10.1111/1541-4337.12322.

[87] J. Li, R. Cha, K. Mou, X. Zhao, K. Long, H. Luo, F. Zhou, X. Jiang, Nanocellulose-based antibacterial materials, Adv. Healthc. Mater. 7 (20) (2018) 1800334, https://doi.org/10.1002/adhm.201800334.

[88] D. Enescu, M.A. Cerqueira, P. Fucinos, L.M. Pastrana, Recent advances and challenges on applications of nanotechnology in food packaging. A literature review, Food Chem. Toxicol. 134 (2019) 110814, https://doi.org/10.1016/j.fct.2019.110814.

[89] E. Jamroz, P. Kopel, J. Tkaczewska, D. Dordevic, S. Jancikova, P. Kulawik, V. Milosavljevic, K. Dolezelikova, K. Smerkova, P. Svec, V. Adam, Nanocomposite furcellaran films-the influence of nanofillers on functional properties of furcellaran films and effect on linseed oil preservation, Polymers 11 (12) (2019) 2046, https://doi.org/10.3390/polym11122046.

[90] K.A. Barnes, C.R. Sinclair, D.H. Watson, Chemical migration into food: an overview, Chem. Migrat. Food Cont. Mater. (2007) 1–14.

CHAPTER 6

Legislation on active and intelligent packaging

Shashi Kiran Misra[a] and Kamla Pathak[b]
[a]School of Pharmaceutical Sciences, CSJM University, Kanpur, Uttar Pradesh, India; [b]Faculty of Pharmacy, Uttar Pradesh University of Medical Sciences Saifai, Etawah, Uttar Pradesh, India

1. Introduction

Food packaging is an amalgamation of science, art and technology that ensures protection of the product during storage, transportation, sale and use. Packaging not only safeguards food product from adverse environmental conditions but also preserves its integrity [1]. Increasing global demands of packed food mandate preservation of quality, safety and integrity of products. Now-a- days, innovative packaging is preferred that can delay adverse effects of the environment and play a key role in food preservation and quality throughout the distribution chain. Although, several packaging terms like active, smart, interactive, clever or intelligent are discussed for innovative packaging but their meanings are different according to use and application. Active and intelligent packaging are considered as smart packaging, that are employed for the extension of product shelf life by stalling the undesirable outcomes (resulting from microbial growth, lipid oxidation and moisture) and monitoring the status of food within the packaging. Both active and intelligent strategies function in collaboration, and are thus acknowledged as smart packaging.

Active packaging involves addition of specific components and freshness enhancers during manufacturing process with aims to extend the quality and shelf life of the food product [2]. These additives are either attached inside the packaging systems or admixed within the packaging materials that function with primary packaging system.

Absorption of gases (oxygen, carbon di oxide), removal of food catalyst (cholesterol and lactose), oxygen scavenging, microbial and temperature control are efficiently executed by active food packaging. Among, all these the oxygen scavenging is the upmost commercially successful application and growing progressively in the meat industry. It includes self-adhesive labels and loose sachet which are added with food product in active packaging system. Ferrous oxide and photosensitive dyes are frequently used as oxygen scavenger and mostly included in the mono- or multilayer polymeric packaging materials, closures and liners of jars and bottles [3]. Moisture absorbers are another active packaging component that are popularly utilized by several food processing industries in the form of moisture absorber pad, sachet and sheets. The tear resistant sachet (permeable) containing desiccants namely, calcium oxide, silica gel, activated charcoal and clays. Food

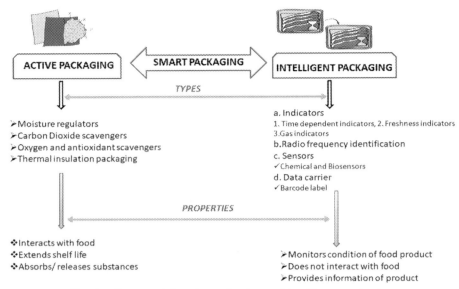

Fig. 6.1 Types and functions of active and intelligent packaging.

industries also use moisture drip absorbent systems (larger sheets, pads and blankets) for liquid moisture absorption in scaled canned food products including fish, seafood and other poultry products. The large moisture absorptive sheets or blankets perform dual action with activated carbon and iron powder that remove odor and scavenges oxygen respectively [4]. Fig. 6.1 portrays types and functions of active and intelligent packaging.

Contrary to the active packaging, intelligent packaging employs integration or combination of sensors in packaging component (primary or secondary) without interacting with food and providesinformation regarding properties and functions to the consumer. It also assures quality, integrity, product safety during storage and delivery [5]. Diverse indicators that sense or monitor altered time-temperature, freshness-ripening, gas sensing dyes, physical shock are displayed by intelligent packaging. In this system, an indicator is attached, printed or incorporated onto the food packaged item. Smart tagsincluding radio frequency identity tag, electronic label and ink printed circuit are placed outside the primary packaging material. For instance, diagnostic indicatorsare used to provide information on food storage temperature, time, gaseous contents (oxygen and carbon di oxide) and hence indicates end use dates [6]. Three important technologies used for intelligent packaging system include Indicators, Data carriers and Sensors [7] and are detailed in preceding text.

1.1 Indicators

This intelligent packaging substance detects the presence or absence of a substance, any reaction that might have occurred or color change of food product. Based upon their function, these areplaced outside or inside the package. Time temperature indicator

(Fresh-Check®, TT Sensor™), freshness indicator (SensorQ™) and gas indicator (UV activated calorimetric indicator) are few subtypes of this category [8].

1.2 Data carriers

These systems ensure traceability and theft protection of product via transmitting information throughout the supply chain. They are fixed within the tertiary packaging. Radio Frequency Identification, RFID (Intelligent Box, Easy2log) and Barcode (One dimensional and 2-dimensional) are frequently used data carriers [9].

1.3 Sensors

These are used to detect and quantify physical and chemical property of food stuff through signal(s). Bio sensors (Toxin Guard™, Food Sentinel®) and gas sensors (Tell-Tab™, Ageless Eye™) are the types of sensor based intelligent packaging system [10].

Few registered active and intelligent packaging [11] and their exclusive applications are listed in Table 6.1.

Table 6.1 Few registered active and intelligent packaging and their exclusive applications.

Active packaging

Principle	Registered name	Food packaging
Low density polyethylene film as moisture absorbent	Activ-Film™	Fruits and vegetables
Ethylene scavenger	PEAKFresh™	Fruits and vegetables
Carbon di oxide emitter	FreshPax®	Processed and precooked food products
Antimicrobial cool bags	Biomaster®	Chilled and frozen products
Oregano-essential oils embedded film coated sachet (antioxidant)	ATOX	Cereal products
Vegetable oil-based phase change materials	Green Box	Perishable food

Intelligent packaging

Self-adhesion pad containing blue dye that monitor temperature change	3M™MonitorMark®	Beverages and meat
Photochromic ink activated through UV light and progressively turns darker (blue) as the ambient temperature increases	OnVu™	Dairy products, fish and meat

Continued

Table 6.1 Few registered active and intelligent packaging and their exclusive applications.—cont'd

Red-ox dye indicator on reacting with oxygen changes color between labeling layers (transparent to blue)	Shelf Life Guard	Meat products
This flexible indicator senses generated toxins produced by species of *Escherichia, Salmonella* and *Listeria* in the packaged food till delivery	Flex Alert	Dried fruits, seeds, coffee beans
The radio frequency identifier tag monitors thermolabile food products during storage and transportation.	Easy2Log®	Dairy products, frozen food, meat and seafoods
This sachet turns pink and blue on absent and present of oxygen in the headspace of food package	Ageless Eye®	Meat products

2. Regulatory aspects in European Union

In the current scenario, different advanced techniques are utilized for the adequate production, distribution and storage of food products globally. Higher demand of packaged foodsnecessitates fulfillment of consumer expectations of food integrity, freshness, safety and quality. To meet such prospects, smart packaging isevolving worldwide to increase the usefulness of packaged foods. For the very first time, Japan introduced active and intelligent packaging in the year 1970 without specific legislations or regulatory guidelines. However, Japanese Food Sanitation Law (1947) and Food Safety Base Law (2003) were implemented by the manufacturers to certify the safety criteria of packaging materials used. Moreover, the guidelines framed by European Union and United States are used as used as reference in any legal situation. Furthermore, the regulatory criteriaof EU and US together with the a given country food law are implemented by Australia, China, New Zealand, Latin America as well, to ensure safe applications of smart packaging [12].

On storage, fresh foods generally release moisture or gas inside the packaging. For instance, the oxygen produced supports mold growth on bakery foods (bread or pizza) surface and causes rancidity of vegetable oils. These instances may challenge organoleptic as well as the food quality of packaged products. The EU legislation is very stringent and allows active and intelligent packaging as long as they comply with the requirements of food products.

In 2003, The European Union Health and Consumer Protection (EUHCP) Commissioner suggested that there must be separate specific legislations for active and

Legislation on active and intelligent packaging

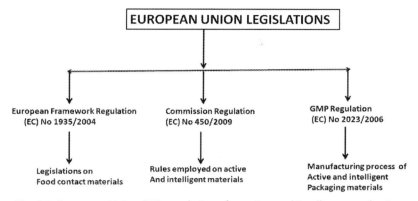

Fig. 6.2 European Union (EC) regulations for active and intelligent packaging.

intelligent packaging. This was proposed to amend the General food law which demonstrated principles for food safety covering complete food chain, Farm to Fork. Since then strict legislation is followed for active and intelligent packaged food materials so that they do not become source of contamination. It is worth mentioning, that in active and intelligent packaging, the inserted or admixed component directly interacts with the food content or head space of the container that may create doubts about the quality, integrity and safety of both the product and container. Advanced packaging materials such as adhesive tag, thermal indicator, moisture bag, printing ink may interact with food products that necessitates revised legislation. On November 23, 2011, modified guidelines on active and intelligent packaging were presented before the Member states of European Union that contained legal aspects for authorization of packaging materials or components. Fig. 6.2 compiles major regulations framed for active and intelligent packaging in the European Union.

The legislation covers following regulations:

2.1 The framework regulation

This part elucidates the legislation or guideline i.e. EC1935/2004, that permits entry of active or intelligent packaged food articles in the European market [13]. Enclosed Article 3, states that the packaged material should not interact or release constituents into the food above permissible limit or change/deteriorate organoleptic parameters that may cause harm to human health [14]. Additionally, Article 4 of the regulationalso details specific requirements that should be complied for selection of packaging components. Article 4 also encloses Good manufacturing practices (GMPs) regulation EC2023/2006 for packaging materials that come in contact with the food substance [15].

A few provisions of Article 4 are mentioned below:

i. Active packaging material should not alter the composition and organoleptic characteristics of packaged food. It should not mask the decay of the food item that can misguide the consumer.

ii. If the active packaging material makes changes in the food composition, then the changes must fulfill the criteria of EU food legislation.
iii. The indicators or active materials that are deliberately added to the food article and get released into the environment of food, should be validated and authorized with EU food safety legislation.
iv. Intelligent indicator or tag should not display the condition of food material which may mislead the users.
v. There must be proper label over active and intelligent packaged material came in contact with food that identifies non-edible parts to the consumer.

The framework regulation (Article 3, EC 1935/2004) focuses on 'functional barrier' which consists of one or more than one layers of packaging material that come in contact with food and prevents migration of active substance toward food item [16]. The legislation permits maximum acceptable migration level of 0.01 mg/kg food for active substance. It is applicable to a category of components and isomers that are related either structurally (having same functional groups) or toxicologically. This migration limit of functional barrier is expressed in concentration on the packaged food label which also should not fall under any of the below mentioned class:

i. Carcinogenic or mutagenic or toxic components that cannot be used as a functional barrier.
ii. Newer technologies that are preferentially based on smaller particles and influence the physicochemical properties are not covered under provision made for functional barrier.

In December 2004, a new community regulation for packaging materials that may possibly come in contact with food materials was proposed by the European Parliament which encompassed Regulation 89/109/EEC. The regulation specifies the basic principles on active and intelligent packaging, advanced technological innovation and legal reference that allows interaction between packaging material and food product. Regulation 89/109/EEC is comprised of following general aspects:

i. Definition and active and intelligent materials listed in the scope
ii. Packaging materials that are designed to be deliberately added as "active" component and absorb substance from foodstuff.
iii. Active substance and materials that contain natural ingredient (wooden barrels) are not recognized as active packaging.
iv. Active substance and materials that modify the organoleptic properties of food products, should comply the Community provisions. For instance, modification of color may mislead product information to consumer, so these substances are not permitted.
v. Labeling should be according to food legislation and appropriately define added active packaging materials.

2.2 Framework regulation (Article 16)

Article 16 includes declaration of compliance that states materials or substance intended for active or intelligent packaging must entertain with written declaration of compliance [17].

This attested form declares to follow legislation or rule applicable to these materials whether or not they come in contact with packaged food. Similarly, written declaration is also applicable for manufacturing of these articles or materials at the time of marketing or sale. In this aspect Annex II of European Commission Regulation (EC No 450/2009) and EU No October 10/2011 detail the related information required by law.

2.3 Commission regulation

Commission regulation covers legislation mentioned in EU No 450/2009 that include definitions or appropriate meanings of active materials and articles, components, release active material or component along with their specifications. According to this legislation active materials are intended for the extension of shelf life and maintaining of packaged food condition. Whereas, releasing active materials are defined as those active substances that are intended to release components into food item [18,19].

The Commission regulation regulates substances or materials utilized for active and intelligent packaging following these provisions:

i. The components or substances meant for active or intelligent packaging must be safe and fulfill both the requirements of Framework regulation legislations i.e. EC No. 1935/2004 and 450/2009.
ii. The substance should be tested for safety protocols designed by The European Food Safety Authority (EFSA) prior to use as active and intelligent packaging.
iii. Release substances from active packaging should follow any limitations mentioned in legislation (authorized food additives) with the food articles.
iv. The passive components or parts (plastic) of active and intelligent packaging substances should follow framework regulation (EU) No.10/2011.
v. For paper and cardboard packaging materials, no specific EU legislation is imposed and existing EU National laws should be applied. Further, intelligent packaging systems which do not come in contact with food item can be separated through functional barrier.
vi. The OML (Overall Migration Limit) from active packaging (releasing) materials can surpass the range mentioned in the EU legislation as long as the migrated substance(s) transferred into the food material complies with the provisions of authorized food additives law. Moreover, the migrated active substance cannot be considered in the calculation of the OML.

2.4 Labeling requirement

European Union food legislation states that all the active or intelligent packaging materials should be well labeled in the perspective of either edible or non-edible. Article 4(d) and article 11–13 of EC No. 450/2009 specify that inserted sachet may be apparent as edible hence; these active packaging materials must be properly labeled as non-edible to avoid any harmful incident [20].

Moreover, the legislation also stresses on providing of complete information and maintain it throughout the packaging chain (manufacturer to consumer) to ensure appropriate use of packaged food item. Similarly, the food article containing antioxidant (active packaging material) that can be released within the food, must contain information such as name of antioxidant, maximum released limit and permitted use. EU labeling provisions shall be implemented for the labeling of active material packaged food articles that may release substance within the food or container. Article 6(4) (a) of Directive 2000/13/EC considers these released substances as active ingredient hence must be treated accordingly. In this consideration, on March 20, 2000, European Parliament states that the food additives are a matter of subject to the labeling responsibilities of the EU Council; thereafter prepared legislations for labeling, packaging, advertising of packaged food articles are to be followed by Member States of EU.

2.5 Specific legislations

Specific legislations were framed to harmonize trade among 27 Member States. As there are different provisions and trade barriers and were not covered by general legislations on food packaging materials. Legislations such as Premarket approval and maximum migration limits were implemented by some countries i.e. Netherland, Italy, Dutch, Germany and France. Newer European Union (Member States) has reframed their national legislations and follows community rules meant for advanced packaging materials. On October 2018, EU parliament sets proposal to reduce usage of plastic containers in packaging industries by 2021. The Member States ambassadors stated directives for innovative packaging technologies and materials that keep packaged food fresh, safe, tasty with extended commercial practice.

Harmonized legislation in European countries is framed for FCM (direct food contact materials) specifically for regenerated cellulose, plastics, ceramics, elastomers and rubbers.

The Directive Regulation No. 78/142/EEC and 1895/2005 are specified for vinyl chloride monomer and epoxy substances containing packaging materials respectively. GMP regulation EC No. 2023/2006 is also applicable on all food contact packaging materials and regulates transfer of material from non-food contact surface to the food article (external surface to internal).

3. Regulatory aspects in India and South East Asian countries (Thailand, Malaysia and Singapore)

India, a continuous growing country with population more than 1.4 billion is specified with diverse zone of climate i.e. extreme wet and dry. Here, climatic conditions drastically vary (temperature, rainfall, dryness and humidity). India is one of the leading nations for the production of dairy, poultry, fishery, tea and sugar substances. Moreover, India is the sixth largest worldwide market of packaging and is expected to show 12.3 % compound annual growth rate in coming five years with 42.7 billion US dollar sales. A study conducted by Tata Strategic Management Group, the growth rate of Indian packaging sector estimated in the year 2015 is progressively surging from 727.09 million US dollar and will observe a substantial pattern for packaged food products, drinking water and pharmaceuticals [21]. The introduction of track and trace devices (2D barcodes, sensors, indicators and RFID) may aggressively hit the Indian packaging market. In India, the food contact legal rules or Food Adulteration Act was enacted in the year 1954 [22].

4. Bureau of Indian standards (BIS) and food safety and standards authority (FSSAI)

BIS sets legal issues for food contact materials in India. The legislation lists maximum permissible limit for colorants and pigments that should be usedwhile selecting contact material (plastic) for food products, drinking water and pharmaceuticals. BIS standards (9845:1998) cover methods of analysis prescribed for overall migrated substance from heat sealable films (single or multilayered), non-sealable films, lids of container and closure in the finished or converted form. The list contains specification for 85 organic pigments, 22 inorganic pigments and 7 dyes. However, purity specification is somehow aligning with framework legislations of Council of European Union. Few amendments are sought with respect to overall migration limits of zinc (0.05%), tighter limit for chromium (250 ppm), antimony (250 ppm) and arsenic (50 ppm). BIS 0171:1986 and 10146: 1982 provide guidelines for aptness of contact plastic and plastic resin such respectively. However, not so strict legislations, unawareness on advanced technology and unskilled users limit acceptance of active and intelligent packaging materials in India. Moreover, Indian market emphasizes more on adequate labeling of the advanced packaged food products with respect to being vegetarian/non-vegetarian, natural/synthetic colors or flavors and language (Hindi and English only).

The demand of active and intelligent packaged food products has raised drastically after pandemic (Covid-19) because people were stocking up much fresh food for long time. FSSAI has strictly prohibited usage of packing materials prepared from recycled plastics or carry bags for the purpose of storing and dispensing of food products. According to FSSAI, more than 95 lakhs tones plastic is wasted per year in which approximately, 38 lakhs tones plastic material ends up in the dumps. In favor of our ecosystem, FSSAI has requested to manufacture bio-plastic or biodegradable plastic materials for the packaging of food products. Furthermore, FSSAI issues new legislation that no food product should be labeled as "Natural" and "Fresh" without fulfilling criteria set by Indian Food Law and permission by FSSAI. Food contact materials such as aluminum, polyethylene terephthalate, polyvinyl chloride and paper are in high demand owing to their sustainability and ease to use. Nanostructured materials as antimicrobial films, oxygen scavengers and gas permeable composites are frequently used to maintain freshness and quality of packaged food items.

Various food regulations on packaging materials are framed in different Southeast Asian countries that focus on prohibitory guidelines for toxic, hazardous, and injurious containers or packaging articles and set limitations on heavy metals or toxic substances [23]. In Thailand, the Food Act of B.E.2522 implements several legislations intended for production, distribution and importation of food packaging materials. The Thai Ministry of Public Health, in year 1979, has issued stringent notifications on specifications of the packaging material intended to be in contact with food item (s). According to the regulation, food packaging material should besafe, previouslyunused, non-contaminated, must not release color and heavy metals into the food content [24]. The Thailand Food Act requires implementation of these three Ministerial Notifications concerned with packaging containers and materials:

i. Ministerial Notification No. 92/2528 specifies the quality and quantities required for food containers those are made up of ceramic, enameled metal and used for infants. The regulation also guides permissible migration limits (not more than 2.5 mg/L) of heavy metals including cadmium and lead leaching from above mentioned small and large deep vessels shaped packaging materials. Although, this regulation (1985) was revised to harmonize international standards.

ii. Ministerial Notification No. 117/2532 was enforced in 1989 that specified regulations on packaging materials for feeding bottles/rubber teat, milk and other liquid preparations meant for children and infants. For instance, the rubber cover and teat should not release heavy metals, nitrosamine (<10 ppm) and color into the food materials. The plastic bottles should be prepared of polycarbonate that could withstand high boiling temperature. Further, Thai FDA regulation is strict on the migration of heavy metals (cadmium and lead) and potassium permanganate (not more than 20 ppm) from plastic milk containers.

iii. Ministerial notification No. 295/2548 was brought by Thai government in the year 2005. It sets standards for 12 types of plastic food packaging containers and their migration testing criteria in different solvent systems namely water/pH < 5, acetic acid/pH < 5, heptane and ethanol. Clause 6 of Notification 295/2548 covers the guidelines for plastic containers to be used for packaging of milk and milk products that should be made up of polyethylene, polystyrene, polypropylene, polyethylene terephthalate. Clause 7 prohibits the application of colored and laminated plastic packaging material for food items. Moreover, this regulation restricts the recycling of food containers that have previously been used to wrap poisonous substance, fertilizer and hazardous things.

The Ministry of Health of Malaysia drives Food Act (1983) and food regulation (1985) for setting regulations for active packaging quite similar to Article 3 of European

Union Framework Regulation. It comprises of safety clause to avoid deterioration of packaged food items. Several sections 27-36 (A) detail legislations and specifications for packaging materials which are as follows:

a. Section 27 prohibits the use of harmful, toxic and injurious packaging materials.
b. Section 28 specifies safety guidelines for packaging, storage and distribution of material containing heavy metals.
c. Section 29 prohibits packaging, sale and recycle of plastic materials containing excess vinyl chloride monomer (not more than 1 mg/kg)
d. Section(s) 32 and 33 set criteria of marking or labeling on active package, container, vessel and appliance containing food which provide information of food. The regulation also makes clear to not place any other material (coins and toys) on the packaged food.

The added active packaging sachet (reduced iron powder) in food article should be sterile and contained appropriate label for the purpose of oxygen absorption. The reduced iron powder (sachet) should be well packed and contain label that claims for non-contamination, non-migration and non-interaction with food. Chemicals such as calcium hydroxide, calcium chloride and iron oxide are allowed to be in the sachet for moisture absorption. Further, caution statement as DO NOT EAT should be mentioned on label.

Singapore follows Chapter 283, Part III of Food Regulations (1988) of the Sale of Food Act. It containsrequisites for containers for packaging of food items. The guideline defines container as any food packaging material which is designed as single item and meant for sale. Part III of Food Regulation mandates following criteria during selection of active package material.

i. The PVC packaged material is restricted if it releases greater than 0.05 ppm monomer (vinyl chloride).
ii. The packaging material that is liable to release teratogenic, carcinogenic or mutagenic components, are restricted for use.
iii. Heavy metals (arsenic, cadmium, lead, antimony etc.) containing packaging materials are not allowed.
iv. Lead piping should not be allowed for the packaging of liquid food products including beer, beverages and cider.

5. Regulatory aspects in US

The term active packaging in United States of America is defined as 'any packaging material or system that shields food content from degradation and contamination by creating a safeguard between outer and inner atmospheric conditions of edible product.' Active packaging materials encompass a variety of active components that are admixed in the packaged food formulation for modification of stability and shelf life of contained food item. For instance, functional barriers (moisture and oxygen scavengers and antimicrobial) are polymer-based packaging films that are intended to absorb unwanted moisture or released oxygen and preserve food from microbial contamination, thus improve stability and quality of packaged food such as bread, vegetables and beverages [25]. Similarly, intelligent packaging materials are defined as the sensors or tags that provide information regarding condition of packaged food to the customer. This facilitates appropriate storage conditions, product integrity and the quality of food item.

6. FDA regulations

The Food and Drug Administration necessitates clearance for packaging food materials that fall under food additives under the laws of FDCA (Federal Food and Drug Cosmetic Act). According to the FDCA, food additive is a related substance that is rationally expected to be a part of food while intake or use. In this way, any component or substance of packaging material that comes in contact of food or migratessignificantly in food, will be considered as food additive. FDCA covers polymer coating, paper, plastic and adhesives films, and bottles as active food contact substances [26]. The substance is treated as unsafe until used according food additive regulation amendment 1958, framed by Food Contact Notification. 21 CFR (Code of Federal Regulation) part 175—186 comprises of food additive regulation to be implemented for active packaging materials. Moreover, the food additives that may reduce the quality of food and supposed to be injurious for human health are not permitted by FDCA [27].

Delaney Clause of FDA focuses that if a food additive is carcinogenic and can induce cancer on ingestion, then it cannot be considered safe and pure according to section 409 (c) (3) (A). Food additives may contain impurities (intermediates, residual reactants, byproducts, manufacturing aids) and should be analyzed and ensured for safety prior to put into use as packaging material. Additionally, if the active component does not migrate from packaging material then also FDA's GMP (Good Manufacturing Practices) confirms its safety and purity. Although, FDA does not provide stringent legislation that decides basic criteria on selection of migrant food additivebut few guidelines based on Ramsey Proposal and Monsanto V. Kennady are followed by manufacturing industries. According to Ramsey Proposal (1969) the active component or food additive would be permitted that migrates not more than 50 ppb (parts per billion) in the packaged food. This food additive regulation would not apply to the components that contain toxic reaction, heavy metals or carcinogen at a level of 40 ppm (parts per million). These migrating components may be relatively highly toxic when mixed with dairy (milk) and beverage (soda) packed products when used by infants and children. This proposal was not officially adopted by the FDA and its standards were accepted scientifically [28]. Further, the Monsanto proposal was an appeal from the FDA Commissioner not to use acrylonitrile polymer in beverages containers as it was unsafe owing to its toxic migrant potential.

7. Threshold of regulation rule (1995)

This rule of FDA exempts non-carcinogenic food additive or active package material from the requirement of premarket clearance. The rule recommends that the food additive or food contact substance is unimportant and needs no special provision. The manufacturer needs to confirm following information:

  i. The added concentration of food additive should not exceed from 0.5 ppb
  ii. If the substance has got clearance from food additive regulation, the dietary amount or concentration after proposed food contact use would not be exceed 1 % of the acceptable or permitted daily intake.
  iii. A stringent acknowledgment from FDA should be released indicating that the presence of food additive or substance in the diet is very less and safe for human health.

8. Regulatory aspects in Brazil, South and Central America

Grupo Mercado Comun (Common Market Group, GMC) initiated the process of harmonizing regulations framed for food contacting packaging materials in the year 1992 [29].

Mercosur is an essential political and commercial alliance of several countries (Brazil and regions of South and Central America) that deals with regulatory scheme for use of food contact materials [30]. On March 26, 1991 Mercosur was recognized by the Treaty of Asuncion del Paraguay to promote free trade among associated members of the countries. Currently, Mercosur follows GMC resolutions related to food contact plastics, metals, glass, ceramics, and lubricants. In Brazil, Mercosurtrade legislation on packaging materials is emerging and emphasizing on interaction of packaging food with/or surrounding environment. This regulation contains framework resolutions both for food contact articles (FCA) and materials (FCM) which are closely similar with established EU legislations and few provisions of USFDA [31]. In Brazil, ANVISA (The National Agency of Sanitation Surveillance) and the Ministry of Agriculture, Livestock and Food Supply are empowered to take decision on import and any contradiction on food contact materials [32]. Mercosur is the largest trading region in South America and contains provinces of Brazil, Argentina, Uruguay, Venezuela and Paraguay. These regions accept GMC Resolution No. 3/92 that cover national legislations on general safety and specified limits for migrants of all FCA/FCM [33].

A few GMC regulations and their specifications are:
  a. GMC Resolutions No. 32/07specifies positive list of food contact plastic additive that are used to prepare food contact plastics.
  b. GMC Resolutions No. 02/12 covers positive list of polymers/monomers and starting materials used for the preparation of food contact plastics and their specifications or limitations.
  c. GMC Resolutions No.32/10 mentions analytical approach and framework analyzing conditions are mentioned.

d. GMC Resolutions No. 27/99 gives specifications for adhesives
e. GMC Resolutions No. 15/10 defines specifications for pigments and colorants

The first two GMC Resolutions are closely related to the EU provisions October 10/2011 and Directive 2002/72/EC. The third one is similar to the Directive 82/711/EEC. The Mercosur Regulation on plastic material additives and polymeric film coating that come in contact with food was implemented and transposed into Member States on July 2019. In the year 2004, European Union introduced provisions and guidelines for smart packaging (active and intelligent) that cover specific methods for determination of suitability of packaging materials. Prior to this many countries utilized traditional materials or substances for smart food packaging. On March 2008 and November 2012, the Positive lists comprising packaging materials and metals for food were released in the resolution RDC NO.17 and 56.

On April 29, 1997, The Ministry of Health of Paraguay framed legislations on food packaging criteria in Decree 17056 by its national law 'Mercosur'. INAN (The National Institute of Food and Nutrition of Paraguay) and INTN (The National Institute of Technology) both are accountable for enforcing the Food contact legislation and publication of technical standards on food contact materials. The legislation (Decree 6115/2011) was revised in February 2011 that contains National Registry of Food Packages materials mandatory in Paraguay. Similarly, Uruguay's Ministry of Public Health and The Uruguay Technological Laboratory both keep an eye over registration and assessment of food contact materials designed through Mercosur Resolution 1994. In Venezuela, the full membership in Mercosur trade was provided on July 2012. The Reglamento Generalde Alimentos, Decree 525 (1959) and Mercosur resolution 82 (2007) were implemented in Venezuela that contain GMP, standardization, quality certification for food contact materials. Presently, 'COVENIN' an industrial standard was enacted that comprises legislations for heavy metal migrants, applications of isocyanate contained films and adhesives in active food packaging materials. SENCAMER, a National Autonomous Standardization, Quality, Metrology and Technical Regulations Services regulate maintenance and certification programs of COVENIN norms.

In Argentina, use of food contact materials is governed by Codigo Alimentario Argentino (CAA). Different entities such as The National Wine Institute regulates packaging material for wine packaging, The National Service of Agricultural Food Health and Quality controls packaging of vegetables, meat and seafoods. The National Food Institute monitors food contact materials registered with SENASA (Servicio Nacional de Sanidad Calidad Agoalimentaria) and packaging materials for beverages and health supplements. Central America and its seven countries (Belize, Costa Rica, Nicaragua, Panama, Guatemala, El Salvador and Honduras) do not have any precise legislation on the active and intelligent packaging for food. The General Health Laws of related countries follows Reglamentos Technicos Centroamericanos (RTCA) provisions for direct food contact packages.

Table 6.2 Different active and intelligent food packaging and their country of origin.

Active and intelligent packaging	Purpose	Products	Country
RFID tags	Product traceability, identification and livestock management	CAEN RFID easy2log™ RT0005ET	Italy
Oxygen sensors	Detection of oxygen inside packing	OxySense®	USA
Gas indicators	Remind time period for usage	Ageless Eye®	Japan
Temperature sensor	Senses temperature of packaged meat and dairy foods	ThinFilm	Norway
Freshness indicators	Detection of oxygen, carbon di oxide, ethylene etc. in packaged fruits	Freshpoint Ripesense®	Israel New Zealand
Light-activated scavenger	Absorption of oxygen via UV light	Zero2™	Australia
Humidity regulator	Humectant	PitChit	Japan
CO_2 absorber	Absorption of carbon di oxide	FreshLock®	USA
Ethylene scavenger	$KMNO_4$ embedded silica oxidizes ethylene into acetate and ethanol	GreenPack®	Japan
Oxygen and Carbon di Oxide scavenger	Scavenge O_2 and CO_2	OXYGUARD™	Japan
Dynamic QR code	Provides complete product information and nearest cycling center.	TetraPAK	India

Table 6.2 compiles some prominent global marketing companies that are key players in the active and intelligent food packaging systems such as Amcor, Ampac, BASF, Avery Dennison, DuPoint, Graphic packaging and many more.

9. Conclusion

Traditional packaging may fail to meet satisfactory standards (safety, quality and cost) for sustainable packaged food products that consumer seek in dairy, meat and beverage articles. Unclear and nonspecific legislations limit the frequent usage of traditional packaging that raise questions on shelf life, quality and safety of food reach to user. Materials involved in active and intelligent packaging must follow the legislations framed by authorized regulatory bodies of concerned regions. The chapter illustrated specific provisions framed by different global provinces engaged in active and intelligent packaging for food. Framework Regulations 1935/2004/EC and 450/2009/EC directed by European Food Safety Authority (EFSA) found more stringent on active and intelligent packaging materials including polymers, inorganic and organic compounds, heavy metals and dyes. However, tracking and complying these legislations will need substantial time and enough capitals that need to be flexible to make ease operations for manufactures.

References

[1] R. Dobrucka, The future of active and intelligent packaging industry, Log Forum. Sci. J. Logist. 9 (2013) 103–110.
[2] R. Ahvenainen, Active and intelligent packaging: an introduction, in: R. Ahvenainen (Ed.), Novel Food Packaging Techniques, CRC Press, Finland, 2003, pp. 5–21.
[3] J.P. Kerry, M.N. O'Grady, S.A. Hogan, Past, current and potential utilization of active and intelligent packaging systems for meat and muscle-based products: a review, Meat. Sci. 74 (2006) 113–130.
[4] M.L. Rooney, Introduction to active food packaging technologies, in: J.H. Han (Ed.), Innovations in Food Packaging, Elsevier Ltd, London, UK, 2005, pp. 63–69.
[5] K.L. Yam, P.T. Takhistov, J. Miltz, Intelligent packaging: concepts and applications, J. Food. Sci. 70 (2005) R1–R10.
[6] D. Dainelli, N. Gontard, D. Spyropoulos, B.E. Zondervan-van den, P. Tobback, Active and intelligent food packaging: legal aspects and safety concerns, Trend. Food. Sci. Technol. 19 (2008) 103–112.
[7] M. Ghaani, C.A. Cozzolino, G. Castelli, S. Farris, An overview of the intelligent packaging technologies in the food sector, Trend. Food Sci. Technol. 51 (2016) 1–11.
[8] L. Roberts, R. Lines, S. Reddy, J. Hay, Investigation of polyviologens as oxygen indicators in food packaging, Sens. Actuators B: Chem. 152 (2011) 63–67.
[9] D. McFarlane, Y. Sheffi, The impact of automatic identification on supply chain operations, Int. J. Logist. Manag. 14 (2003) 1–17.
[10] S.Y. Lee, S.J. Lee, D.S. Choi, S.J. Hur, Current topics in active and intelligent food packaging for preservation of fresh foods, J. Sci. Food Agric. 95 (2015) 2799–2810.
[11] E. Drago, R. Campardelli, M. Pettintao, P. Perego, Innovations in smart packaging concepts for food: an extensive review, Foods 9 (2020) 1628.
[12] General Food Law: regulation (EC) No. 178/2002 of the European Parliament and of the Council of 28 January 2002, laying down the general principles and requirements of food law, establishing the European Food Safety Authority, and laying down procedures in matters of food safety, OJL 31 (2002) 1.
[13] Regulation (EC) No. 882/2004 of the European Parliament and of the Council of 24 April 2004 on official controlsexercised to ensure compliance with the feed and food law, animal health, and animal welfare, Corrigen. OJ L191 (2004) 1.
[14] EFSA Guidelines on submission of a dossier for safety evaluation by EFSA of active or intelligent substances present in active and intelligent materials and articles intended to come into contact with food, EFSA J. 1208 (2009) 2–11.
[15] EFSA Scientific Committee, Scientific opinion on exploring options for providing advice about possible human health risks based on the concept of Threshold of Toxicological Concern (TTC), EFSA J 10 (2012) 103.
[16] Regulation (EC) No. 1935/2004 of the European Parliament and of the Council of 27.10.2004 on materials and articlesintended to come into contact with food and repealing Directives 80/590/EEC and 89/109/EEC, OJ L338 (2004) 4.

[17] Commission Regulation (EC) No 450/2009 of 29 May 2009 on Active and Intelligent Materials and Articles Intendedto Come into Contact with Food of L135, 2009, pp. 3–11.
[18] Council Directive 82/711/EEC of 18 October 1982 laying down the basic rules for testing migration of the constituents ofplastic materials and articles intended to come into contact with foodstuffs, OJ L297 (1982) 26.
[19] Council Directive 85/572/EEC of 19 December 1985 framing the list of simulants to be used for testing migration of the constituents of plastic materials and articles intended to come into contact with foodstuffs, OJ L372 (1985).
[20] B. Magnusona, I. Munro, P. Abbot, N. Baldwin, R. Lopez- Garcia, K. Ly, L. McGirr, A. Roberts, S. Socolovsky, Review of the regulation and safety assessment of food substances in various countries and jurisdictions, Food Addit. Contam. 30 (2013) 1147–1220.
[21] R. Shinde, R. Victor, K. Shanthanu, J. Subramanian, Active and Intelligent Packaging for Reducing Postharvest Losses of Fruits and Vegetables, first ed., Postharvest Biology and Nanotechnology, 2018.
[22] L.S. Lee, K.D. Fiedler, J.S. Smith, Radio frequency identification (RFID) implementation in the service sector: a customer-facing diffusion model, Int. J. Prod. Econ. 112 (2008) 587–600.
[23] K.S.M. Naim, Food labeling regulations in South asian association for regional cooperation (SAARC) countries: benefits, challenges and implications, TURAF 3 (2014) 196.
[24] D. Pereira, N. Teixeira, I. Zin, R. Gonçalves, N. Melo, Active and intelligent packaging: security, legal aspects and global market, Braz. J.Dev. 6 (2020) 61766–61794.
[25] J. Heckman, Food packaging regulation in the United States and the European Union, Regul. Toxicol. Pharmacol. 42 (2005) 96–97.
[26] D. Dainelli, Global legislation for active and intelligent packaging materials, in: J.S. Baughan (Ed.), Global Legislation for Food Contact Materials, Woodhead Publishing, Cambridge, 2015, pp. 183–199.
[27] Z. Fang, Y. Zhao, R.D. Warner, S.K. Johnson, Active and intelligent packaging in meat industry, Trend. Food Sci. Technol. 61 (2017) 60–71.
[28] L.S. Gold, E. Zeiger, Handbook of Carcinogenic Potency and Genotoxicity Databases, CRC Press, Florida, 1997.
[29] A. Ariosti, MERCOSUR Legislation on Food Contact Materials, 2018, https://doi.org/10.1016/B978-0-08-100596-5.21879-9.
[30] H. Soto-Valdez, Latin American food contact legislations, in: Food Contact Legislation for Packaging in Emerging Markets, PIRA, San Francisco, Leatherhead, 2006, p. 19.
[31] Global Agricultural Information Network Report-Brazil," a Report by the Foreign Agricultural Service of the, U.S. Department of Agriculture (USDA), 2016.
[32] B. Magnuson, I. Munro, P. Abbot, N. Baldwin, R. Lopez-Garcia, K. Ly, L. McGirr, A. Roberts, S. Socolovsky, Review of the regulation and safety assessment of food substances in various countries and jurisdictions, Food Addit. Contam. Chem. Anal. Control Expo. Risk Assess. 30 (2013) 1147–1220.
[33] P. Silva, A. Da Rocha, Perception of export barriers to Mercosur by Brazilian firms, Int. Mark. Rev. 18 (2001) 589–611.

CHAPTER 7

Market demand for smart packaging versus consumer perceptions

Madhur Babu Singh[a,b], Prashant Singh[b] and Pallavi Jain[a]
[a]Department of Chemistry, SRM Institute of Science and Technology, Delhi NCR Campus, Ghaziabad, Uttar Pradesh, India;
[b]Department of Chemistry, Atma Ram Sanatan Dharma College, University of Delhi, New Delhi, India

1. Introduction

Food safety, defined as the pledge that food products will not give rise to any harm if eaten as intended, is a global concern that has an influence on consumer health in both developed and developing countries. As a consequence, strengthening food safety is an ongoing and critical exploration for governments around the globe, especially as consumer concerns rise [1]. Meanwhile, buyers are wanting better levels of quality and taste, with words like "clean label" and "minimal processing" gaining momentum. Traditional techniques of assuring food safety, such as the inclusion of preservatives and heat processing, would not fulfill the expectations of today's consumers [2].

Effective food packaging methods are an alternate technique for guaranteeing customers of a food item's safety. Packaging delivers extended of functionality and satisfaction to consumers as a ubiquitous feature in current consumption patterns. Packaging elements are often used in the food industry to preserve product quality, reduce product loss, improve transportation and storage, and create market distinctiveness [3]. While traditional or passive food packaging has offered preservation for food items, challenges in transportation and user orders have prompted substantial research into new food packaging strategies. These packaging approaches, which are commonly mentioned as "smart packaging" (SP), include both active and intelligent packaging strategies.

Protection, communication, convenience, and containment are all goals of traditional food packaging. The packaging is used to protect the product from degradation led by external environmental circumstances such as heat, light, moisture, pressure, microbes, gaseous outrush, and so on. It also offers the user more simplicity of use and time savings, as well as a variety of product sizes and forms [4]. Traditional packaging materials that come into touch with food must be as inert as feasible in order to be safe. Smart packaging solutions, such as active and intelligent packaging ideas, are depended on the helpful reaction between the package surrounding and the food in order to assure food security [5].

Consumers' buying choices are influenced by packaging, which is one of the most essential product features. Consumer perceptions toward the new generation of packaging are a useful source of information for manufacturers when building marketing strategies for new product design and positioning on the market. "At a time when new technologies and possibilities are generated by the present economy, it is precisely the inventive customers that build the market for new brands and new items, first by demonstrating their usage in front of imitators and then by popularizing their attitudes," says the author [6]. Traditional packaging still dominates the market, but newer packaging materials, packaging architectures, and packaging technologies are also being employed more frequently. Food items' use-by dates are extended due to advancements in packaging, while better storage procedures and cold chains allow for longer transportation. New-generation packaging's properties and qualities can highlight a product's distinctiveness and reliability, and so attest to its quality.

Most industry experts anticipate that active packaging, the next generation of food packaging, will be the norm in the coming years. Active systems are anticipated to account for 35% of the value of the advanced packaging sector, which is around 5% of the entire packaging industry value [7]. In spite of the fact that active packaging represents just a tiny fraction of total package sales, there are indicators that this sector may see considerable development in the next years [8,9].

The primary goal of packaging is to preserve a goods from deterioration effected by exposure to and use in the external atmosphere. Furthermore, goods packaging is a powerful marketing tool for communicating with customers. It comes in many different of forms and sizes, and as a user interface, it offers users comfort and simplicity of use. Traditional packaging, on the other hand, is no longer adequate due to rising customer demands, increased goods complexity, and now a days, national and global measures aimed at supporting a circular economy and reducing production and manufacturing carbon footprints. To meet a range of new customer demands, creative packaging with greater function is also necessary. Offering meals prepared having fewer preservatives, items which satisfy greater regulatory standards, and packaging that enables cradle-to-grave identification and so protects against litigation are just a few examples. Furthermore, smart packaging aids to accept higher national and global food security laws and also protects from possible dangers of food bioterrorism in the context of globalization. Packaging is used to safeguard the goods, it may also be utilized by the company to promote their marketing pitch and increase sales. A nice package aids consumers in recognizing and distinguishing items. Packaging is used to facilitate transportation and ensure product safety. Packaging aids in product differentiation from competing brands. Companies must comprehend how consumers are influenced during the purchasing process. They must also comprehend what variables impact purchasing behavior and the importance of packaging features in the purchasing decision process of customers. Market research assists businesses in developing the "proper" packaging for a product, as well as the packaging aspects that may be important to customers. Companies may take advantage of this knowledge in a thoughtful plan to provide the appropriate products and services to the right customers on time. Consumers react to packaging depending on prior

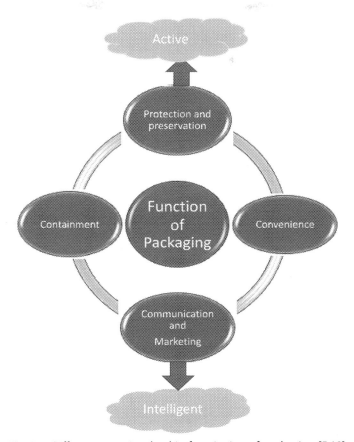

Fig. 7.1 Different steps involved in functioning of packaging [5,10].

knowledge, learned responses, and personal preferences. As a result, package components such as forms, colors, sizes, and labels may impact customers' favourable responses. Packaging is frequently the last impression a customer or consumer will have of your items before making a final purchase decision, so it's essential making sure it's working as hard as it can to get that sale, whether that's through imagery, brand values, product functionality, or sheer innovation (Fig. 7.1). Packaging may be used to enhance value in a variety of ways.

2. Smart packaging

"Smart packaging is a whole package alternative that tracks and responds on variations occurred in goods or its surroundings and on the other side respond on these variations (active)." Chemical sensors or biosensors are utilized in smart packaging to regulate food standards and safety from manufacturers to consumers. Smart packaging uses a range of

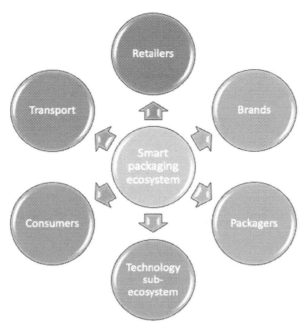

Fig. 7.2 Things needed for smart packing solution [11].

detectors to regulate food standards and security, such as finding and identifying freshness, pathogens, leaks, CO_2, O_2, temperature, time and pH level. Specific smart packaging systems' functionality varies and is dependent on the goods being packaged, such as food, drinks, medications, or many sorts of health and home supplies. Things required in smart packaging are shown in Fig. 7.2.

3. Active packaging (AP)

The modern customer is dissatisfied with traditional packaging's "passive" protective function and lack of information about the product and how it is preserved. As a result, in the 1980s, research on alternative, active packaging began, which affected product quality and safety. The package, the product, and the environment are interconnected in active packaging, also known as interactive packaging. In contrast to typical packing materials, the active ingredients utilized to cause the product's use-by date to be extended while maintaining its superior quality while interaction with the inside atmosphere [12]. Aside from product protection, active packaging also serves as a shield against the effects of external stimuli. Interaction with the packaged product is their primary method of action. The relationship between the goods and its packaging is critical since it increases the product's storage life or enhances its sensory characteristics. The purpose of AP is to improve the storage of products in the packet by combining multiple approaches such

Table 7.1 Name of a few moisture absorbers commonly employed in food packaging [17,18].

Sorting	Materials used as a moisture absorbing
Inorganic	Silica gel, natural clay, $CaCl_2$, $MgCl_2$, $AlCl_3$, LiCl, CH_3COOK, $CaBr_2$, $Ca(NO_3)_2$, $ZnCl_2$, P_2O_5, activated alumina, CaO, BaO, NaCl, KCl, K_2CO_3, NH_4NO_3, bentonite, $(NaPO_3)_6$
Organic	$C_6H_{14}O_6$, $C_5H_{12}O_5$, fructose, ammonium carboxymethyl cellulose, Na-carboxymethyl cellulose, K-carboxymethyl cellulose, triethanol amine monoethanolamine carboxymethyl cellulose, $HN(CH_2CH_2OH)_2$
Polymer build	Starch copolymers, PVA, absorbent resin
More synthesized	Acrylamide synthesis attapulgite, starch-grafted sodium poly acrylate, acrylamide synthesis attapulgite, diatomaceous earth

as temperature regulation, oxygen removal, moisture management, and the inclusion of chemicals such as CO_2, natural acids, sugar, and salts [13]. Delay oxidation in muscle meals, monitored the respiration rate in horticulture goods, microbiological germination, and moisture migration in dry goods are just a few of the advancements made possible by active packaging. Active packaging also uses coating, micro-perforation, lamination, co-extrusion, and polymer mixing to alter selectivity in order to change the air proportion of gaseous components within the container [8]. Various applications of Active packaging in food industry are listed in Table 7.2. Active elements are delivered to this form of packaging in two different ways (i) they are placed in little bags within the packaging (ii) they are delivered directly to the packaging material [6]. Various types of active packaging are used in the food industry as shown in Fig. 7.3.

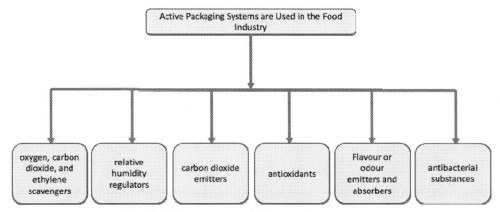

Fig. 7.3 Different types of Active packaging system.

3.1 Oxygen scavengers (OS)

Nowadays OS are among the highest extensively utilized active packaging techniques for foods. The introduction of O_2 in a packaging hastens food's oxidative degradation. The proliferation of erobic microorganisms, the formation of bad flavors and odors, color variation, nutritional drop, and general shelf-life balance of muscle foods are all aided by oxygen. As a result, controlling O_2 concentration in food packaging is critical for restricting the pace of such food spoiling processes. Although O_2-sensitive foods can be wrapped in modified atmosphere packaging (MAP) or vacuum packing, these methods will not totally exclude O_2. The O_2 that enters the packing film is not eliminated by the method. Quality changes in oxygen-sensitive foodstuffs might be reduced by using oxygen scavengers, that absorb the remaining oxygen after packing [14]. The oxygen absorbent are engineered to lower oxygen content in package headspace to less than 100 ppm. Fe powder oxidation, $C_6H_8O_6$ oxidation, photosensitive dye oxidation, saturated fatty-acid oxidation, immobilized yeast on solid substrate, and other methods are used in commercially accessible oxygen scavengers [15].

3.2 Moisture scavengers

Excess moisture in food packing must be controlled to avoid microbiological development and hazy film formation. The collection of moisture inside the packaging will become more evident if the packaging has low water vapor penetration. Intake of fresh produce, temperature variations, or leaking of tissue moisture from fresh-cut flesh are the most common causes of excessive water production inside packaged foods. Extra moisture within the packaging encourages the growth of germs and mold, decreasing the standard and shelf-life of the items. Moisture scavengers actively collect and retain water molecules from their surroundings [16]. Drying agents lower humidity levels in the atmosphere by absorbing moisture from the surroundings via both physiochemical adsorption. Moisture scavengers including such silica gel, molecular sieves, natural clay, CaO, $CaCl_2$, and modified starch are the most efficient technique to manage excess water collection in food packaging with a strong barrier toward water vapor. Because it is safe and non-corrosive, silica gel has been the most extensively utilized moisture scavenger. Physically moisture adsorption happens when silica gel absorbs moisture, and chemical moisture adsorption happen when calcium chloride absorbs moisture. Moisture scavengers are the most widely utilized moisture scavengers in packaged foods. Different types of moisture absorbers were given in Table 7.1.

3.3 Ethylene scavengers

Ethylene is a colourless and odorless pure unsaturated hydrocarbon. It is a hormone generated by plants that regulates its growth phase, respiratory speed, somatic embryogenesis, seed germination, and basic development and growth. On the one side, its

Table 7.2 Food applications for active packaging [17,19].

Active packaging types	Type of food	Exception
O_2 scavenger	Cooked meat products	Discoloration protection
	Grated cheese, bakery products	Mold growth protection
	Fruit and vegetable juices	Vitamin C retention, browning protection
	Seeds, nuts, and oils; quick fat supplements, cooked snacks; and dried beat items	Rancidity protection
Moisture scavenger	Seeds, strawberries, mushrooms, fresh seafoods, maize, tomatoes	Prolonging one's life by preserving moisture content, moisture condensation in the packing is reduced, good influence on appearance, browning or blemish decrease
Ethylene absorber	Climacteric fruits and plant edible	Reduced ripening and infirmity, which improves standard and extends shelf-life
Antioxidant releaser	Fresh fatty fish and meat; fat-containing instant powders; seeds, nuts, and oils; fried products	Improvement of oxidative stability
CO_2 releaser	Fresh fish and meat	Microbiological shelf life is extended, and the head space capacity of modified environment packaging is reduced.
Anti-microbial packaging system	Vegetable, bakery items, meat, dairy goods, fresh food, frozen foods	Bacterial growth suppression or retardation, shelf life extension

approach led to guiding an adequate ripening stage which leads to the result of making fresh products for consumers, but on the next side, the rate of change of improvement and diminishment of chlorophyll results to end up causing a decline of standard and shorten the shelf-life of perishable goods in the period of post-harvest storage [19]. During the ripening period, climacteric fruit and vegetables create a large quantity of ethylene, aldehydes, and many more gases, that helps to accelerate the process. Active technologies that reduce ethylene in the package environment can help to mitigate negative impacts on these goods [20]. Climacteric fruit ripening can be triggered by as little as

1 ppm of ethylene in the packaging. KMnO$_4$ (4%–6%) based on neutral matrices like alumina or silica gel is the most extensively utilized ethylene scavenger [18]. Because of its toxicity, this molecule is not incorporated onto areas that come into touch with food; instead, it is incorporated in minerals or nanoparticles to improve its scavenging capacity, and it is normally packaged in permeable pouches. Additionally, potassium permanganate based scavengers are accessible as tube filters, blankets, and films.

3.4 Carbon dioxide emitter (CO$_2$)

Some foods produce CO$_2$ as a result of rotting and respiratory responses. The CO$_2$ must be evacuated from the package to avoid the goods from rotting and the packaging from being damaged. The Strecker decomposition process involving sugars and amines, for example, roasted coffee can result in 15 atm of dissolved CO$_2$. CO$_2$ scavengers might help in this situation. O$_2$ and CO$_2$ scavenging sachets are being utilized to prevent oxidation-related flavor alterations and absorb CO$_2$ [21]. CO$_2$ permeability is 3–5 times greater as compared to oxygen permeability in some polymeric films. CO$_2$ must be continually created in such instances to maintain the required level in the packing. CO$_2$ generating devices are utilized in the packing of fresh meat, poultry, fish, and cheese. O$_2$ scavengers and CO$_2$ releasers are utilized simultaneously in certain foodstuffs when the size and look of the package are important. This prevents the packaging from collapsing owing to O$_2$ absorption [22].

Carbon dioxide acts as a microbial germination inhibitor in a packaging system. As a result, a CO$_2$ generation system may be thought of as a supplement to O$_2$ scavenging. CO$_2$ has a three to five times greater penetration throughout most plastic films over oxygen, thus it should be continually generated to keep the correct concentration inside the package [23].

3.5 Flavor or odor emitters and absorbers

Volatile chemicals like aldehydes, amines, and sulfides which collect within the packaging as a result of food deterioration can be effectively scavenged [24]. While transporting heterogeneous loads, flavor scavengers restrict cross contamination of unpleasant odors. Morris created odor-proof packaging for the shipment of durian fruit. The packet is composed of a smell-resistant polythene, such as PET or polyethylene of appropriate width, as well as a valve for the flow of breathing gases and an odor-absorbing packet consisting of charcoal and nickel combination. By adding acidic substances like citric acid into polymers, the generation of volatile amines as a result of protein breakdown in fish muscle may be eliminated. Other non-food aromas, such as taints, can be prevented by using high barrier packing materials [8].

3.6 Antioxidants

Antioxidants are commonly employed as food additives for increasing lipid oxidation durability and shelf life, especially in dry-foods and foods that are susceptible to oxygen. Sometimes antioxidants might put in plastic films to help preserve them from deterioration by stabilizing the polymer. Antioxidant levels in polymeric films are known to decline over storage owing to oxidation, as well as diffusion from the polymer's core to its surface, following by evaporation. The waxed paper has been employed as a container for antioxidant emission in the grain sector in the United States [25]. The benefit of containing antioxidants inside packaging material outweighs the benefit of using them directly in food compositions. As a result, the majority of antioxidant systems are produced as packets, pads, or tags, or are included into packaging monolayer or multilayer materials [26].

4. Intelligent packaging (IP)

IP is a sort of SP which is commonly employed in the food, drink, and drug companies. Despite the fact that intelligent packaging is linked to the food sector, it has no straightforward impact on the product. They have the capacity to transmit the packed product's situations, and they do not interface with the goods. The purpose is to keep track of the goods and provide updates to the consumer. Intelligent packaging, according to the European Commission, is "materials and items that record the state of packed food or the conditions encompassing the food" [27]. The introduction of new IP that can continuously supply data status of both food items and packaging integrity allows it to function as a safe and efficient distribution chain, reducing food waste and avoiding unnecessary transportation and logistics [5]. Intelligent packaging systems use indicators, sensors, and data carriers as its key technology [7].

4.1 Indicators

Consumers are given evidence of the existence or non-presences of a material, the range of a relation among 2 or more compounds, or the quantity of a particular items or category of products through indicators [28]. The most common way to convey these information is through quick visual changes, such as varying color intensity or dye diffusion along the indicator geometry [29]. Despite the wide range of indicators, they may all be categorized into three types: time-temperature indicators, freshness indicators, and gas indicators geometry [30].

4.2 Data carrier

Data carrier gadgets, often termed as automated identification gadgets, improve the efficiency of information transfer throughout the food supply network, which benefits food safety and quality [31]. Data carrier technologies, in particular, are not meant to offer data on the quality state of food, but rather to automate, track, stop theft, or guard against counterfeiting. Barcode labels and radio frequency identification (RFID) tags, which fall under the of class simplicity intelligent technologies, are the most essential data carrier gadgets in the food packaging sector. Barcodes have already been broadly utilized in large-scale retail businesses from the start to help with stock control, stock reordering, and checkout. For the time being, one-dimensional barcodes have now been created, however, they have restricted data store ability. Radio frequency identification systems, on the other hand, are the most powerful data transmission devices. An RFID technique having of three basic parts: a tag, which is made up of a microchip attached to a tiny antenna; a reader, which sends out radio signals and accepts answers from the tag; and middleware, which links RFID hardware to corporate operation [32,33].

4.3 Sensors

The most potential and novel innovation for future intelligent packaging systems is sensors. A sensor is a device that includes electronics for management and interpretation, as well as an interlinkage network and software [34]. By sending an indication for the identification or assessment of a chemical or physical attribute for the gadget to reacts, a detector can identify, find, or increase energy or matter. Temperature, humidity, pH, and light exposure are all measured using the most common conventional sensors [35]. Furthermore, the requirement to monitor food quality and package integrity has sparked a rising interest among researchers in the use of biodegradable and sophisticated sensors in intelligent packaging. The first part of a sensing element is a receptor, that is typically made up of a particular coating that serves as a sampling area for identifying the presence, action, proportions, or level of a particular chemical sample matrix via layer adsorption, resulting in the reconfiguration of a particular property of the coating.

5. Consumer perception

If you want people to remain long-term customers, then company must earn their trust. Greater than half of the globe's population is prepared to pay as much for items from trusted brands. And 75% of customers would only buy from companies they trust completely. You might need smart packaging to offer consumers with detailed that verifies that company product is exactly what you say it's real and that it is created in the manner you claim. Consumers can now view geographical data by the virtue of smart packaging to learn where such goods was created, how long it took to transport, as well as other details.

We live in an emerging "smart world," as we have undoubtedly heard it thousand times. Anything nowadays can be "smart," even your refrigerator to your pen, provided you are ready to pay the proper amount. A few of these devices are, admittedly, a little strange. It may also seem a little foolish when you hear the terms "smart" and "packaging." Smart packaging aids in the creation of a digital link among customers and merchants [36]. A smart packaging may have a significant impact on a company's success. Therefore, the corporation must be ready to go the extra mile and spend more money on packing. More people will be willing to purchase their products on a regular basis, social media presence will increase, and return on investment will be substantial. It is advisable to go one step further and invest in intelligent packaging. Smart packaging helps you to build a strong relationship with the customers and create purchase habits in the consumers that will benefit both the business and its well-being. Consumers considered smart packaging extremely beneficial for a number of reasons, ranging from practical utility to luxurious emotional and social ones, and the exact was observed for purchasing obstacles. Investigating these explanations in greater depth, as well as how wider psychological moderating factors and socio-demographic control variables influence the thought of various worth and barriers, can lead to an excellent understanding of the factors that influence the buying of foods increased with SP [37]. Customers respect honesty and authenticity, in addition to simplicity or in other terms, consumers desire to know that the item on the label matches the goods inside. Companies must constantly guarantee that their packaging makes them appear credible in order for this to work. This may be accomplished by ensuring that the package is manufactured of the best possible quality materials. The industry's character should be reflected in the packaging. Packaging should be recyclable and inventive if a firm is green and current [38]. Customer purchasing behavior is directly influenced by purchase intention and consumer perception. The packing material's grade may preserve goods, make it more appealing, make it more desirable, and give it a favourable image. The consumer is drawn to the product by a beautiful backdrop, color, form, good labeling, arrows, icons, and tiny (smart) packaging [39]. It is also suggested that advertising and business actually pay attention to the importance of effective packaging. They will be the reason of product failure if they utilize, adopt, launch/introduce inadequate packaging. As a result, marketing managers must concentrate on packaging standards and create a plan that considers variables and dimensions of marketing in product packaging (Fig. 7.4) [40].

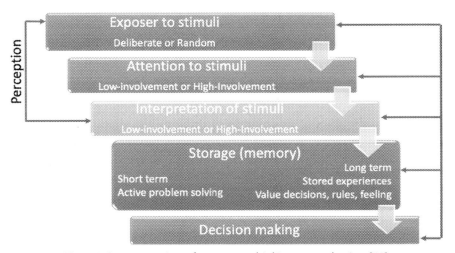

Fig. 7.4 Representation of consumer thinking on packaging [40].

6. Conclusion

Packaging is now becoming extremely important in today's competitive environment, where almost every company strives for victory in its area, that can only be achieved via the use of new packaging strategies that entice people to acquire their goods. The color of the packaging aids consumers in distinguishing their preferred brands, and it aids businesses in catching the eye and curiosity of consumers. As a result, color, together with other packaging components, makes the marketing offer more appealing and distinguishes it from competing items. In addition, creativity in packaging design aids in the retention of all product details and attributes in the memory of customers. This also aids in the categorization of their preferred items. To promote customer trust in the safeguard of wrapped food, intelligent technologies be properly labeled. Administrating considerations, like potential impacts on person health, variations in the content and sensory profiles of foods, and feasible contamination movement, must all be considered by packaging producers, particularly those devices deliberated to be put within the packet.

In past few years, several smart packaging methods have emerged, that are being incorporated into packaging system to satisfy the needs of the food logistic network. The food business can benefit from adopting appropriate packaging techniques to lengthen shelf-life, enhance quality, increase safety, and give details regarding the product. Research upon these smart packaging technologies may lead to enhancements to the current system. Smart packaging has unexplored potential in the future to provide customer advantages and convenience.

References

[1] H.-M. Lam, J. Remais, M.-C. Fung, L. Xu, S.S.-M. Sun, Food supply and food safety issues in China, Lancet 381 (2013) 2044–2053.

[2] T. Li, K. Lloyd, J. Birch, X. Wu, M. Mirosa, X. Liao, A quantitative survey of consumer perceptions of smart food packaging in China, Food Sci. Nutr. 8 (2020) 3977–3988, https://doi.org/10.1002/fsn3.1563.

[3] N.D. Steenis, E. van Herpen, I.A. van der Lans, T.N. Ligthart, H.C.M. van Trijp, Consumer response to packaging design: the role of packaging materials and graphics in sustainability perceptions and product evaluations, J. Clean. Prod. 162 (2017) 286–298, https://doi.org/10.1016/j.jclepro.2017.06.036.

[4] K.B. Biji, C.N. Ravishankar, C.O. Mohan, T.K. Srinivasa Gopal, Smart packaging systems for food applications: a review, J. Food Sci. Technol. 52 (2015) 6125–6135, https://doi.org/10.1007/s13197-015-1766-7.

[5] K.L. Yam, P.T. Takhistov, J. Miltz, Intelligent packaging: concepts and applications, J. Food Sci. 70 (2005) R1–R10.

[6] A. Barska, J. Wyrwa, Consumer perception of active and intelligent food packaging, Probl. Agric. Econ. 4 (2017) 138–159.

[7] M. Ghaani, C.A. Cozzolino, G. Castelli, S. Farris, An overview of the intelligent packaging technologies in the food sector, Trends Food Sci. Technol. 51 (2016) 1–11, https://doi.org/10.1016/j.tifs.2016.02.008.

[8] A.L. Brody, B. Bugusu, J.H. Han, C.K. Sand, T.H. McHugh, Scientific status summary, J. Food Sci. 73 (2008) R107–R116, https://doi.org/10.1111/j.1750-3841.2008.00933.x.

[9] S.Y. Lee, S.J. Lee, D.S. Choi, S.J. Hur, Current topics in active and intelligent food packaging for preservation of fresh foods, J. Sci. Food Agric. 95 (2015) 2799–2810, https://doi.org/10.1002/jsfa.7218.
[10] S.D.F. Mihindukulasuriya, L.-T. Lim, Nanotechnology development in food packaging: a review, Trends Food Sci. Technol. 40 (2014) 149–167, https://doi.org/10.1016/j.tifs.2014.09.009.
[11] M. Armstrong, F. Lazio, D. Herrmann, D. Duckworth, Capturing Value from the Smart Packaging Revolution, Deloitte Insights, 2018.
[12] J. Wyrwa, A. Barska, Innovations in the food packaging market: active packaging, Eur. Food Res. Technol. 243 (2017) 1681–1692, https://doi.org/10.1007/s00217-017-2878-2.
[13] D. Restuccia, U.G. Spizzirri, O.I. Parisi, G. Cirillo, M. Curcio, F. Iemma, F. Puoci, G. Vinci, N. Picci, New EU regulation aspects and global market of active and intelligent packaging for food industry applications, Food Control 21 (2010) 1425–1435, https://doi.org/10.1016/j.foodcont.2010.04.028.
[14] L. Vermeiren, F. Devlieghere, M. van Beest, N. de Kruijf, J. Debevere, Developments in the active packaging of foods, Trends Food Sci. Technol. 10 (1999) 77–86, https://doi.org/10.1016/S0924-2244(99)00032-1.
[15] J.P. Kerry, M.N. O'Grady, S.A. Hogan, Past, current and potential utilisation of active and intelligent packaging systems for meat and muscle-based products: a review, Meat Sci. 74 (2006) 113–130, https://doi.org/10.1016/j.meatsci.2006.04.024.
[16] K.K. Gaikwad, S. Singh, A. Ajji, Moisture absorbers for food packaging applications, Environ. Chem. Lett. 17 (2019) 609–628, https://doi.org/10.1007/s10311-018-0810-z.
[17] A. Ozcan, New approaches in smart packaging technologies, Int. Symp. Graph. Eng. Des. (2020) 21–34, https://doi.org/10.24867/GRID-2020-p1.
[18] K.K. Gaikwad, S. Singh, Y.S. Negi, Ethylene scavengers for active packaging of fresh food produce, Environ. Chem. Lett. 18 (2020) 269–284, https://doi.org/10.1007/s10311-019-00938-1.
[19] S. Yildirim, B. Röcker, M.K. Pettersen, J. Nilsen-Nygaard, Z. Ayhan, R. Rutkaite, T. Radusin, P. Suminska, B. Marcos, V. Coma, Active packaging applications for food, Compr. Rev. Food Sci. Food Saf. 17 (2018) 165–199, https://doi.org/10.1111/1541-4337.12322.
[20] K. Sadeghi, Y. Lee, J. Seo, Ethylene scavenging systems in packaging of fresh produce: a review, Food Rev. Int. 37 (2021) 155–176, https://doi.org/10.1080/87559129.2019.1695836.
[21] J.H. Han, J.D. Floros, Casting antimicrobial packaging films and measuring their physical properties and antimicrobial activity, J. Plast. Film Sheeting 13 (1997) 287–298, https://doi.org/10.1177/875608799701300405.
[22] K.A. Mane, A review on active packaging: an innovation in food packaging, Int. J. Environ. Agric. Biotechnol. 1 (2016) 544–549, https://doi.org/10.22161/ijeab/1.3.35.
[23] P. Suppakul, J. Miltz, K. Sonneveld, S.W. Bigger, Active packaging technologies with an emphasis on antimicrobial packaging and its applications, J. Food Sci. 68 (2003) 408–420, https://doi.org/10.1111/j.1365-2621.2003.tb05687.x.
[24] B.P.F. Day, Active packaging of food, in: Smart Packaging Technologies for Fast Moving Consumer Goods, John Wiley & Sons Ltd, 2008, pp. 1–18, https://doi.org/10.1002/9780470753699.ch1.
[25] T.P. Labuza, W.M. Breene, Applications of "active packaging" for improvement of shelf-life and nutritional quality of fresh and extended shelf-life foods 1, J. Food Process. Preserv. 13 (1989) 1–69, https://doi.org/10.1111/j.1745-4549.1989.tb00090.x.
[26] C. Vilela, M. Kurek, Z. Hayouka, B. Röcker, S. Yildirim, M.D.C. Antunes, J. Nilsen-Nygaard, M.K. Pettersen, C.S.R. Freire, A concise guide to active agents for active food packaging, Trends Food Sci. Technol. 80 (2018) 212–222, https://doi.org/10.1016/j.tifs.2018.08.006.
[27] M. Latos-Brozio, A. Masek, The application of natural food colorants as indicator substances in intelligent biodegradable packaging materials, Food Chem. Toxicol. 135 (2020) 110975, https://doi.org/10.1016/j.fct.2019.110975.
[28] S. Kalpana, S.R. Priyadarshini, M. Maria Leena, J.A. Moses, C. Anandharamakrishnan, Intelligent packaging: trends and applications in food systems, Trends Food Sci. Technol. 93 (2019) 145–157, https://doi.org/10.1016/j.tifs.2019.09.008.
[29] S.A. Hogan, J.P. Kerry, Smart packaging of meat and poultry products, in: Smart Packaging Technologies for Fast Moving Consumer Goods, John Wiley & Sons Ltd, 2008, pp. 33–59, https://doi.org/10.1002/9780470753699.ch3.

[30] J. Kerry, P. Butler, Smart Packaging Technologies, 2008.
[31] D. McFarlane, Y. Sheffi, The impact of automatic identification on supply chain operations, Int. J. Logist. Manag. 14 (2003) 1–17, https://doi.org/10.1108/09574090310806503.
[32] A. Sarac, N. Absi, S. Dauzère-Pérès, A literature review on the impact of RFID technologies on supply chain management, Int. J. Prod. Econ. 128 (2010) 77–95, https://doi.org/10.1016/j.ijpe.2010.07.039.
[33] P. Kumar, H.W. Reinitz, J. Simunovic, K.P. Sandeep, P.D. Franzon, Overview of RFID technology and its applications in the food industry, J. Food Sci. 74 (2009) R101–R106, https://doi.org/10.1111/j.1750-3841.2009.01323.x.
[34] B. Kuswandi, Y. Wicaksono, A. Abdullah, L.Y. Heng, M. Ahmad, Smart packaging: sensors for monitoring of food quality and safety, Sens. Instrum. Food Qual. Saf. 5 (2011) 137–146, https://doi.org/10.1007/s11694-011-9120-x.
[35] M. Vanderroost, P. Ragaert, F. Devlieghere, B. De Meulenaer, Intelligent food packaging: the next generation, Trends Food Sci. Technol. 39 (2014) 47–62, https://doi.org/10.1016/j.tifs.2014.06.009.
[36] S. Chaudhary, The role of packaging in consumer's perception of product quality, Int. J. Manag. Soc. Sci. Res. 3 (2014) 17–21.
[37] E. Young, M. Mirosa, P. Bremer, A systematic review of consumer perceptions of smart packaging technologies for food, Front. Sustain. Food Syst. 4 (2020). https://www.frontiersin.org/article/10.3389/fsufs.2020.00063.
[38] P.R. Smith, J. Taylor, Marketing Communications: An Integrated Approach, Kogan Page Publishers, 2004.
[39] K.L. Keller, Building strong brands in a modern marketing communications environment, in: Evolution of Integrated Marketing Communication, Routledge, 2013, pp. 73–90.
[40] R.K. Singh, The effect of packaging attributes on consumer perception, Int. J. Innov. Res. Multidiscip. F. 4 (2018) 340–346.

CHAPTER 8

Metal packaging for food items advantages, disadvantages and applications

Nadia Akram, Muhammad Saeed, Asim Mansha, Tanveer Hussain Bokhari and Akbar Ali
Department of Chemistry, Government College University Faisalabad, Faisalabad, Punjab, Pakistan

1. Introduction

Since the birth of the man, the food is the primary need. As it is very delicate item, it cannot be stayed for longer without any protection measures, hence there is always the need to store and preserve food. As the world has turned as a global village, the transportation of this food in every corner of the world is also a primary reason to pack it intelligently. The demand of food packaging has increased all over the world. There are several items which have been used as food packaging materials, such as paper, polymer and plastic, wood and metals [1–4]. The metal packaging has played a significant part in recent years. One of the important use is in the form of cans for processed food, fruits, vegetables, and drinks. There are several metals which have been used as food packaging materials including; aluminum (Al), stannous or tin (Sn), Sn free steel (SFS), and rustles steel (RS) commonly known as stainless steel. These metals provide hard food packaging such as cans and flexible food packaging in the form of foil and bags. The market share of metal food packaging was estimated as 430 billion cans in the year 2020 [2,5–8]. Majority of this share is used in soft and hard drinks. Hence it is also estimated that with increase in the alcoholic beverages demand, the metal packaging will also rise rapidly. However, both alcoholic and non-alcoholic beverages use the metal packaging cans for protection. Majority of the cans are made up of Al, SFS or RS. Due to malleability of metals, the metal packaging is available in various shapes and sizes however, the spherical lids are commonly used in packaging cans [9–12].

The food packaging with metallic materials is associated with health and environmental conditions. It has both advantages and disadvantages which must be considered while making metallic packaging a choice. As a matter of fact metals are better in providing the shield toward light. These provide good gas barrier properties. These are resistant to moisture and have good shelf life. The metals also have the high tolerance against thermal transitions making them an ideal candidate for the transportation of the food materials all over the world. Moreover, they maintain the shapes and the structures as the packaging materials which plays an important role to ensure the safety of the food items for short and long distances equally [13–16].

Although the metallic packaging is very efficient in all aspects, there are certain concerns associated with this packaging. The global warming is playing an important part in deteriorating the efficacy of metallic food packaging. The ejection of CO_2 is a major component which is not only emitted by the metallic plants but these are responsible to spoil the food items much earlier than their shelf life as well [17,18]. Other toxic chemicals released from the metallic food packaging are also hazards for food items as well. There are continuous efforts to maintain the quality of food in metal packaging. Apart of the high costs of metal packaging, the disposal and recycling is also an important issue associated with these packaging materials. The metallic wastes produced by the individuals along with paper and plastic waste is a main source of environmental pollution, with the growing population, the percentage of the waste is alarmingly increasing as well. In order to develop a better understanding of how these metals are contributing in the food processing industry, it is important to discuss the types of various metals involved in this industry [19–22].

1.1 Types of metal packaging for food industry

The followings are the important type of metallic packaging in food industry as shown in Fig. 8.1;

1.1.1 Alloys of iron (steel)
Steel is a combination of iron (Fe) in the form of alloys. As Fe is the central atom, it is bonded by carbon in the form of a lattice. The percentage of carbon may vary up to 2%. The prime reason to include the carbon is to provide strength and fracture resistance.

1.1.2 Stannous or tin (Sn) plate
One of the major food packaging material is the Sn free steel. This material is also used for the cooking of food items all over the globe. In this packaging the base of steel is immersed with Sn. Previously the hot dipping method was used for the coating of Sn

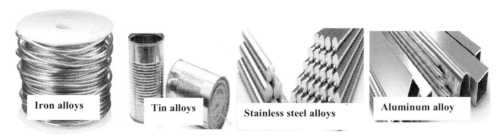

Fig. 8.1 Types of metals used in food packaging industry.

which has been replaced with electroplating method now. The thickness of the coating can be varied on both sides making it an electrolytic Sn plate. For the manufacturing of this electrolytic plate: the steel base in covered up with various thickness (usually thin layer) of Sn, the plate is thermally treated up to 270 °C followed by quick water quenching. This develop a $FeSn_2$ layer. The Sn and Chromium oxide surface layer is produced by the process of chemical passivation with the help of sodium dichromate. The layer produce protection against atmospheric factors. The coating of these materials with oil also helps to cope up with environmental factors. There are several advantages to use Sn plates as these are cost effective, heavier than Al, good recyclability, and can prevent the penetration of moisture along with gases as well. The good thermal stability is also responsible to tolerate the thermal fluctuations. These are also good materials to withstand long time storage of food items.

1.1.3 Sn free steel (SFS)

The SFS resembles to Sn plate however the flow of Sn is avoided in this case. For this plate the electroplating of chromium is followed by coating with oil making it more prone toward corrosion in acidic conditions as there is no protective Sn layer. However, it is very cheap as compared to tin coatings hence used as lids of cans, vacuum lids for glass containers as well. It is also used for small storage or bulk storage of food quantity as well. It is also an excellent material for transportation of food material as well. The recycling is also convenient for these packaging materials. The corrosion is prevented by the polymer coating including polyaniline and polypyrolle. The corrosion can also be prevented by the coating of fluoropolymers as well. The polypropylene and polyethylene terephthalate can also be used to avoid corrosion. The polymer coated steel provides protection against moisture contents as well.

1.1.4 Stainless steel

Another important alloy of Iron which provides excellent corrosion and chemical resistance is the stainless stee. One of the major component of this alloy is the chromium (Cr). The Cr develops a layer of Cr_2O_3 in the presence of atmospheric O_2 resulting in the prevention of corrosion. This makes it a good material for the food processing and packaging. Based on the crystallinity of the stainless steel materials it is available in three forms such as; Austenitic, ferritic and martensitic. Among these forms austenitic is commonly used for food packaging. As compared to Sn and Al it is expensive however for long term storage, transportation it is ideal [7].

1.1.5 Aluminum

Another important metal used for food packaging is Al. The process is carried out at 175 °C. The PPTs of $Al(OH)_3$ are white solids. It is expensive than coated steels. It is hard to mold or weld due to which it is preferably used as a single piece of storage material. Due to low density it is used as packaging of sea foods, beverages and pet foods. The strength of Al can be increased by adding manganese. It is an excellent foil and lamination material. The recycling of Al is also convenient as compared to other metallic packaging items [9].

1.2 Shapes of metal packaging

There is a single piece, two piece and three piece metal packaging items. The two piece items contains the main body along with the lid however, for a three piece item there are two lids at the end with a central body. Its manufacturing can be done at any diameter with the height adjustment as well. Hence there is a huge range in the specification of the materials which can be adopted according to requirement of food packaging. Welding is desired for meals industries, which now no longer overcomes the worries of soldering however additionally it reduces the steel utilization. After the formation of spherical hole structure in the form of cylinders, necking and flanging for cans are performed. Necking typically denotes the decreasing in size of ends in the form of flawless cylindrical frame to concave ends consequently providing steel utilization reduction. One of the remedy known as beading of meals cans offers round curves to the spherical shape. There is another term known as flanging which use outer knobs for the adjustment of lids. Two piece cans had been a primary revolution in this connection which are; commercial, sterile and feature excessive printing region. Two piece cans are about 35% lower in weight and offer higher integrity. For the processing of two-piece cans there are two important processes known as drawn and wall ironed (DWI) and drawn and redrawn (DRD). United states developed a DWI can in 1958. The manufacturing procedure included stamping of round discs which were developed from oily metallic sheets, uniform thickness was maintained during the entire process [23–26]. The thickness for base of DWI is more than stretched and thin cans. Some typical shapes of metal food packaging is shown in Fig. 8.2.

Fig. 8.2 Some typical shapes of metals used in food packaging industry.

1.2.1 Aluminum foil
The aluminum foil is found in the form of thin films with ultimate level of purity (99%). The size of the thickness varies in several microns. Commercially this material was introduced in earlier 20th century in US for the packaging of sweets and candies. In the first step of its manufacturing, Al foil is casted in the form of square blocks and the oxides are removed. Desired thickness is achieved through hot and cold rolling. During the process of rolling, the Al foil gets hard, the problem can be solved by the process of annealing. Aluminum is particularly desired due to the fact that it is of low density malleable material.

1.2.2 Aluminum collapsible tubes
Collapsible tubes are elastic materials for food packaging which are hygienic as well. They offer various applications. It is developed by extrusion process followed by annealing and enameling which increase its value. It is primarily used for dairy products. Presently, accessibility of inexpensive replacements has significantly decreased its usage.

1.2.3 Aluminum bottles
The bottle cans of Al were prepared in 2000. It was a brewing company of Japan which introduced these bottle. These were made through extrusion process. These were developed in various shape and size at lower prices.

1.2.4 Laminated and metallized films
Laminated materials are developed by the adhesion of two substrates. The adhesion may be carried out by adhesive or the temperature. For the lamination of Al foil, a variety of paper material have been used based on aqueous and solvent based adhesives. The functionality of the materials has also been improved by the use of waxes in order to enhance the barrier properties. Both plain and printed lamination has been used for food packaging whereas, the thickness has also be controlled in microns for these materials. Although the impurities of other metals can be tolerated to some extent however, the contents of As, Cu, Fe and Pb should be under the limit of 2 ppm, 30 p.m., 70 ppm and 20 ppm respectively. The physical deposition on aluminum can also be performed by the metallization process. In this process the Al is heated to vaporize which are allowed to deposit on the substrate and condensed to develop layer which may be of nano size [27–30].

1.2.5 Retort pouches
Another important packaging of Al is retort pouch which may be in the form of multilayers. These are flexible foils and act as barrier. In a typical design of retort pouch outer layer is made up of PET, the central layer is made up of Al, nylon and PP due to air, moisture, light, abrasion barrier. These even work at lower thickness. In these retort

pouches Al may be replaced with other materials of EVOH and PVDC etc. These materials have also developed ecofriendly materials.

1.2.6 Metal containers

Metallic containers are particularly used for industrial scale packaging and storage. These are mainly used for transportation. Both processed and semi processed food items. These containers have large capacity to carry food items usually of several liters. The typical container contains a lid and a central body whereas, the processing may be of any type according to the need of the food items. Usually the unit is made up of a single sheet it is helpful against the corrosion.

1.2.7 Metal lids

Metal lids are used to seal the various packaging materials. The lid may be in the crown shape. It may be metallic or plastic in nature. The lids are used for the protection and decoration purpose of the materials. All these lids are used for crewing to close the containers, it is always considered to make the lids which can screw the containers, the containers can be closed by vacuum.

1.3 Advantages of metal packaging

The metal packaging is being used since ages for food packaging however, historically it was not considered significant primarily because it was always over looked in comparison to abundantly available and economic materials with easy transportation as well. Although the cost of the original metal is high for such packaging however the reusability adds in its money saving quality a positive impact. The practicality can be helpful to understand the diverse advantages of metal food packaging in a broader aspect.

1.3.1 Product protection

Some products require to be stored in dark and prevent from sunlight. Whenever aluminum or steel packaging is used it is opaque and prevent sunlight from penetrating in the product. Moreover, metal is durable and can keep the inside contents away from damage.

1.3.2 Durability

The damages and breakdown of materials during packaging for a long time results in the damaging of food items as well. The damaging may be caused by moisture, breakdown of Al and other components. Metals are used for long time storage items.

1.3.3 Sustainability and long shelf life

The one of the amazing facts related with the metal food packaging is the strength and durability of metallic food cans. This not only keeps the food sustained for longer time the can itself shows significant shelf life as compared to plastic, polymers or other packaging materials. The bursting of the food confined in the cans is not feasible endorsing the reliability of this material. The little damage is expected to metal packaging due to high strength of the materials. The stacking of metal cans improves the storage duration and in a large aspect the business associated with this market. In addition to these qualities the quality of the used material for cans manufacturing enhance the quality of the seasonal food items making them fresh and edible for long time. As the small containers are usually packed in the large containers, the durability of the packaging increases and the food can be supplied in its best form to the customers. The process is not very simple, each step is linked with sanitization of cans and hermetic closures to keep the food in good condition for long time.

Most forms of metallic goods can be recycled easily. Al and steel are commonly used substances in metal packaging hence a lot of recycled materials can be obtained.

1.3.4 Light weight metallic food packaging

Some styles of metallic packaging, specifically Al weigh plenty much less than different materials. For example, a median six-percent of Al beer cans weighs considerably lower than a median six-percent of glass beer bottles. The lower weight decrease transportation cost and improves the export quality of the food items.

1.3.5 Customer's attraction

Numerous customers are searching out approaches to lessen their carbon footprint and to stay extra sustainable. Since metallic packaging is widely recognized to be smooth to recycle and recognized to have fewer dangers to humans and earth than polymers and plastic, the use of metals to enhance the merchandise allow to hook up with customers who need green environment. Currently, extensive type of inventory metallic packaging is provided to consumer [31–34].

1.4 Disadvantages of metal packaging
1.4.1 Metal corrosion

The biggest disadvantage of the metallic food packaging materials is the corrosion, which is purely destructive. It involves an electrochemical reaction and all metal packaging consisting of Sn, Fe, Pb food cans can be affected by the corrosion. The vulnerability of corrosion increases in the presence of moisture where the metals dissolve and produce ions, these ions produce salts on the surface of packaging body which act as anode. The packaging materials where different metals are involved, the process takes place at different potential level. The release of the hydrogen atom develop cathode via Spontaneous reaction. There are several types of corrosion which may occur including stress corrosion which cause the cracking process. It arise due to stress at different areas and develop localized corrosion. The chlorides and the sulfur dioxide or other components may result in this type of corrosion. The sulfur containing compounds are also responsible

Table 8.1 Impact of alloys on corrosion resistance.

Metals for alloys	Impact on corrosion
Cu	Reduction in corrosion of Al packaging
Mn	Increase in corrosion resistance
Mg	Good corrosion resistance
Zn	Reduction in resistance in case of acid and alkali media
Cr	Increase in corrosion resistance
Fe	Reduction in corrosion resistance

of sulfur blackening. Another type of corrosion is the pitting corrosion, in this corrosion the iron is rapidly dissolved. The foods containing chlorides also damage the food packaging specially Al and cause holes in the metals. Another type of corrosion is known as filiform which is external and occurs under the conditions of temperature 20–35 °C and humidity in the range of 60–95%. The interaction between food and metal packaging drastically affect the shelf life and human health. Apart from the above mentioned factors, several other factors such as labels containing acidic and alkaline medium, improper air circulations in the storage places, low quality metallic coating, leakage of food products and humidity are the biggest sources of damaging of metallic packaging [35,36]. The tendency of metal packaging against corrosion resistance is shown in Table 8.1.

1.4.2 Sightlessness of contents

One of the disadvantage of the metal packaging is the sightlessness of the contents. Although the closed containers ensure the storage of the food items however the contents cannot be seen hence any damage during the transportation or storage cannot be seen unless the packaging is reopened. Hence this serious disadvantage can only be observed after the food item reach to the customer. It is difficult to prepare the transparent packaging of metals hence making it a serious problem of food damaging. This also results in the reduction in the purchase demand and provide a chance of other packaging materials such as plastics to be used in the packaging materials.

1.4.3 Storage issues

The metal packaging requires special storage places during the transportation and long term storage. Also usually the metallic food packaging items are very specific which limit the use of metallic packaging. As the metal containers are stiff in nature they cannot be bend or mold and hence requires specific places for the storage as well. This usually lead to the wastage of space as compared to original contents.

1.4.4 Protection and decoration of metallic cans

The majority of metallic food packaging require protective coating such as Sn plate is not inert toward environment. This results in easy corrosion of the Sn packaging and results in food rotting. Although all type of food is affected by this however, the soft fruits and juices are usually affected by this type of packaging. The acidic environment also results in the food damage. Hence these metal packaging require the additional coating on packaging to prevent it from these damages. There must be coating of lacquer on the metallic food containers to effectively hinder decomposition of the interior of the food cans and to protect the food items from microbial or environmental decomposition. This results in the increase in the cost of these containers and also increase the time of preparation of these containers making it non convenient for the easy access. On the other hand the lacquer developed on the surface as the protective coating usually harden and leave the residues causing the food rotting as well. Similarly, there coatings may also be dissolved in the liquid food items making it hazardous for the consumers. Like the internal coatings, the external coatings can also absorb moisture and results in early decomposition.

1.4.5 Health issues with metal packaging

There is a major disadvantage of metal food packaging associated with health. It involves two process including migration and the second is interaction. The term migration is the exchange of coating and packaging constituents in to the food items or from the food items to the metal packaging. The migrant species includes; Sn, bisphenol A (BPA), Pb, Al, Cr and other metal impurities. The term interaction is associated with the physical interaction, chemical interaction or microbiological interactions associated with food packaging. The interactions depend on various factors such as composition, temperature, pH, humidity, the material of the container etc. The interaction also results in corrosion, roughness, damage on the surface and staining.

1.4.6 Environmental concerns of metal packaging

A very alarming disadvantage associated with metal packaging is the collection of the municipal waste which has increased with the consumption of the processed food. The use of all the metals as food packaging raw material is not equal. Some of them has limited use such as Fe, Al and Sn. Due to various magnetic properties, the separation and recycling is convenient. A number of approaches have been for energy retrieval based on recycling and combustion with the aim to reduce the waste of metal packaging and to avoid its hostile impacts. The complete disposal of a Sn can decays in 12 months however it may take up to 10 years for complete decomposition. The metal packaging plants pose the threat of carcinogenicity to the workers and the additives release significant quantity of hydrocarbon. It is also the source of global warming [31−33,35].

1.5 Metal packaging applications

The various types of foods items are considered secure for metal packaging as shown in Fig. 8.3. The detail is given below;

1.5.1 Milk products

The primary job of packaging is to foresee weakening variables, for example, light, dampness, oxygen and microorganisms from influencing the time span of usability of milk items. Polyethylene packaging is used for pasteurized milk. The milk variations such as ultra-heat treated (UHT) milk, dry milk popularly known as condensed milk are stuffed in metal based drums and jars or multifaceted bundles. These packaging are mostly made up of Al. These packaging enhance the shelf life of the materials for a considerable duration. The fat oxidation is a very serious issue in dairy items. The packaging provide a shelter against the oxidation and bacterial attack. The storage capacity of each edible item vary according to nature of the food item. The bread, oil, fats, creams, margarine, cheese all be carried in different metal packaging for long time storage and for the purpose of transportation. The metallic paper packaging such as Al foil packaging allows very small quantity of light to pass on making Al foil a good packaging material

Fig. 8.3 Types of foods used for metal storage.

especially for dairy product. The Al lids are also used for the fermented food such as yogurt. The metallic coated films of PET are also utilized for the storage of dairy products. The metallic packaging is also used for the packaging of condensed milk [35–38].

1.5.2 Beverages
Another applications of metallic food packaging is the soft drinks packaging. The soft drink contains sugar in carbonated water. The carbonated drinks contain dissolved carbon dioxide hence the metal packaging should be corrosion resistant. The cans may be in the form of single piece and its lid hence making it two piece. Whereas, the three piece system is also used for the packaging of beverages.

1.5.3 Fruits and vegetables
The metallic food packaging is also used for fruits and vegetables which is also used for protection purpose as well. The food processing may be treated thermally for storage conditions. The activity of metallic packaging depends on bio components. The vegetable, rich in nitrate components such as spinach, radish and green beans corrodes the metallic packaging. Other food items containing sulfur such as garlic and onion also act in the same manner. Some of the food items are rich in Zn and can result in black staining.

1.5.4 Flesh products
There are numerous food items consisting of meat products which can be packed in the metallic containers. The meat items can be freshly packed or packed in processed form. The variety of meat also vary from sea food; crab, shrimp, prawn and other products. The storage of food also vary from short storage time to long storage time. The processing of food items is carried out in various oil which is also packed in metallic cans to have better shelf life. Tuna fish is stored in sulfur coated containers with transparent PET coating. Al wrapping for meat items is also reportedly very effective.

1.5.5 Bakery and confectionary products
The metallic packaging is also used for Al foil to wrap the bakery items in the form of chocolate block, biscuits, cookies and crackers etc. Along with Al the LDPE is also used for the packaging of bakery products for short and long term storage. Often the packaging is flushed with Nitrogen gas to remove oxygen which is also an effective method to improve the storage life of food items.

1.5.6 Coffee and tea
The Sn plate cans are used for the packaging of coffee which offers barricade to loss of volatiles. It also acts as barrier to emit pressure developed by CO_2. It also offers good covering for instant coffee however the flexible aluminum is also used for this purpose.

1.6 Conclusion

Although the polymer and plastic industry is the major packaging industry however the metallic food packaging has a special place in packaging industry as well. There are numerous metals such as copper, aluminum, steel which are currently used to store and transport the food. The metal packaging is selectively used for approximately all type of food packaging. There are several advantages and disadvantages associated with the metallic food packaging as well. However, the strict regulations are required in order to ensure the safety of food items and to prevent the harmful impurities from metallic packaging into the food items and vice versa. The metal packaging is associated with health and environmental concerns directly.

References

[1] J. Bayus, C. Ge, B. Thorn, A preliminary environmental assessment of foil and metallized film centered laminates, Resour. Conserv. Recycl. 115 (2016) 31–41.
[2] P. Beigl, S. Salhofer, Comparison of ecological effects and costs of communal waste management systems, Resour. Conserv. Recycl. 41 (2) (2004) 83–102.
[3] P.E.M. Bernardo, J.L.C. Dos Santos, N.G. Costa, Influence of the lacquer and end lining compound on the shelf life of the steel beverage can, Prog. Org. Coat. 54 (1) (2005) 34–42.
[4] J. Bindu, C.N. Ravishankar, T.S. Gopal, Shelf life evaluation of a ready-to-eat black clam (Villorita cyprinoides) product in indigenous retort pouches, J. Food Eng. 78 (3) (2007) 995–1000.
[5] B. Boelen, H. den Hartog, H. van der Weijde, Product performance of polymer coated packaging steel, study of the mechanism of defect growth in cans, Prog. Org. Coat. 50 (1) (2004) 40–46.
[6] R. Catala, M. Alonso, R. Gavara, E. Almeida, J.M. Bastidas, J.M. Puente, N. De Cristaforo, Titanium-passivated tinplate for canning foods, Food Sci. Technol. Int. 11 (3) (2005) 223–227.
[7] S. Cvetkovski, Stainless steel in contact with food and beverage, Metall. Mater. Eng. 18 (4) (2012) 283–293.
[8] R. Diaz, M. Warith, Life-cycle assessment of municipal solid wastes: development of the wasted model, Waste Manag. 26 (8) (2006) 886–901.
[9] K. Ertl, W. Goessler, Aluminium in foodstuff and the influence of aluminium foil used for food preparation or short time storage, Food Addit. Contam. Part B 11 (2) (2018) 153–159.
[10] T. Geens, T.Z. Apelbaum, L. Goeyens, H. Neels, A. Covaci, Intake of bisphenol A from canned beverages and foods on the Belgian market, Food Addit. Contam. 27 (11) (2010) 1627–1637.
[11] M.S. Jellesen, A.A. Rasmussen, L.R. Hilbert, A review of metal release in the food industry, Mater. Corros. 57 (5) (2006) 387–393.
[12] S.P. Joshi, R.B. Toma, N. Medora, K. O'Connor, Detection of aluminium residue in sauces packaged in aluminium pouches, Food Chem. 83 (3) (2003) 383–386.
[13] J.H. Kang, F. Kondo, Bisphenol A migration from cans containing coffee and caffeine, Food Addit. Contam. 19 (9) (2002) 886–890.
[14] Y. Kim, B.A. Welt, S.T. Talcott, The impact of packaging materials on the antioxidant phytochemical stability of aqueous infusions of green tea (Camellia sinensis) and yaupon holly (Ilex vomitoria) during cold storage, J. Agric. Food Chem. 59 (9) (2011) 4676–4683.
[15] J.M. Kim, I. Lee, J.Y. Park, K.T. Hwang, H. Bae, H.J. Park, Applicability of biaxially oriented poly (trimethylene terephthalate) films using bio-based 1, 3-propanediol in retort pouches, J. Appl. Polym. Sci. 135 (19) (2018) 46251.
[16] J. Lange, Y. Wyser, Recent innovations in barrier technologies for plastic packaging—a review, Packag. Technol. Sci. Int. J. 16 (4) (2003) 149–158.
[17] E. Leivo, T. Wilenius, T. Kinos, P. Vuoristo, T. Mäntylä, Properties of thermally sprayed fluoropolymer PVDF, ECTFE, PFA and FEP coatings, Prog. Org. Coat. 49 (1) (2004) 69–73.

[18] J.Z. Li, J.G. Huang, G.R. Zhou, Y.W. Tian, Y. Li, Study on the growth mechanism of electrolytic chromium coated steel (ECCS), Adv. Mater. Res. Trans. Tech. Publ. 154 (2011) 663−666.
[19] K.J. Maheswara, C.V. Raju, J. Naik, R.M. Prabhu, K. Panda, Studies on thermal processing of tuna-a comparative Study in tin and tinfree steel cans, Afr. J. Food Agric. Nutr. Dev. 11 (7) (2011) 5539−5560.
[20] A.K. Mallick, T.K. Srinivasa Gopal, C.N. Ravishankar, P.K. Vijayan, Polymer coated tin free steel cans for thermal processing of fish, Fish. Technol. 43 (1) (2006a) 47−58.
[21] A.K. Mallick, T.K. Srinivasa Gopal, C.N. Ravishankar, P.K. Vijayan, Canning of rohu (Labeo rohita) in North Indian style curry medium using polyester-coated tin free steel cans, Food Sci. Technol. Int. 12 (6) (2006b) 539−545.
[22] C. Mannheim, N. Passy, A.L. Brody, Internal corrosion and shelflife of food cans and methods of evaluation, Crit. Rev. Food Sci. Nutr. 17 (4) (1983) 371−407.
[23] K. Marsh, B. Bugusu, Food packaging—roles, materials, and environmental issues, J. Food. Sci. 72 (3) (2007) R39−R55.
[24] S.F. Mexis, K.A. Riganakos, M.G. Kontominas, Effect of irradiation, active and modified atmosphere packaging, container oxygen barrier and storage conditions on the physicochemical and sensory properties of raw unpeeled almond kernels (Prunus dulcis), J. Sci. Food Agric. 91 (4) (2011) 634−649.
[25] A. Montanari, C. Zurlini, Influence of side stripe on the corrosion of unlacquered tinplate cans for food preserves, Packag. Technol. Sci. 31 (1) (2018) 15−25.
[26] M.L.M. Saraiva, J.L. Lima, P.C. Pinto, Sequential injection fluorimetric determination of Sn in juices of canned fruits, Talanta. 79 (4) (2009) 1100−1103.
[27] E.M. Munguia-Lopez, S. Gerardo-Lugo, E. Peralta, S. Bolumen, H. SotoValdez, Migration of bisphenol A (BPA) from can coatings into a fatty-food simulant and tuna fish, Food Addit. Contam. 22 (9) (2005) 892−898.
[28] G.O. Noonan, L.K. Ackerman, T.H. Begley, Concentration of bisphenol A in highly consumed canned foods on the US market, J. Agric. Food Chem. 59 (13) (2011) 7178−7185.
[29] T.E. Norgate, S. Jahanshahi, W.J. Rankin, Assessing the environmental impact of metal production processes, J. Clean Prod. 15 (8−9) (2007) 838−848.
[30] U.S. Pal, M. Das, R.N. Nayak, N.R. Sahoo, M.K. Panda, S.K. Dash, Development and evaluation of retort pouch processed chhenapoda (cheese based baked sweet), J. Food Sci. Technol. 56 (1) (2019) 302−309.
[31] J. Pasqualino, M. Meneses, F. Castells, The carbon footprint and energy consumption of beverage packaging selection and disposal, J. Food Eng. 103 (4) (2011) 357−365.
[32] L. Piergiovanni, S. Limbo, The protective effect of film metallization against oxidative deterioration and discoloration of sensitive foods, Packag. Technol. Sci. 17 (2004a) 155−164.
[33] E. Pongracz, The environmental impacts of packaging, Environ. Conscious. Mater. Chem. Process 2 (2007) 237.
[34] M. Ramos, A. Valdés, A. Mellinas, M. Garrigos, New trends in beverage packaging systems: a review, Beverages 1 (4) (2015) 248−272.
[35] M.A. Shah, S.J.D. Bosco, S.A. Mir, K.V. Sunooj, Evaluation of shelf life of retort pouch packaged Rogan josh, a traditional meat curry of Kashmir, India, Food Packag. Shelf-Life 12 (2017) 76−82.
[36] Y. Uematsu, K. Hirata, K. Suzuki, K. Iida, K. Saito, Chlorohydrins of bisphenol A diglycidyl ether (BADGE) and of bisphenol F diglycidyl ether (BFDGE) in canned foods and ready-to-drink coffees from the Japanese market, Food Addit. Contam. 18 (2) (2001) 177−185.
[37] N.M. Vasava, P. Paul, S. Pinto, Effect of storage on physicochemical, sensory and microbiological quality of gluten-free gulabjamun, Pharma. Innov. J. 7 (6) (2018) 612−619.
[38] M. Xie, W. Bai, L. Bai, X. Sun, Q. Lu, D. Yan, Q. Qiao, Life cycle assessment of the recycling of Al-PE (a laminated foil made from polyethylene and aluminum foil) composite packaging waste, J. Clean Prod. 112 (2016) 4430−4434.

CHAPTER 9

An approach of smart packaging for home meals

Uzma Hira and Muhammad Husnain
School of Physical Sciences (SPS), University of the Punjab, New Campus, Lahore, Punjab, Pakistan

1. Smart packaging approach

The demand for food increases day by day, so as to fulfill the needs of increasing population and therefore food packaging technology has been increasing massively worldwide. Packaging of food material also exists in nature e.g., egg, banana and orange etc., so we can say humans get the idea about packaging from nature. Customer requirements for the preservation and the better quality of home meal cause the change in food packaging. The word "Smartness" covers the huge number of methods which include in the preservation of home meal products [1]. Old traditional methods are not scientifically approved methods, therefore we have to move forward from traditional methods to scientific methods.

1.1 Purposes of smart packaging for home meal products

The purposes of smart packaging can be described as follows:
1) Protect the meal from external environment like:
    - Light
    - Heat
    - Moisture
    - Pressure
2) Emission from gases
3) Microorganisms
4) Prevent food spoilage
5) Increase the products attributes
6) Maintain the quality of products
7) To meet the food supply chain

Nowadays, Customers prefer fresh and preserved food having good tastes and quality. All the people, producers, retailers and consumers are interested in the quality of packaged food. Fig. 9.1 is showing that intelligent and active packaging are considered to be a part of smart packaging. Active packaging deals with the types of chemicals that are released or absorbed from the meals and intelligent packaging deals with the condition of packaging [2]. Intelligent Packaging and Active packaging have been introduced to fulfill the requirements of consumers because from some previous years demand for packaged food increased.

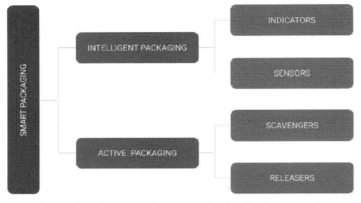

Fig. 9.1 Classification of smart packaging for home meals.

1.2 Aim of active packaging

There are following aims and objectives for active packaging approach for home meals:
- Moisture regulation
- Ethylene removal system
- Carbon dioxide scavenger
- Antimicrobial active packaging
- Anti-oxidant
- Thermal insulation

1.3 Aim of intelligent packaging

The intelligent packaging will be described in the coming part of this chapter. These are following aims for intelligent packaging for home meals:
- Use of indicators
- Chemical sensors
- Biosensor
- Data carriers
- Monitors the conditions
- Shelf life

The traditional methods demand that materials that are used in the packaging for home meal products should be inert but now in a smart packaging approach, scientists use different types of scavengers, preservers, indicators and sensors in order to protect the meal, its taste, its color and all essential elements of food. The approach of smart packages is rapidly increasing in research areas nowadays. With the help of new technologies like nanotechnology and advanced molecular biology, smart packaging is becoming popular [3].

2. Current need of smart packaging for home meal

In the era of technology when people do not have time to talk face to face and whatsapp, Skype and cellular phone calls are made for some time owing to the active lifestyles. The consequence of the altering attitudes and active lifestyles result is that individuals do not have enough time for homemade food and moreover cooking at home is a time taking process. There are also many other reasons including time limitations, absence of cooking expertize, growing number of members for a single family, troublesomeness in preparing special diets with care for babies and elderly people, and currently lock-down conditions owing to the Covid-19 global epidemic, customers require easy and ready to made food instead of going outside in restaurants/hotels or cooking foods at home [1]. All food chain companies are highly interested in providing high quality dishes or whole meals that may easily and swiftly replace home cooked food [2]. The meal solution gives the rushed consumer additional options while still allowing them to enjoy a wonderful dinner. There are a growing variety of production/distribution systems available to consumers with the goal of partially or completely replacing homemade meals. Homemade-food-replacement (HFR) is a type of easy meal that can be made away from the house for home intake/feeding. HFR furthermore comprises food solution, manufactured foods, prepared meals and easy accessibility of food [2]. The HFR single or multiple meal portion serving container currently comprises starches (carbohydrates), plant and meat proteins and vegetables [2]. Currently, HFR is easily available in numerous parts of the world. This convenient meal can be purchased easily not only from cafeterias, bistros, restaurants, supermarkets, departmental and grocery markets, but also in HFR specific markets, stalls, cabins and through foodstuff ordering with various virtual mobile phone apps.

The increasing reputation of HFR is owing to the ease to make food and moreover not more time is needed for cooking. The market value of HFR items is progressively growing throughout the globe. According to the "Research and Market", 2020 report in terms of value that worldwide ready to made meal goods shop is anticipated to enlarge at a the rate of annual growth (RAG) of ∼ (6.83 %) within 2019–2025 and it is assessed to be valued around US$ 156.807 billion at the end of 2025 [3]. Europe is predicted to have the biggest market share over the coming years due to the rise in the work-load of working class people of the developed nations such as France, United Kingdom and Germany. The high living standard and established economy of customers are anticipated to increase marketplace prices in the European and also in the Northern American areas. The Asia-Pacific (APAC) area is projected to produce maximum RAG due to the varying customer routines, rising number of mid-class individuals in the underdeveloped nations and numerous manufacturers are introducing assorted fresh HFR goods in the countries like: China, India and South Korea. According to the Statista Research, 2021, the rate of yearly transaction of HFR goods in the South Korea was assessed to be around 3.0 billion United State dollars in 2019 and it increased up to US$ 3.4 billion in the year 2020 [4].

The customers are impressed by several factors, when they are buying HFR products and the factors include quality of food (flavor, freshness of product and valuable nutritional contents), packaging style, affordable prices, availability, and delivery service quality. Numerous investigations have highlighted that HFR goods characteristic is the main feature that customers ponder during buying these products [5,6]. As far as food quality concerns, it will be determined by considering several aspects including: (1) the value of food ingredients, (2) fresh materials were considered or not and (3) which approach and materials used for food handling and packing. The all packing strategies, honesty and protection measure features have a crucial impact on buyer's HFR choice [5]. The precise packaging material with good features and design can provide a high degree of product quality and food safety. Consequently, smart packaging technology is subtly applied in the HFR manufacturing. The aim of the current book chapter is to deliver contemporary inclinations and advanced research on packaging products involved in the HFR packaging technology. It is essential to know the critical prospects of manufacturing and customers' concern about the HFR marketplace. The required materials properties, smart packaging features, and design will be presented in this chapter. Available advanced innovative smart packaging characteristics, like easy to peel off, microwavable packaging and intelligent/smart packing techniques will be debated in detail. Lastly, the effect of packing technology inclinations, evolving innovative technologies and worldwide problems on food products will be debated for upcoming advances in this field.

3. Different HFR goods and essential smart packaging features

A great diversity/variety of HFR goods offer customers the choice of half-done or whole cooking at home. As it is already reported in the literature that HFR goods are divided into four (See Fig. 9.2) groups depending on the nature of manufacturing actions taken by commercialized HFR technology [2,7].

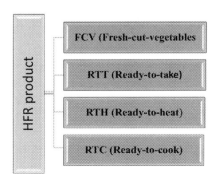

Fig. 9.2 Classification of different HFR products.

3.1 Fresh cut vegetable (FCV) goods

The fresh cut vegetable (FCV) products have been processed, cleaned/washed, peeled off, cut up, and shredded and chopped before packing and can be taken without cooking them. Moreover, HFR products are distinguished into long- and small lifespan goods. Several HFR goods with diverse lifespans have different packing qualities and resources. As a result, it is very essential for both manufacturers and producers to contemplate the qualities and protection of the packing materials, including: corporeal, blocking, and instinctive characteristics (like, honesty & fortitude) and potential passage/composite development that have an effect on the odor and flavor of the packed products during lifetime of a food. Table 9.1 shows classification of HFR goods with primary requirements of packing according to buyer/purchaser consumption. Processed/treated vegetables produce physiological pressures in the still alive cut tissues of the plants and cause worsening of quality and decrease shelf life than freshly whole vegetables. In order to extend the lifespan and reduce the exterior yellowing of these goods, improved atmospheric packing within great honesty and anti-microbial activity is commonly prerequisite [8].

Table 9.1 Explains different classes of ready-made products, examples, and properties of material.

Classification of HFR products	Example of HFR products	Properties of material needed for packing
Fresh cut vegetable (FVC) Ready to take (RTT) Ready to heat (RTH) Ready to cook (RTC)	Salad with fresh fruits and vegetables/Mixed salad Fresh fruit juices Food delivery/Take away products/Bento lunch boxes Sandwiches/Burgers/Pizzas/Tortilla Freshly manufactured sideways dishes (e.g., guacamole, salsa, kimchi and salad dressings) Raw oysters/fishes "sashimi" Frozen and chilled pizzas, noodles, pastas, nuggets, fried rice and dehydrated pastas item Instant noodle, rice, pasta Marinated and frozen raw	Improved atmospheric packing, sterile and high integrity packing, antimicrobial packing and easy to open packing Acid resistant characteristics High temperature and grease/oil lubricant confrontation wrapping High temperature and grease/oil lubricant confrontation wrapping Acid resistant properties, better taste barrier packing, good integrity packing Improved atmospheric packing, packaging with high honesty Ovenable/Microwaveable packing, packing with heat

Continued

Table 9.1 Explains different classes of ready-made products, examples, and properties of material.—cont'd

Classification of HFR products	Example of HFR products	Properties of material needed for packing
	chicken, beef and mutton with vegetables chilled uncooked fish cut with bread crumbs and sauces, Frozen stewed vegetables Chilled seafood paellas, pastas, fried rice and wheat breads	resistant characteristics, Easy to open packing, Moisturize and high gas barrier packing Moisturize and high gas barrier packing, Anti-oxidant packing High sturdiness with puckering resistant packing characteristics

3.2 Ready to take (RTT) goods

The second category of HFR products is ready to take (RTT) goods and it can be directly obtained from take away as a main course and that products can be eaten as bought. For freshly manufactured RTT, for example takeout meals, food baskets, and junk foodstuff items such as: pizzas, burgers, and tortillas, grease, heat and resistant properties are main parameters for packing and containers. Though pasteurized RTT products as wrapped side dishes (guacamole, salsa, kimchi, and salad dressings) normally need good taste and gas-barrier packing with high reliability (See Table 9.1).

3.3 Ready to heat (RTH) goods

The third category of HFR items is ready to heat (RTH) goods, which require little heating for complete cooking. This is the easiest procedure to just heat the products without adding ingredients in it. You just need to heat 10 min in a frying pan or 15 min in a conventional oven and less than 10 min in a microwave oven. The materials that are appropriate for rejoinder and re-heating in the microwave ovens, simmered aqueous medium, and traditional furnaces are required for this category of food items [9].

3.4 Ready to cook (RTC) goods

The last category of HFR products is ready to cook (RTC) and it requires substantial heating (approximately greater than 15 min in a frying pan, greater than 20 min in a traditional oven, or greater than 10 min in a microwave oven) before eating. Most of the items in this category are frozen goods, which need perforation resistant characteristics and sturdiness to endure a hefty drop load at low temperature. Antioxidants with oxygen and light barrier packing features are important parameters for lengthening/increasing the shelf life of food goods, mainly meat/processed meat food items [10]. Furthermore, moisture resistant characteristics with leak proof packing can stop decomposition and dryness of frozen products with packaging during storing [11].

4. Impact of smart packaging in HFR technology

4.1 Why smart packaging is necessary in HFR industry

The packing of food products is an important parameter for keeping the freshness, taste and safety of foodstuff throughout the supply process. Packaging process preserves food and allows HFR goods to travel securely and wholesomely over long distances from place of production to point of usage. The packaging industry has reported that losses of food items are < 1 % during the distribution process [12]. Packaging creates delicate collaboration between the products and packaging process not only fulfills desires of the consumer but it also meets the needs of the producer and the supplier. Therefore, the packing industry is important for food safety as well as for the ease of buyer. Not only employees and people who live alone but also the aged people are a significant customer category. Apart from working class or busy individuals living away from their homes, the category of old people is becoming an important and huge customer category to benefit the HFR industry [13]. The requirements and desires of these consumers must be taken into account by HFR producing companies.

4.2 Required characteristics of smart packaging for HFR industry

It is a need of consumers of the modernized localities that good design of HFR packaging have been established to satisfy their desires. Therefore HFR packaging must have the following characteristics:

- Packaging that is aseptic and has a high level of integrity.
- Packaging that is antimicrobial.
- Packaging that is simple to open e.g., juices straight from the press.
- Packaging with a high level of integrity and acid resistance e.g., RTT delivered food/main meals from take away.
- Packaging that is resistant to oil and heat e.g., burger/sandwich/tortilla/pizza.
- Packaging that is resistant to acid e.g., side dishes that have been freshly cooked (i.e., salad dressing, salsa, kimchi and guacamole.
- Packaging with high coherence e.g., "Sashimi" of raw oysters and raw fish Packaging with a modified environment, pizza (frozen and chilled), noodles, spaghetti, and fried rice.
- Packaging that can be microwaved or baked e.g., pasta recipes made using dehydrated pasta.
- Packaging that is resistant to heat e.g., soups (canned or instant) or broth.

- Packaging that is simple to open, e.g., rice, pasta, and instant noodles.
- Packaging with a high gas and moisture barrier, e.g., raw beef and poultry that has been chilled and marinated with vegetables, frozen raw fish, breaded and dipped in sauces.
- Packaging that is antioxidant.
- Consumers have gained additional convenience from the HFR market, when packaging technology has been combined with sophisticated/innovative and biodegradable packaging materials and functional packaging.
- The above mentioned complementary characteristics of packaging materials along with food safety have an important influence on customers' buying choices [14].

5. Smart packaging technologies available in HFR industries

The different HFR products have different shelf lives, so there would be a need for different technologies for packing these products. The barrier, mechanical and physical properties along with the compound formation or migration which can directly affect the taste and smell of the enclosed product should also be considered by the packaging companies. Therefore, depending upon these features different packaging technologies are introduced (See Fig. 9.3) for the formation of more effective and useful packaging of the products.

5.1 Easy to exposed packing (EEP)

The comprehensive/considerable literature survey suggested that the main customer objection regarding food packing was the damages/harms connected within foodstuff

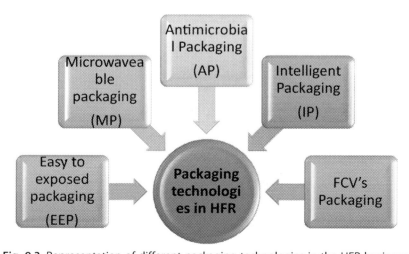

Fig. 9.3 Representation of different packaging technologies in the HFR business.

packing during opening, besides certain kinds of packing that need extra gears for opening of foodstuff packaging are very problematic to access [15]. Easy to expose packaging (EEP) permits buyers to just open food packaging without using some gears. According to technical description, EEP normally needs approximately 6–20 N/15 mm continuous amounts of force through using only hands to open packaging and without damaging the packing material [16]. The most demanding feature from the buyers is the easy opening packages along with the seal. At first, the packages with zippers became very famous which were initially introduced for the meat product. The food companies then realized that the buyers do not want zippers on the small packages, as these packages are not in use for a long time. But these zipper packages are in demand now, as these are easy to use and provide long-term safety and freshness to the food enclosed [17]. The bags that can be shrunk by applying heat are also used to protect the products by sealing them in closed conditions. The seals can be opened by applying a certain amount of force [18].

EEP can be categorized into two different classes according to the action taken for opening:

5.1.1 Easy to open for rigid packing

The action is taken by using right-handed individuals in EEP for rigid packing. For example: bottles, cups, trays and cans.

5.1.2 Easy to peeling for flexible packing

The action is required by using both hands in easy to peel for flexible packing, For example: bags, sachets, pouches and flexible thermo shaped packing [19].

Currently, according to the technological strategy, two main technological classes have been proposed for EEP.

5.1.3 Easy peel sealant

This technique is usually used for sausages, dairy products, microwaveable foods, and freshly cut fruits, meat, and frozen food products [19]. Different essential parameters are considered while using this technique, such as: the seal-substrate adhesion, type of adhesion, tear velocity, temperature, tear tab and seal seam designs, sealing contour, and peel angle [16].

The cohesive peeling is generally manufactured through dry amalgamation of base sealed polymer (which is known as sealant layer) with another mismatched polymer material, which results in the formation of numerous tiny islands in the peal sealing sheet. A semi-crystal-like and high isotactic thermoplastic (polybutene-1) material will be fabricated through the polymerization process of 1-butene and adding Ziegler Natta as a catalyst. Polybutene-1 can be utilized as a mismatched polymer layer to manufacture tiny islands, and the polymer base sealing layer is polyolefin (low density poly-ethylene, lined low density poly-ethylene, and poly-propylene). The bond between these two polymer

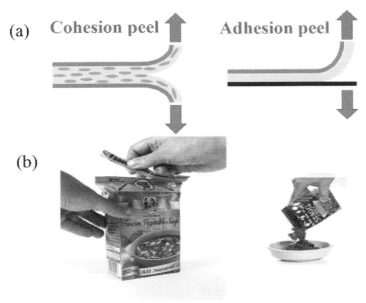

Fig. 9.4 Representation of different peel mechanisms (A); Pictures showing the opening process for Hormel Tetra Recart™ packaging (B).

layers is lower than the cohesive nature within the mismatched polymeric island and the polymeric base medium. During the opening process of the packaging, the cohesive devastation arbitrarily spreads sideways and via polymeric medium, as represented in Fig. 9.4A. The sealing sheet is detached through a flat/even and clean peeling surface, which provides blanching of the stopper with addition of integral interference mark [16]. Commonly, cohesive peelings are employed for supple sacks and bags. Though, adhesive peelings are categorized through a parting between a substratum medium and cover layer, like polymeric trays, saucers, and void membrane packing. The parting process takes place along sideways of the substrate and surface region of the sealant sheet during peeling of packaging materials, though every surface region remains undamaged [20]. The peel off power rests on the apparent energy of the wrapped external side and setting developed for the close up process.

5.1.4 Laser perforation technology

Normally, laser perforation technology (LPT) is utilized for nonmetals, like laminated plastics and paper based packaging [21]. For instance, the Tetra Recart is a carton that is formed by the laser perforation technique, this carton is designed to maintain the sterile environment of the product in it. The line perforated by laser action on the carton helps in tearing it easily, no specific tool is required to open it up as displayed in Fig. 9.4B. LPT comprises six layered packing materials, which is made of laminated polymers, foils and paper board materials. This Tetra Recart packaging excellently bears the severe countering procedure. It is usually employed for packaging wet shelf stable items, like meat sauces, beans, rice, soups, different oat and barley cereals and RTT foods. Moreover, LPT is extensively used in foils, plastic-coated bags, pouches and caps. Split outlines are developed through two different ways: (1) inscribing and (2) creating a lined puncture/perforations. The laser beam makes its passage via the outside plastic film region without producing any kind of imperfections. At that point, it will be engrossed through the bottom film, and it is known as a highly absorbing sheet, beforehand it will be mirrored through the internal aluminate foiling coating and without disturbing blockade characteristics [22].

5.2 Microwaveable packaging (MP)

In the 21st century, people prefer the food items that can be directly microwaved or cooked directly from their frozen form. These types of products require special packaging which cannot explode in the heat of the microwave and cannot fracture at low temperature in freezers. It is necessary for the packaging to bear the temperature fluctuations from freezing temperature ($\sim 20\ °C - 0\ °C$) to the higher temperature of the microwave for cooking (71 °C–105 °C). Usually, these types of microwaveable frozen products are packed in tight containers along with a lid (See Fig. 9.5) [23].

The daily use of microwave ovens has tremendously gained attention in the industry for food sterilization as well as in the homes for heating, defrosting and cooking food goods, particularly RTH and RTC items [24]. The heating variation between traditional and microwave ovens for the food goods are revealed by the modification in the rate of crispness and frying. The conventional ovens use hot air produced via the electrical heating system/Bunsen burner to cook foodstuff in the closed compartment. The food heat equilibrates up to the oven compartment heat, and it varies from (48 °C–260 °C). The

Fig. 9.5 Microwaveable packaging for HFR products.

conducting heat of this nature is transported throughout the foodstuff from outer to the inner side; consequently, the outer part has the warmest region and temperature attains prerequisite value for food crispness. A heat-resistant characteristics is a key demand for packing materials. On the other hand, microwave ovens work on the principle of non-conducting behavior of the foodstuff. The targeted molecules (polar-type) in the foodstuff only absorb microwave radiation and can be excited, whereas the surrounding atmosphere in the microwave oven has low temperature [25].

Though, packaging technology uses two different techniques to overcome the above mentioned problems:

5.2.1 Susceptor method

Crust formation and the surface browning in the heated product (fries, pizza, sandwiches, pies, etc.) are achieved by using the susceptors. Susceptors are specially designed materials that efficiently convert electromagnetic radiation into heat. The susceptor fabrication requires vaporization of the thin layer of metal with thickness less than hundred nonmetric range, which is done on robust polymeric layers like poly-ethylene tere-phthalate (PET) as described in Fig. 9.6. This polymeric layer is translucent to microwave radiation. If the metal is aluminum which is vaporized on PET film, more electromagnetic radiation will get absorbed because of the increase in the external resistivity from 20 Ω^{-2} to 400 Ω^{-2}. The exposed surface of food products in the microwave will receive this produced heat through the conduction process. The crisping and browning process needs 150 °C temperature, and these susceptors can work efficiently at high temperatures up to 260 °C [26]. Therefore, it can be used efficiently in crisping and browning the products. These susceptors are used by well-known companies for the packaging process of food goods.

5.2.2 MicroRite method

MicroRite is another technology that is used to maintain the energy of microwaves and in the improvement of packaging rigidness. It is simply a paper board plate with the lamination of a polyester film deposited with Al, as represented in Fig. 9.7A. The layer of aluminum on the MicroRite trays helps in the redistribution of the energy in the depth of frozen products with even heating [27]. This working principle can be applied to the

Fig. 9.6 Susceptor method for microwave heating.

An approach of smart packaging for home meals 155

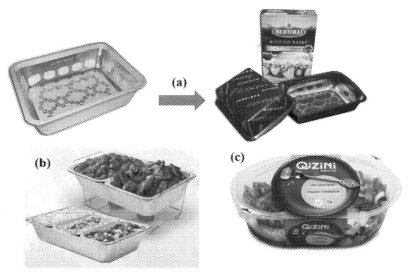

Fig. 9.7 A MicroRite technology on the thin decorated Al coated tray [27] (A); many fold portion boxes coated with foiling sheets (B); and microwave shielding packing from Shieltronics company [28] (C).

number of containers requiring high-temperature treatment, as it just requires the lamination of the aluminum sheet on the compartments sections for the food that need high heat, as represented in Fig. 9.7B. It might accelerate the cooking procedure in the designated tray, on the other hand food in other sections heats gradually [29]. Instead of uniform heating of all over sections of the box, one section in the box is secured from heating through micro-wave insulating material. "Shieltronics" a highly reputed establishment with patent license for protected microwave expertize, and this company also has established micro-wave guarding packing and developed different HFR items. Fig. 9.7C is representing that a multiple sectioned injection shaped polypropylene (PP) plate full with fresh or incompletely cooked poultry food in one translucent section and FCV are placed in mildew categorized section. An aluminum foil tag performs as a protection cover for a wrapped meal during the microwave process and does not change the flavor and consistency of the packaged food items [28].

5.3 FCV product packaging

The FCV items intake has increased due to the growing trend of healthy living style among communities. FCV products comprise fresh vegetables and fruits slightly processed through washing, peeling, and chopping or cutting. Assorted vegetables salad, ready vegetables for soup preparation, or RTT fruits are the key food items in this class, which are usually sold as fresh food items. Though, nonstop metabolic action in the food items result in quality reduction owing to the enzyme reactions, transpiration and respiration. Modified atmospheric packaging (MAP) methods are broadly employed for FCV goods to increase shelf life through decreasing the rate of respiration, ethylene (C_2H_4) bio-synthesis, water content loss, and browning of cut surfaces [30]. The main criteria for MAP is that an improved atmosphere can be generated either inactively via right penetrable packing materials or energetically through removing a quantified gas mixture along with permeable/penetrable packaging items [31].

Passive mode of MAP utilizes a gas-like penetrable sheet, wherein a required ambiance created automatically by nature due to the food items respiration mechanism and diffusion of different respiratory gases take place via packaging sheet. Fig. 9.8A shows mechanism of respiration in the presence of oxygen (aerobic respiration) of the fresh cut vegetable items and it is comprised of decomposition of complex large molecules (carbohydrates, fatty acids and proteins) into simple molecules, in the

Fig. 9.8 The mass transferring in perforated FVC packaging through food respiration (A); Macro perforated box film for freshly packed strawberries (B); and micro scale punctured polymeric bags for already cleaned vegetables (C).

course of this process of respiration energy, CO_2 and H_2O delivered. The respiration method uses O_2 for a chain of enzyme-based procedures [32]. Consequently, the basic demanding characteristic of FCV products packaging is H_2O vapors and gas penetrability, which can be obtained either through the polymeric medium itself or via perforation. Though, most of the FCV items are passed through the packaging process with polymeric products (like: PE, PP and PET), which might not permit H_2O vapors and gas transport used for sufficiently good atmospheric conditions. Therefore, a film with extra perforation is utilized to allow the food item to "respire/inhale" [33]. Fig. 9.8B and C are representing variation of different perforations depending on their sizes i.e., macro perforations and their size is greater than 200 μm in diameter and micro perforations around 50—200 μm in diameter size. A perforation size essential to be optimized dependent on the oxygen intake and carbon dioxide manufacturing rates for these gases [34]. Single or numerous pores not only increase the gases and water vapor exchange rate but also decrease the bad smell generation from anaerobic respiration and bacterial growth connected with wetness condensation, which results in a higher lifespan of fresh vegetables. A permeable pore sheet might allow an entrance of bacteria's or other microorganisms into a sealed cartons during moisture handling conditions [35]. Consequently, a good MAP method had been established and used for impermeable wrapping. A better modified atmospheric packaging method is grounded on the displacement of head space respiratory gases in the packaging material through purging a gas mixture or utilizing absorber/emitter gases to create the required atmospheric conditions [33]. To decrease respiratory process, transpiration products, and ethylene generation, environments with low oxygen around 1%—5% and higher carbon dioxide amounts ~ 5%—10% have been used for FCV packaging. The high carbon dioxide amount can impede production of both Gram + ive & −ive bacterial strains; though, lower oxygen quantity on the inner side of FVC packaging may stimulate anaerobic metabolism process, which causes bad taste and fermentation mechanism [36]. Hence, the amount and stoichiometric ratio of particular gases blend in the packing desires to be selected sensibly and smeared practically, which depends on the food item nature and storing conditions, as summarized in Table 9.2. Moreover, absorbing/emitting gases, particularly carbon dioxide, oxygen and ethylene, may be added inside food packaging in order to regulate quantities of these gas compositions. Recently, "punctured mediated-modified-atmospheric packing" (PMMAP) method replaced the non-perforated traditional MAP method for many fresh vegetables [37]. The atmospheric level of gases obtained by means of out-of-dated MAP method are inadequate to guarantee a long lifespan and value of food through storing the process in typical polymer films. The application of the PMMAP method can enhance the replacement rate of respiratory gases and water vapors via the food packaging wall.

Table 9.2 Adapted atmospheric storing approvals for chosen fresh cut vegetable food items at zero to 5 °C. Repeated after taking approval from the Elsevier [33].

Food items	Atmospheric (O_2%)	(CO_2%)	Food items	Atmospheric (O_2%)	(CO_2%)
Broccoli	2, 3	6–7	Cubed cantaloupe	3–5	6–15
Shredded cabbage	5–7.5	15	Cubed honeydew	2	10
Shredded sticks or sliced carrots	2–5	15–20	Sliced apple	<1	—
Sliced or diced onion	2–5	10–15	Sliced orange	14–21	7–10
Zucchini	0.25–1	—	Sliced pear	0.5	<10
Cut or whole-peeled potato	1–3	6–9	Sliced persimmon	2	12
Sliced mushrooms	3	10	Cut kiwifruit	2–4	5–10
Chopped green leaf lettuce	0.5–3	5–10	Sliced peach (storing at zero °C)	1,2	5–10
Chopped red leaf lettuce	0.5–3	5–10	Sliced tomato	3	3
Chopped or shredded iceberg lettuce	0.5–3	5–10	Sliced strawberry	1, 2	5–1
Cleaned spinach	0.3–3	8–10	Arils (seed coating) pomegranate	—	15–20

5.4 Antimicrobial packaging (AP)

Most food products especially meat items are considered as the effective media of growth for microorganisms like bacteria. The growth of bacteria and other microorganisms on the food leads to changes in food texture, discoloration, and flavor change, which produces the chances of illness and also reduces the lifespan of the food goods [38]. The quality of meat products is worsened because of the wastage and the infection-causing microbes. This type of contamination is usually caused by improper handling during the process of packaging. So, to avoid this situation many different techniques and antimicrobial agents are being used [39].

One of the most effective techniques to avoid microbes in food products is antimicrobial packaging. This technique increases the shelf life of food along with its quality by slowing down the growth of bacteria and other microorganisms, prolonging their lag phase [40]. The most effective agents which are used to reduce the bacterial spoilage of food are ClO_2, organic acids, peptides, sulfites, antibiotics, CO_2, and nitrites (See Fig. 9.9) [41,42]. The natural antibacterial agents are thyme, clove, garlic, cinnamon, rosemary, oregano, etc., which are directly obtained by the plants [43–45].

Antimicrobial coatings are prepared by coating, surface modifying, coating, or immobilizing the antimicrobial agent on the materials used for packaging [46]. These agents are introduced directly in the packaging for more effective results.

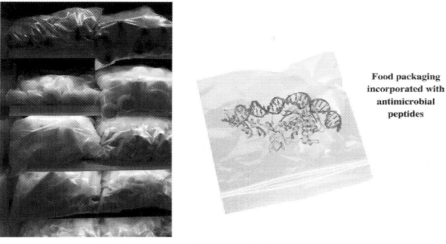

Fig. 9.9 Antimicrobial packaging for home meal.

There are two types of agents:
➢ **Heat sensitive e.g., enzymes**
These antimicrobial agents are added into the material by non-thermal processes to maintain the antimicrobial activities of the prepared film. The process of solvent compounding, electro spinning, and casting are extensively used for this purpose [47].
➢ **Thermally stable**
These agents are introduced in the packaging by different processes i.e., extrusion, injection molding, and extrusion [48,49].

In some packaging, the antimicrobial agents are used in the form of multilayers, where every layer performs its specific function. As in the three-layer film, the outer layer acts as a barrier as it stops the outside movement of active components, a middle layer which acts as a matrix firmly holds these active components, and the third and innermost sheet monitors diffusion rate of the agents [50]. The process of diffusion of active components in multiple film layers is considered extra intricate than simply in solution.

5.5 Intelligent packaging (IP)

In this packaging field, intelligent packaging technology is considered as one of the effective newly emerging processes of packaging. This technology is developing, as all its characteristics are not completely known. Besides this point, this technology is still used for providing quality, traceability, and safety to the packed food products, along with being convenient for the consumer. This packaging controls and examines the surrounding of food and also its condition [51]. In other words, this technology is the communication of the packaging system which promotes external and internal environmental communication and improves the internal condition of packed products [52]. This technology works on different principles as compared to the active packaging system because the intelligent packaging system aims to provide complete evidence about the state of a foodstuff by connecting all the food supplying chain participants which are the producers, the retailers, and the users/consumers [53].

This packaging technique also indicates the current condition of packed food whether it's fresh or expired. In intelligent packaging, different temperature indicators (See Fig. 9.10) are also induced e.g., MDI (microwave doneness indicator) which shows the food temperature, and TTI (time-temperature indicator) which shows the overall history of temperature for packed food. This packaging also checks the effectiveness of the active packaging system [21]. This packaging shares the informative data to the supplying chain while active packaging takes action for packing products [36]. Both of these packaging technologies can be used at the same time for making a smart and well-organized packaging system [33].

Timestrip® is a United Kingdom based company and it uses technological procedures that monitor a food item shelf life through a pictorial watch to substitute a perished mold on the item. Fig. 9.11A displays that the pictorial watch will continue its function by discharging liquid of food rank color into the permeable film through pushing a trivial basin. It screens the intervening time of unfastened or cold/chilled products retained within a cooled environment throughout a year. The current technology is feasible not only for frozen/chilled RTH and RTC food items but it is also applicable for fresh meats and seafood products [55]. The "Keep-it" Technology of Norwegian based company was established by the Life Science department of Norway University. It continuously controls

Fig. 9.10 Intelligent packaging with different temperature indicators.

An approach of smart packaging for home meals 161

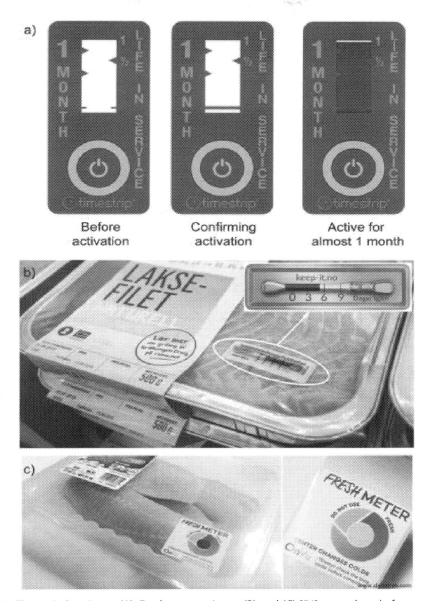

Fig. 9.11 Timestrip® pointers (A); Freshmeter pointers (B) and (C) [54]; reproduced after permission.

temperature (T) as a function of time (t) and indicates the real lasting lifespan of food items like raw fish, minced meat and poultry products as shown in Fig. 9.11B [54]. "OnVu™ (Israel-based company: fresh point quality assurance Limited)", is an irreversible and nonpoisonous "T" and "t" gauge and it is extensively applied for frozen goods. The ink color varies from colorless to a blue color due to photo-chromic solid state reactions under ultra violet (UV) light radiation. Fig. 9.11C is representing, when undesired temperature observed by food goods, the ink changes from blue to colorless form with passage of time, and the customer can match with a mention shade published on the box. This technological advancement was industrialized by "Israel-based company: fresh point quality assurance Limited" and German "Bizerba Group" [56].

6. Different smart packaging approaches for different products

6.1 Innovative packaging approach of meat & poultry goods

Customers prefer fresh red meat. Usually, the pieces of meat are placed over the tray and enfolded with paper through a porous oxygen membrane. Poultry products are also wrapped with oxygen. The quantity of oxygen maintained by 'Modified atmosphere packaging' technology. Its value for meat is approximately 70%—80% oxygen and 20%—30% carbon dioxide and for poultry products is 65%—75% nitrogen and 25%—35% carbon dioxide [57]. MAP is an important technology and is used to maintain the proper conditions of poultry and meat items.

The purpose of oxygen is to preserve the muscle's pigments, nitrogen is used to provide the inert environment and carbon dioxide is used to prevent the production of bacteria. Oxygen scavengers are mostly used to maintain the quality of these goods [8].

6.2 Innovative packaging of fish and sea-food food items

Fish and other sea-food meals are one of the favorite dishes of many people but due to busy lifestyles people demand or order the cooked packaged sea-food. The sea-food departments or industries are trying to make good quality products. The quality of sea-food is one of the difficult tasks for industries [9]. The reason is that a large number of varieties of sea animals exist. Each type of sea animal has its own properties. Both types of smart packaging (intelligent & active) show their vital part in the packaging of sea-food. TTI, MAP, RFID and FQI techniques are used to fulfill all requirements of seafood packaging.

6.3 Smart packaging of fruit and vegetable goods

The smart packaging of fruit and vegetable goods is most critical and difficult from all other packaging. Generally, highly controlled conditions are required for fresh fruit and vegetables. Actually, both fruit and vegetable goods have living tissues, they continuously intake O_2 and discharge CO_2 [8]. Many factors control the process of respiration. Therefore, highly controlled conditions are required. The major conditions for packaging the fruits and vegetables are …

- Humidity
- Temperature
- Mechanical Damage
- Gas composition

Variation in these conditions can deteriorate the fruit and vegetable [10].

6.4 Smart packaging of beverage products

Beverage industry also used smart packaging technology. In order to attract the customer, different types of fragrance substances, aromas and scents are used by different beverage industries. Usually children do not like the milk for this purpose, companies use a lot of flavoring agents in order to make the milk tasty like chocolate, banana and strawberry, etc. [58]. In the beverage industry different types of smart packaging are approached by using following ways:
- Nutrient releasing packaging
- Flavor releasing packaging
- Gas releasing packaging
- Pro-biotic releasing packaging
- Enzyme releasing packaging
- Odor removal packaging
- Thermo chromic technology
- Tamper-proof packaging

7. Smart packaging trends for home meals

Recent smart packaging inclinations are comparatively influenced by many other inducing trends of the societies, like purchaser living style, health related concerns, recyclability, packaging distribution, and advanced emerging knowledge (such as: artificial intelligence (AI), 3D printing electronics, and IOT (Internet of Things)) [59]. Consequently, the improvement of packaging technology and smart approach is necessary to fulfill the desires of both industry and buyers. The modern-day customers have inadequate time or skills to cook food at home, they want HFR food to have a good taste, high nutritional values, safe, and good quality [60]. To fulfill the customer desires of slightly processed meals with good quality of commercialized goods, moreover development of innovative non heated foodstuff handling advances (like pulsated electric field, higher hydrostatic pressure, UV light and ionization radiations, etc.) has been under attention among nutritionists, scientists, food producing companies and fast food chains, and customers to substitute or combine with traditional thermal food goods processing [61]. The variety of suitable packing materials is critical for consumers, therefore, innovative technologies are required for the process of pre-packed goods. The smart food packaging materials require to be supple to permit the electronic beam/radiation to penetrate from packaging goods or to endure inflexible powers, whereas stabilizing the characteristics and functionalities of wrapping products afterward curing process [62].

Moreover, the vegetarian and organic food products market are gaining more popularity in the developed countries due to increasing health concerns [63]. The market players are trying to meet the demands of vegetarian consumers. Food packaging with high honesty, like barrier rejoinder bags and vacuum packaging is a prerequisite to improve customers' concern about an adulteration of food products. It can offer whole control of contents and safeguard that their HFR products remain pure without addition of any non-organic materials contamination from the point of packaging to the time of cooking and heating [64].

Recently, the "circular economy" term has obtained great attention especially in the drink and food industries. Nowadays the main concern of HFR and smart packaging technology is to promote the principle of 3R (reuse, recycle and reduce). Packaging Producing companies are acknowledged to develop products with considering parameters like ecofriendly design, recyclable/biodegradable and prolonging the shelf life of food goods in their design-Based concept [65]. For example, a Coca Cola Beverage Company introduced the cold drink bottles manufactured with recoverable and recyclable oceanic plastics recovered from the beaches of Mediterranean Sea [66]. The industries have introduced Di-ethylene polymerization through Microwave E Technology (DEMETO), the formerly used poly ethylene terephthalate plastic materials were disintegrated into their pure monomeric units (i.e., ethylene glycol & terephthalic acid). These monomeric units again combine through the polymerization process into PET with improved plastic quality, which can be utilized again for packing of beverage products. The discussed recycling procedure reveals outlook for locked hoop economical features for plastic products.

Concerning the aquatic toxic waste from plastic packing technology, compostable/recyclable packing might be a feasible key to stop plastic left-over from getting into an aquatic ecosystem. Though, the existing worldwide recycling technology for plastic wastes is not appropriate, which might cause problems for the waste management system. The application of recyclable plastic packing or prohibition on polymer products can be considered a good approach for current problems. Many nations are trying to make an act/law for leftover controlling of plastic materials. A huge class of customers around the globe are willing to spend for sustainable food packaging products [67].

In recent years the appearance of smart-phone applications and powerful fridges offers exciting prospects to intermingle with customers through different technologies such as (1) sensor connection, (2) radio-frequency identification (RFI) stickers and (3) near-field communication (NFCM) labels as a smart packing. People can envisage the value and lifespan of packed food goods through digital information by using smartphone applications [68]. The above mentioned labels can regulate CO_2 and O_2 quantities, NH_3 and moisture contents of the packaging. Therefore, customers can envisage the lifespan of a wrapped product. As presented in Fig. 9.12A, the packing materials need 2 parameters, (a) metallic structure and (b) RFI/NFCM labels. The mobile phone applications obtain data with smart tags for the temperature and warming time, while the metallic structures of the packing bring and transfer the heat waves to the packaging. A Norwegian (Thin Film Electronic ASA) company established a temperature sensor (completes practical standards) with 3D printing techniques, as shown in Fig. 9.12B. Therefore a less expensive and varying temperature checking tag can monitor the lifespan of consumable

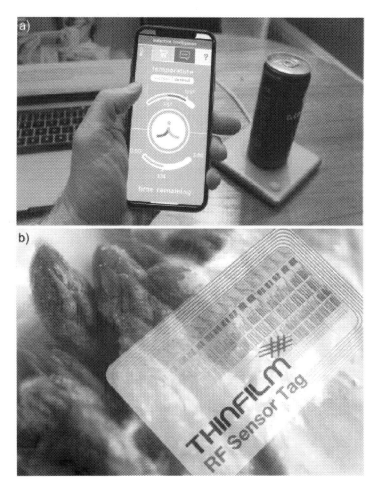

Fig. 9.12 Food warming equipment [69] (A); different temperature devices formed by "Thin Film Electronic, ASA company" [70] (B). *(Produced with permission.)*

products [70]. While the IOT and 3D printings are comparatively new exciting concepts and undeveloped technologies in the food business, the use of these technologies is anticipated to surge in the coming era.

During COVID-19 pandemic time, food buyers are more anxious about contact to corona viruses through just touching an object [71]. To reduce the risk of touch during food shopping, the vendors can offer "QR/NFC" tags on the packaging to develop a touch-less procurement environment. Buyers may attain the required evidence from their smartphone applications before making a buying choice. Therefore technological advancement might decrease probabilities of infective corona virus's dispersal. Moreover, emerging technologies, like worker robotic systems for packaging, drone delivery systems and automatic processes, can overcome the human to human direct contact. Consequently, the corona or other viruses are slightly practical for surfaces of the packaging products.

8. Consumer benefits of smart packaging

The consumers are getting following benefits by using smart packaging approach for home meals:
- Active and intelligent smart approaches introduce the innovation in the packaging of home meal products.
- The quality of home meal products can be checked by sensors and indicators.
- Home meal products analysis by smart packaging technique saves time and cost.
- With the help of smart packaging home meal products are less wasted.
- Smart packaging helps to meet the demand of consumers [11].
- We can use different types of techniques to guarantee the quality of goods.
- Smart packaging approach for home meals helps us to increase the lifespan of products.
- Help us in the supply chain of food and other eatable products.
- Smart Packaging improves authentication and traceability.
- With the help of smart packaging techniques such as FTIR, QR and NFC etc. Customers by scanning may see tutorial videos and all the information about the product directly from the website.
- Packaging material design in such a way that in smart packaging may be reusable for other purposes [12].

9. Issues related to the smart packaging

There are also some issues related with small packaging approach for home meals and few of them are listed below:
- Intelligent and active packaging may contain different types of chemicals, some are released and absorbed, so may contaminate the food material.
- Consumer does not rely only on the appearance, freshness and taste. Quality assurance is also one of the major requirements of customers.
- People prefer fresh and good quality products. Many retailers demand this type of packaging which is totally inert to the home meal products.
- Before the implementation of smart packaging producers need to show the fabrication of all sensors and detectors.
- No doubt smart packaging is an advanced technique for home meal products but mostly smart packaging materials are not environmentally friendly because many batteries, sensors and detectors have not been recycled.
- The implementation of smart packaging is a complex procedure so it requires regular checks and balances. This also needs a lot of time and resources.

- Due to complex procedures, customers may confuse and mislead by procedure.
- In active packaging except few techniques are not applicable for liquid home meal products.
- Active/intelligent packaging increases the cost of food products and moreover, active/intelligent packing is limited [13].

10. Conclusion

Variations in the customer living style around the world have increased demand for HFR and approach of smart packaging for home meals. Numerous investigations have demonstrated that not only the quality of foodstuff but also packing methods conclusively affect consumer's choice. Innovative smart packaging methods are providing new prospects for the HFR industry through offering several remunerations for both the food market and customers. For instance, easy to open packing and microwave susceptor approaches have been considered by customers owing to their efficacy and comfort to use. Antimicrobial packaging and MAP techniques under the category of active packaging have gained great consideration owing to their potential to increase the food items life period and also improve protection of food products. From a large-scale commercialization point of view, active packaging method is extensively in use by consumers, whereas intelligent packaging method is progressively increasing for the frozen food goods. To advance a next generation smart food packaging approach for home meals, other correlated trends must be considered. Current worldwide environmental concerns comprising maritime pollution and weather fluctuation have shifted the packaging trends in the direction of the use of recyclable, ecofriendly and biodegradable green materials and modern customers also want to decrease the global ecological effect. The arrival of printable electronic gadgets and smartphone applications has inspired the smart packaging market. Customers can obtain valuable digital statistics, such as quality, actual lifetime of food product and nutritional assessment, through smart tags on the packaging. Especially, evolving technologies like drone delivery, or robotic systems, NFC and QR-codes can decrease human to human contact and make a touch less atmosphere, which endorses customer assurance in procuring. Though the key restrictions of price and technology incorporation are quite present, more advanced investigations can contribute to a broader acceptance of these smart packaging knowledge for home meal in the coming future.

References

[1] K.-I. Kim, et al., Quality characteristics of beef by different cooking methods for frozen home meal replacements, Korean J. Food Sci. Anim. Resour. 35 (4) (2015) 441.
[2] A.I.A. Costa, et al., A consumer-oriented classification system for home meal replacements, Food Qual. Prefer. 12 (4) (2001) 229–242.
[3] Research and Markets, Global $156+ Billion Ready Meals Market to 2025, Research and Markets, 2020.

[4] Statista Research, Sales of Home Meal Replacements in South Korea from 2010 to 2020 Statista Research, 2021.
[5] S. Kim, K. Lee, Y. Lee, Selection attributes of home meal replacement by food related lifestyles of single-person households in South Korea, Food Qual. Prefer. 66 (2018) 44–51.
[6] S. Thienhirun, S. Chung, Consumer attitudes and preferences toward cross-cultural Ready-To-Eat (RTE) food, J. Food Prod. Mark. 24 (1) (2018) 56–79.
[7] J.A. Park, Y.S. Jang, Differences in consumer purchasing biases of home meal replacement by gender and involvement: target to university students, Culin. Sci. Hosp. Res. 24 (7) (2018) 63–73.
[8] D. Rico, et al., Extending and measuring the quality of fresh-cut fruit and vegetables: a review, Trends Food Sci. Technol. 18 (7) (2007) 373–386.
[9] S.J. Eilert, New packaging technologies for the 21st century, Meat Sci. 71 (1) (2005) 122–127.
[10] M. Wrona, F. Silva, J. Salafranca, C. Nerín, M.J. Alfonso, M.A. Caballero, Design of new natural antioxidant active packaging: screening flowsheet from pure essential oils and vegetable oils to ex vivo testing in meat samples, Food Control 120 (2021) 107536.
[11] D.S. Lee, Active packaging, in: D.W. Sun (Ed.), Handbook of Frozen Food Processing and Packaging, CRC Press, Florida, 2016, p. 826.
[12] K. Sonneveld, What drives (food) packaging innovation? Packag. Technol. Sci. 13 (1) (2000) 29–35.
[13] L.M. Duizer, T. Robertson, J. Han, Requirements for packaging from an ageing consumer's perspective, Packag. Technol. Sci. 22 (4) (2009) 187–197.
[14] A.R. Raheem, P. Vishnu, A.M. Ahmed, Impact of product packaging on consumer's buying behavior, Eur. J. Sci. Res. 122 (2) (2014) 125–134.
[15] C. Caner, M.A. Pascall, Consumer complaints and accidents related to food packaging, Packag. Technol. Sci. 23 (7) (2010) 413–422.
[16] S. Sangerlaub, K. Reichert, J. Sterr, N. Rodler, D. von der Haar, I. Schreib, et al., Identification of polybutene-1 (PB-1) in easy peel polymer structures, Polym. Test. 65 (2018) 142–149.
[17] T.J. Rourke, T. Rourke, Enhancement of cooked meat quality & safety via packaging, Proc. Recipr. Meat Conf. 54 (2001).
[18] D.A. Busche, G.R. Pockat, T.A. Schell, Easy Open Heat-Shrinkable Packaging, Google Patents, 2009.
[19] A. Liebmann, I. Schreib, R.E. Schlozer, J.P. Majschak, Practical case studies: easy opening for consumer-friendly, peelable packaging, J. Adhesion Sci. Technol. 26 (20) (2012) 2437–2448.
[20] M. Nase, L. Großmann, M. Rennert, B. Langer, W. Grellmann, Adhesive properties of heat-sealed EVAc/PE films in dependence on recipe, processing, and sealing parameters, J. Adhesion Sci. Technol. 28 (12) (2014) 1149–1166.
[21] A. Stepanov, E. Saukkonen, H. Piili, Possibilities of laser processing of paper materials, Phys. Procedia 78 (2015) 138–146.
[22] F. Gaebler, E. Büchter, CO2 laser system enhances food packaging: a reliable and viable solution for perforation tasks, Laser Technik J. 7 (5) (2010) 39–41.
[23] J.-M. Su, P. Georgelos, Freezable/microwaveable Packaging Films, Google Patents (2011).
[24] N.V. Olsen, E. Menichelli, O. Sorheim, T. Naes, Likelihood of buying healthy convenience food: an at-home testing procedure for ready-to-heat meals, Food Qual. Prefer. 24 (1) (2012) 171–178.
[25] V. Meda, V. Orsat, V. Raghavan, Microwave heating and the dielectric properties of foods, in: K.K.M. Regier, H. Schubert (Eds.), The Microwave Processing of Foods, Woodhead Publishing Ltd, Cambridge, 2017, pp. 23–43.
[26] M. Celuch, W. Gwarek, M. Soltysiak, Effective modeling of microwave heating scenarios including susceptors, in: International Conference on Recent Advances in Microwave Theory and Applications, IEEE, Jaipur, India, 2008, pp. 404–405.
[27] C. Hine, Seafood is served, in: Paper, Film and Foil Converter, (Paper, Film and Foil Converter Website: Paper, Film and Foil Converter Magazine) vol. 79, 2005, p. 46.
[28] T.H. Bohrer, Shielding and field modification—thick metal films, in: M. Lorence, P. Pesheck (Eds.), Development of Packaging and Products for Use in Microwave Ovens, Woodhead Publishing Ltd, Cambridge, 2009, pp. 237–266.
[29] ft.org/news-and-publications/food-technology-magazine/issues/2008/october/features/novel-ideas-in-food-packaging, 2008.

[30] O.J. Caleb, Modified atmosphere packaging technology of fresh and fresh-cut produce and the microbial consequences—a review, Food Bioproc. Technol. 6 (2) (2013) 303—329.
[31] R. Ahvenainen, New approaches in improving the shelf life of minimally processed fruit and vegetables, Trends Food Sci.Technol. 7 (6) (1996) 179—187.
[32] M.A. Rojas-Graü, G. Oms-Oliu, R. Soliva-Fortuny, O. Martín-Belloso, The use of packaging techniques to maintain freshness in fresh-cut fruits and vegetables: a review, Int. J. Food Sci. Technol. 44 (5) (2009) 875—889.
[33] M. Oliveira, M. Abadias, J. Usall, R. Torres, N. Teixido, I. Vinas, Application of modified atmosphere packaging as a safety approach to fresh-cut fruits and vegetables—a review, Trends Food Sci. Technol. 46 (1) (2015) 13—26.
[34] Z. Hussein, O.J. Caleb, U.L. Opara, Perforation-mediated modified atmosphere packaging of fresh and minimally processed produce-A review, Food Packag. Shelf Life 6 (2015) 7—20.
[35] M. Scetar, M. Kurek, K. Galic, Trends in fruit and vegetable packaging—A review, CJFT 5 (3—4) (2010) 69—86.
[36] J.R. Chen, A.L. Brody, Use of active packaging structures to control the microbial quality of a ready-to-eat meat product, Food Control 30 (1) (2013) 306—310.
[37] B. Salemi, N. Sedaghat, M.J. Varidi, S.M. Mousavi, F.T. Yazdi, The combined impact of calcium lactate with cysteine pretreatment and perforation-mediated modified atmosphere packaging on quality preservation of fresh-cut 'Romaine' lettuce, J. Food Sci. 86 (3) (2021) 715—723.
[38] K. Biji, et al., Smart packaging systems for food applications: a review, J. Food Sci. Technol. 52 (10) (2015) 6125—6135.
[39] J. Kerry, M. O'grady, S.J. Hogan, Past, current and potential utilisation of active and intelligent packaging systems for meat and muscle-based products: a review, Meat Sci. 74 (1) (2006) 113—130.
[40] D.D. Jayasena, C. Jo, Essential oils as potential antimicrobial agents in meat and meat products: a review, Trends Food Sci. Technol. 34 (2) (2013) 96—108.
[41] P. Suppakul, et al., Active packaging technologies with an emphasis on antimicrobial packaging and its applications, J. Food Sci. 68 (2) (2003) 408—420.
[42] Y. Zhao, et al., Recent development in food packaging, a review, J. Chin. Inst. Food Sci. Technol. 13 (4) (2013) 1—10.
[43] A.A. Abou-Arab, Heavy metals in Egyptian spices and medicinal plants and the effect of processing on their levels, J. Agric. Food Chem. 48 (6) (2000) 2300—2304.
[44] S.F. Hosseini, et al., Development of bioactive fish gelatin/chitosan nanoparticles composite films with antimicrobial properties, Food Chem. 194 (2016) 1266—1274.
[45] H.-J. Yang, et al., Antioxidant activities of distiller dried grains with solubles as protein films containing tea extracts and their application in the packaging of pork meat, Food Chem. 196 (2016) 174—179.
[46] C. Véronique, Bioactive packaging technologies for extended shelf life of meat-based products, Meat Sci. 78 (1—2) (2008) 90—103.
[47] P. Appendini, F.S. Hotchkiss, Review of antimicrobial food packaging, Innov. Food Sci. Emerg. Technol. 3 (2) (2002) 113—126.
[48] K.J. Cooksey, Antimicrobial food packaging materials, Trends Food Sci. Technol. 2001 (8) (2001) 6—10.
[49] S.-Y. Sung, et al., Antimicrobial agents for food packaging applications, Trends Food Sci. Technol. 33 (2) (2013) 110—123.
[50] S.A. Onaizi, S.S.J. Leong, Tethering antimicrobial peptides: current status and potential challenges, Biotechnol. Adv. 29 (1) (2011) 67—74.
[51] European Parliament, Commission regulation no. 1935/2004 of 27 October 2004 on materials and articles intended to come into contact with food and repealing directives 80/590/EEC and 89/109/EEC, Off. J. Eur .Union 338 (2004) 4—17.
[52] K. Yam, Intelligent packaging to enhance food safety and quality, in: Emerging Food Packaging Technologies, Elsevier, 2012, pp. 137—152.
[53] D. Restuccia, et al., New EU regulation aspects and global market of active and intelligent packaging for food industry applications, Food Control 21 (11) (2010) 1425—1435.
[54] A. Fagerstrøm, N. Eriksson, V. Sigurðsson, What's the "thing" in internet of things in grocery shopping? A customer approach, Procedia Comput. Sci. 121 (2017) 384—388.

[55] B. Kuswandi, Y. Wicaksono, A. Abdullah, L.Y. Heng, M. Ahmad, Smart packaging: sensors for monitoring of food quality and safety, Sens. Instrum. Food Qual. Saf. 5 (3) (2011) 137–146.
[56] A.P.D.R. Brizio, C. Prentice, Use of smart photochromic indicator for dynamic monitoring of the shelf life of chilled chicken based products, Meat Sci. 96 (3) (2014) 1219–1226.
[57] Y.J. Jang, W.G. Kim, I.-S. Yang, Mature consumers' patronage motives and the importance of attributes regarding HMR based on the food-related lifestyles of the upper middle class, Int. J. Hosp. Manag. 30 (1) (2011) 55–63.
[58] M. Mathlouthi, Food Packaging and Preservation, Springer Science & Business Media, 1994.
[59] E. Poyatos-Racionero, J.V. Ros-Lis, J.L. Vivancos, R. Martinez-Manez, Recent advances on intelligent packaging as tools to reduce food waste, J. Clean. Prod. 172 (2018) 3398–3409.
[60] Y.J. Jang, W.G. Kim, I.S. Yang, Mature consumers' patronage motives and the importance of attributes regarding HMR based on the food-related lifestyles of the upper middle class, Int. J. Hosp. Manag. 30 (1) (2011) 55–63.
[61] N. Chotyakul, P. Rungpichayapichet, Commercialization of high pressure processed foods: a consumer choice for quality and safety products, Sci. Eng. Health Stud. 12 (3) (2018) 139–148.
[62] P. Juliano, T. Koutchma, Q.A. Sui, G.V. Barbosa-Canovas, G. Sadler, Polymeric-based food packaging for high-pressure processing, Food Eng. Rev. 2 (4) (2010) 274–297.
[63] S. Jeske, E. Zannini, E.K. Arendt, Past, present and future: the strength of plant-based dairy substitutes based on gluten-free raw materials, Food Res. Int. 110 (2018) 42–51.
[64] P.G. Creed, The potential of foodservice systems for satisfying consumer needs, Innov. Food Sci. Emerg. Technol. 2 (3) (2001) 219–227.
[65] B. Geueke, K. Groh, J. Muncke, Food packaging in the circular economy: overview of chemical safety aspects for commonly used materials, J. Cleaner Prod. 193 (193) (2018) 491–505.
[66] Coca-Cola Company, Introducing a World-First: A Coke Bottle Made with Plastic from the Sea, Coca-Cola Company, 2019.
[67] D. Eriksson, Sustainability in Packaging: Inside the Minds of Global Consumers, McKinsey & Company, 2020.
[68] M. Vanderroost, P. Ragaert, J. Verwaeren, B. De Meulenaer, B. De Baets, F. Devlieghere, The digitization of a food package's life cycle: existing and emerging computer systems in the logistics and post-logistics phase, Comput. Ind. 87 (2017) 15–30.
[69] Inductive Intelligence, L. L. C., Inductive Intelligence, 2017.
[70] Thin Film Electronics ASA, Thinfilm Builds First Stand-Alone Sensor System in Printed Electronics, 2013.
[71] R. Martin-Neuninger, M.B. Ruby, What does food retail research tell us about the implications of coronavirus (COVID-19) for grocery purchasing habits? Front. Psychol. 11 (2020) 1448.

CHAPTER 10

Perspective and challenges: intelligent to smart packaging for future generations

Sri Bharti[a], Shambhavi Jaiswal[a], V.P. Sharma[a] and Inamuddin[b]

[a]CSIR-IITR, Lucknow, Uttar Pradesh, India; [b]Department of Applied Chemistry, Zakir Husain College of Engineering and Technology, Faculty of Engineering and Technology, Aligarh Muslim University, Aligarh, Uttar Pradesh, India

1. Introduction

Smart and Intelligent packaging refers to packaging that may accomplish "intelligent" functions such as tracking product information, sensing an attribute of the packaged food or its immediate environment such as pH or temperature, and communicating it to the user, such as a manufacturer or consumer. This packaging system with inbuilt sensor technology is used with foods, pharmaceuticals, automotives, personal care and many other types of products. It is anticipated that the smart packaging market will be unevenly distributed in the geographical scenario, but overall it will grow by US $38,662.0 million by 2030 with a CAGR of 5.5% [1,2].

The Food Business Operators (FBO) and manufacturers need to overcome challenges of techno feasibility, circular economy for viable mass-production costs. We need to address the complexity of integrating smart devices into latest packaging processes for security, data privacy, and confidence on depicted information on labels, quality enhancements, sustainability etc.

Moreover, packaging aids in the extension of shelf life, the control of freshness, the display of quality information, the improvement of safety, and the enhancement of user comfort. This enables the capacity to measure or detect product functionality, the transportation environment, or the atmosphere within the package, which aids in the prevention of food spoilage. These systems also improve the product's properties, such as flavor profile, palate, and fragrance; ensure the integrity of the seal; and actively track environmental and product changes, as well as test the authenticity of the package. Packaging manufacturers use smart and intelligent packaging techniques to keep nutritional value and product freshness at a reasonable rate. Traditional packaging methods, such as canning, are being overtaken by novel technologies such as smart active packaging and changed atmosphere packaging. It is also useful for logistics management, premium pricing, waste minimization, and quality control.

Smart packaging has recently received a lot of attention because of the numerous ways it is profoundly changing not only the consumer product packaging industry but also the packaging industry as a whole. So, what exactly do we mean by "smart packaging"? **Smart packaging refers to emerging packaging techniques to improve consumer and commercial functionality while also making product and business data more accessible and trackable** [3–5].

The rapidly increasing development of advanced packaging solutions in the food processing and pharmaceutical industries, growing consumer concern about food waste reduction and management, and rising demand for smart and operational packaging are the factors that drive the growth of this packaging industry. Besides that, the expanding e-commerce market and soaring industrialization, combined with effective consumption for superior logistics and supply chain management, are expected to provide lucrative opportunities for market participants. Smart packaging is more important than ever. As a result of the pandemic, technologies such as near-field communication (NFC) and QR (quick response) codes have become more widely used and are supported by the majority of iOS (iPhone Operating System) and Android devices [6]. Where almost every retail product category, including toys, cosmetics, games, and clothing, uses "connected packaging." A QR code is printed on the product packaging in connected packaging applications, or an NFC label is present on the inside. Consumers can use a mobile device to activate this code and receive informational content. Systemic reviews of varied perceptions of food packaging have been studied by Young et al.

2. Connected and smart packaging: past, present, future ...

Intelligent packaging is not a new concept in the packaging industry. It is frequently used to track inventory ranging from food to personal care products (PCP) from place of its generation/production to point - of - sale. They can be barcodes, RFID (Radio Frequency Identification) tags, QR codes, or electronic chips. It is also used outside of the supply chain to provide consumers with product information or to access sales promotions (Fig. 10.1).

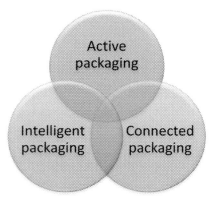

Fig. 10.1 Representation of smart packaging with reference to active, intelligent and connected packaging.

Customers can use their smart phones to find out not only which farm their ingredients come from, but also the distance traveled and storage time, thanks to smart packaging. Aside from authenticity information, a handful of companies are attempting to incorporate practical information into their packaging, such as usage instructions. This linked packaging creates a new channel of communication, and thus advertising, between a brand and a consumer. It is possible to offer competitions, games, and advertisements for other products to consumers using this mode of digital communication [6]. Intelligent packaging is also beneficial to the environment. There is an environmental impact: for example, it is possible to know if the expiry date is approaching and thus minimize waste; it is also possible to transmit relevant data on the method of recycling the packaging. One of the most important aspects of smart and connected packaging in the future is that it provides a positive experience for consumers by providing them with transparent information. It gives brands a better understanding of their customers and it's a great tool for the environment, as it can be used to reduce waste and encourage recycling (Fig. 10.2).

For goodness's sake, connected content is becoming more accurate and resourceful, particularly in the food industry, where a package can now send a notification if a product is out of date. Smart packaging can now be used in the healthcare sector to remind patients to take their medications and to remind them of expiry dates. It is anticipated that the results provided by connected packaging will become more customized and personalized. Due to retrieval data storage, the information and experience contained in the package will be tailored to each individual customer. Every customer will have a more connected experience with their purchased product.

RFID sensors will also become more prevalent in the market. It will inform the package about the current state of the product, which will be especially useful in the food

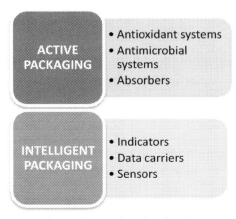

Fig. 10.2 Outline of the driving forces that lead to the development of smart food packaging.

industry. These sensors may transmit data over extremely long distances. It can, for example, notify a farmer that his or her products have been exposed to inclement weather on their way to the point of sale and will thus be unfit for consumption. To provide results based on customer actions and requirements, more powerful AI and AR will be used. Augmented reality is a technology that will turn packaging into a more interactive experience. These experiences can assist brands in demonstrating their values, products, and commitments [7,8]. Due to the obvious current health situation, we are all scanning QR codes found on restaurant tables and store entrances on the spur of the moment, a gesture that we will undoubtedly remember in the future. As a result, now is the time to incorporate intelligent and connected packaging into one brand strategy!

3. Bio-based plastics

Bio-based plastic products are materials that biodegrade in industrial facilities over a set period of time under different composting conditions. As a result, bio-based plastics have a very vast growing market and application range, and as a result, they are becoming more popular in research and economics. Such a type of plastic is produced from biomass and degrades gradually over time. The scarcity of fossil resources, as well as related environmental issues, has fueled the development of an innovative bioeconomy as well as the evolution from fossil-based plastics to bio-based plastics. The conventional, petroleum-based wide selection accounts for approximately 99% of the world's plastic, and much of it will deteriorate for centuries. Avocado seeds are among the feedstocks used to make bio-plastics. Although these products' current exchange rates are marginally higher than any of their petroleum-based equivalents, with rapid manufacturing and a community emphasis on green chemistry, manufacturing costs are expected to drop, making them a more interesting and sustainable alternative than traditionally produced plastics. Despite the fact that this product's market is still in its infancy, research and innovation attempts are being made to encourage the sector. Several polymers, such as Poly Lactic Acid, which is polyester, Poly Hydroxyalkanoates, Bio Poly Trimethyl Terephthalate, Bio Propanediol, and several others, are on the horizon. Certain plastics derived from renewable materials, such as biomass, can save up to 40% of overall energy in power generation when compared to their petrochemical counterparts [9–11].

4. Cutting edge advancement

The advancement in packaging due to selective compositions/specifications, designs, texture, etc., has drastically changed the outlook of the packaging industry. Packaging has become such an important part of the product and brand that it is frequently used as a differentiator in consumer purchase decisions. As a result, brands and retailers are investing more time and money in developing innovative packaging solutions. Packaging companies, material providers, converters, and even industries which are indirectly connected to or outside of packaging are developing hundreds of new technologies aimed at providing packaging advantages to attract buyers. One of the primary objectives is to capture the consumer's attention by conveying the brand message through graphics or frames. Package appearance can influence purchase decisions, whether it is through shelf "squeeze" with an entirely new shape or format in a segment or through packaging graphics [12].

Today's packaging industry does not rely solely on plastic and paper; many other materials, such as seaweed packaging, cornstarch packaging, recycled cardboard and paper, air pillows made of recycled materials, biodegradable packaging peanuts, eco-friendly plastics and recycled plastics, and so on, are used as an innovative methodology in the packaging industry, which is widely accepted around the world [11].

4.1 Innovative materials for sustainable packaging

Sustainable packaging has become even more widespread and innovative. It is the creation and use of recyclable or reusable packaging created from speedily renewable resources or materials. Packaging designers and manufacturers are constantly looking for new innovations for the improvement of their packaging designs that will be safer for mankind, more efficient than the prior one, and, most significantly, greener. This exercise benefits the environment and its ecological footprint [13].

- **Packaging materials with micro-patterns:** Micro-patterning is a new technology that transforms the surfaces of cups. This modern technology not only improves the hold, solace, and handling of hot beverages, but it also slows condensation creation and enhances material characteristics without introducing new chemistry. Market demands necessitate greater and more significant technological advancements, leading to increased digitalization on multiple levels. We are seeing an increase in demand for more sustainable packaging as a result of each of the factors mentioned, as well as more regulatory requirements forcing companies to consider new alternatives. The frequency at which technologies change continues to pose a challenge to industries. Introducing advancements into products and packaging will become a standard and a continuous evolutionary process, alongside the changing regulatory implications and restrictions, in order to remain relevant and efficiently effective. With this, researchers can see that the industry will continue to propagate its own need for revolutionary design, function, operation, and technology, based not only on the advancements in the examples above, but also on the continued innovation in packaging throughout all spectrums.

- **New sustainable coatings with an aqueous barrier**: New sustainable coatings improve any fiber product by preventing moisture from penetrating and potentially contaminating foods and provide alternatives to laminated structures that, rather than those handled with new sustainable coating materials, cannot be recycled. This is a significant differentiating component of this new technology in sustainability practices.

- **Printing technology based on moulded fibers:** High-resolution four-color graphics can now be applied directly to the surface of moulded fiber packaging thanks to new technology. Potential advantages include visually appealing graphics and improved packaging sustainability as there is no need for extra outer packaging or wrap with a label on it at a low per-unit cost and a higher production rate with fewer time constraints.
- **Micro-Fibrillated Cellulose (MFC)** is a one-of-a-kind fiber derived from plant waste that is used to sustainably enhance and lighten fiber products. This technology enables advantages such as reduced substance with retained effectiveness, a resource stronger and tougher than glass or carbon fiber and yet lighter in weight, enhanced crack resistance, and efficient rheological aid. It also serves as an oxygen and moisture barrier. With only minor advancements of the material to emerging substrates, users can significantly improve effectiveness. For example, a 1% improvement in product performance can result in a 15%–20% improvement in product performance. Although it is derived from crops rather than trees, it has a much lower cost structure than some other micro-fibrillated cellulose goods.
- **Chill Buddy: lightweight insulation material-** Chill Buddy is a lighter weight, long-lasting insulation that may be comfortably used in warm or cool environments. The temperature-controlling packaging material outperforms traditional sealed blanket technology and is a flexible, environmentally friendly alternative to expanded polystyrene (EPS) foam. Retail stores are leveraging modern technology for wrappings over skid loaders, roll cages, and pallets. These shrouds keep products chilled for more than 3.5 h and are especially useful in retail locations with limited refrigeration interiors and when products have a massive turnover.

5. Packaging types

5.1 Active packaging

Active packaging methods are being formed with the objective of elongating food shelf life and boosting the physiological, chemical, or biological aspects that alter the relationships between the package and product to achieve the desired result. One of the most frequently detected mechanisms is depleting oxygen from the pack or product, which can even be stimulated by an external cause like UV light. Active packaging is commonly classified under two categories: sachets and pads that are arranged within packages; and active components that are linked into packaging materials [14,15].

5.1.1 Radio Frequency Identification (RFID)

Tags based on RFID are expected to be in greater usage for packaging of food items. These tags are a sophisticated form of data carrier capable of identifying and tracking a product. Reportedly, they are preferred to monitor high-value goods and livestock. A sender transmits out using a radio signal in order to read data from an RFID tag in a typical system. After that, the data is converted into digital form for analysis. RFID tags are equipped with a microchip that is connected to a single antenna. This allows tags to be analyzed from a distance of at least 100 feet or more in costly tags and the distance is reduced to 15 feet in less costly tags. RFID tags have the potential to provide much more than a traditional bar-code. An RFID tag can be estimated at a higher rate and it can tag multiple numbers of things at the same time. RFID tags can store temperature and humidity records, as well as nutritive benefits and cooking instructions. These tags could be coupled with a biosensor and a timer that stores data that can be microbiological and bacteriological data along with time and temperature. RFID technology is still in its infant stages in the food system. The majority of food science issues revolve around simple applications like identification and tracking, which must be developed before more complicated applications [8].

5.1.2 Freshness indicators
Chilling food products in the refrigerator may cause changes like microbiological growth and metabolism, which result in pH change, the appearance of noxious substances, rancid odors and flavors, and slug formation. We need to know about analysis for water vapor, oxidation, microbial growth, enzymatic activity etc to save time, money and prediction methods for food packaging.

Food spoilage, the formation of toxins with harmful biological reactions, or stains may be caused by oxidation of lipids and colorants. These freshness indexes are focused on intelligence, a concept that refers to changes in both sections. A freshness indicator depends upon the quality of a food product directly by monitoring the freshness of a food product using metabolic pathways from microbial activity or chemicals from oxidations as information.

5.1.3 Enzyme-based time-temperature indicator
Chilling food products in the refrigerator may cause changes like microbiological growth and metabolism, which result in pH change, the appearance of noxious substances, rancid odors and flavors, and slug formation.

Food spoilage, the formation of toxins with harmful biological reactions, or stains may be caused by oxidation of lipids and colorants. These freshness indexes are focused on intelligence, a concept that refers to changes in both sections. A freshness indicator depends upon the quality of a food product directly by monitoring the freshness of a food product using metabolic pathways from microbial activity or chemicals from oxidations as information.

5.1.4 Emerging technologies: sustainability of the food supply system
The security of food and challenges to feeding the increasing population are critical issues for policy planners. The high per capita consumption, inequality of wealth, weaknesses in

the public distribution system and environmental issues compel us to devise packaging solutions to mitigate the weaknesses. Demographic and environmental factors are to be taken into consideration prior to the drafting of policies and procedures for supplying sufficient calories and nutrients without any adulterations or contaminants.

The application of refillable containers may help to reduce the transportation and manufacturing costs, supply chain intricacies besides reducing the potential of viral or micro biological diseases. The chitosan or similar edible packaging may protect food from multiple mechanisms, inhibiting the respiratory activity of bacteria by blocking the oxygen supply in packaging solutions for takeaways epically in fruits and vegetable sectors.

Sustainable development goals (SDG: zero hunger) also discuss food and its safety and hygiene. We need to have increased awareness among the stake holders for effectively using the new emerging technologies with special are and knowledge. Conventional packaging systems have few limitations, particularly with regard to shelf life, food safety, and consumer expectations. Several of the customers want to have the packaging material to keep their food fresh for a longer duration and experience the freshness of the food for them. With increasing industrialization, changing lifestyles and preferences for food varieties, there has been an increased demand for packaged and processed food. Smart sensors and stickers are being added to the advanced packaging solutions to represent the next generation of technology helping us monitor the condition of the products. For the safety of packaged food, we are experiencing increased competitiveness, a paradine shift in consumer behavior and digital interaction. In brief, containment, preservation, communication, distribution, and facilitation are the business approaches in view of future demands. In view of regulatory requirements and growing interest in safety, the state of the art and technologies help in the prevention of the growth of microorganisms and pathogens and prevent the transport of contaminants with appropriate responses to external stimuli. We require new packaging techniques that are widely applied to significantly support this demand. The post consumer recycled resin (PCR) may be used for making packages for electronics, retails, sheets and films from the surrounding areas and create recycled packaging materials. This is packaging system sensor solutions from primary to end of line. We need to develop networking and educational opportunities for packaging education and training [1,12,16].

The Sustainable Packaging Alliance (SPA), a professional organization for sustainable packaging, describes sustainable packaging solutions as having the primary goal of creating a world where all packaging is sustainably sourced, intended to be safe and efficacious throughout the lifecycle of a product, must meet market performance standards, including financial aspects, is entirely made with renewable energy, and therefore is recycled proficiently to provide an essential resource for flora and fauna.

The food supply chain involves usage of polymeric LLDP and HDPE woven sacks. The Bureau of Indian standards have set the guidelines IS: 14887:2014 for packaging of food grains. The life cycle analysis of the raw material used in fabrication of sacks has been conducted. It was has been determined that PP woven bags are environmentally friendly. The production processes and reusability facilitate comparatively less pollution to the environment. The limitation of non-biodegradability persists.

5.2 Intelligent packaging
5.2.1 Absorbing system

$$C_6H_{12}O_6 + 6O_2 = 6CO_2 + 6H_2O + \text{heat}$$

The traditional role of packaging is to shield the product from the elements, i.e., chemical as well as physical changes. Increased consumer demand for low-quality processed foods and additive-free products, as well as food safety concerns, have prompted the food packaging industry to seek new innovative packaging technology in which the enhanced package makes a significant contribution to the product's quality and safety while also functioning as a protective barrier. The majority of these specifications can be met by active packaging, a newer technology. The concept of active packaging has changed the function of the package from passive preservation to an active contribution to the protection of food quality and safety (Fig. 10.3).

Moisture vapor gathers inside the packaging as a result of the moisture humidity difference between the produce's surface and the surroundings, as well as vapor perspiration from respiration, which can saturate the headspace and compress the inner surfaces of the packaging and the produce's surface. The commodity's shortened moisture may ultimately serve as a breeding ground for bacterial activity and deterioration. The monitored absorption of humidity from the packaging headspace by a moisture absorbing material can maintain an appropriate level of unsaturated moisture content within the packaging. Depending on the requirements, different kinds of moisture absorbing films are available to consumers, such as $CaCl_2$, KCl, in which sorbitol is for fast absorption and bentonite

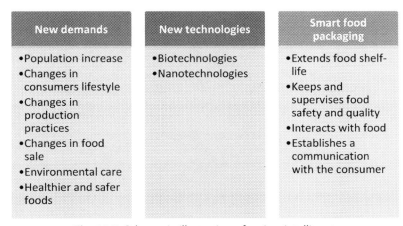

Fig. 10.3 Schematic illustration of active, intelligent.

is for slow absorption. A preferred absorption rate is obtained by combining multiple absorbers.

CO_2, which is usually produced by the metabolic activities of fresh fruits and vegetables, can cause oxidative responses. If left unchecked, it can build up to toxic levels within the packaging. As a result, CO_2 discharge is advantageous in terms of quality. The most well-known CO_2 absorbers are calcium hydroxide, zeolite, activated charcoal, and magnesium oxide.

Although some oxygen is consumed for the regular metabolism of respiration in fresh produce, absorption is not usually targeted in fresh produce packaging. In some cases, oxygen absorbers can be incorporated within the package or attached to the polymeric membrane as sachet plugs or self-adhesive labels to reach the optimum atmosphere as soon as possible.

5.2.2 Releasing system

Several gas-phase particles, including the ethylene antagonist 1-methylcyclopropene (MCP, C_4H_6) as well as the microbial agent chlorine dioxide (ClO_2), are impactful at preserving the freshness of fruits and vegetables and are frequently made available within the packaging so that they are widely obtainable in the packaging headspace. When delivered in antimicrobial sachets or films, high volatility antimicrobials aid in the protection and preservation of the produce's microbiological quality. When fresh produce is exposed to ethylene, it overripens and promotes microbial activity, resulting in much shorter lifespan. The discovery of the ethylene inhibitor MCP adds a new method for delaying ripening and damage to fruits and vegetables, thereby extending shelf life.

At different concentration levels ranging from 2.5 nL/L to 1 L/L, MCP inhibits the effects of ethylene in a broad array of fruits, vegetables, and floriculture crops at different concentration levels. The emission of MCP gas from cyclodextrin powder is the most frequently used approach for MCP treatment. Moisture must be present as a trigger for the cyclodextrin gas to encounter the fresh produce-containing space. For something like the release of MCP in the produce packaging, sachet and film forms have been examined.

5.2.3 Antimicrobial packaging

Active packaging is a form of antimicrobial packaging in which antimicrobial compounds are released. The majority of readily accessible antimicrobial methods are based on antimicrobial migration, though several systems rely on the inherent antimicrobial applications of the packaging material's alerting surface. The indirect contact framework only allows for the use of volatile antimicrobials and does not allow for antimicrobial delivery in liquid or solid form.

Antimicrobials can move or be transported via gas-phase diffusion, desorption, and evaporation. Volatile antimicrobials have the advantage of effortlessly testing asymmetrical food layers via void space or streams. With the increased awareness and demands for preserving the quality, prolonging the shelf life of food items and sustainability give rise to the exponential development of antimicrobial packaging solutions. The application of nano textiles or advanced films may help to increase efficiency of novel biobased packaging systems. In view of this highly efficient anti fungal anti bacterial, nano particles or anti oxidants may help to develop green polymers with impregnation of antimicrobial agents. ClO_2-releasing packing materials are one useful indirect system [17–19].

5.2.4 Edible coatings

Edible coatings have traditionally been applied on the surface of fresh products to improve physical appearance, reduce bruising during collection, processing, and shipping, reduce the weight of the product, and serve as a carrier for different active compounds. Among the most commonly coated commodities are apples, tomatoes, turnips, cantaloupes, cucumbers, grapefruits, avocados, bell peppers, lemons, parsnips, eggplants, passion fruit, peaches, rutabagas, limes, melons, oranges, squash, pineapples, pumpkins, sweet potatoes etc. The three types of edible macromolecules used as adhesives are polysaccharides, lipids, and proteins. The characteristics of the coatings may depend on the composition and value added properties for attaining critical beneficial effects [10].

One of the recent steps in the evolution and application of additives that can increase the safety and efficiency of fresh-cut products is that they are widely accepted. Edible additives may help to increase the shelf life of fresh-cut produce by minimizing its moisture content, solute flow of migrants, as well as gas exchange, respiratory rate, oxidative reaction rates, and can also minimize or suppress physiological disorders [2].

Innovative packaging includes active and intelligent packaging systems. They offer innovative and unique convenience, quality, and food safety options. According to some experts, the next generation of packaging technology will include nanotechnologies that will enable the incorporation of new compounds such as gas scavengers and unique antimicrobials into packaging applications.

Low-cost electronic device development will, however, contribute to the innovation of active and intelligent packaging. As our society evolves, customer expectations will continue to grow. As new technologies enter the market, the use of active and intelligent packaging is likely to increase in popularity. Innovative packaging will become more common in active and intelligent systems. Possibly, active and intelligent packaging will supplant traditional packaging completely. According to the upcoming trends, less packaging will be present, and what is present will be more immersive.

5.2.4.1 Inference

The packaging industry has immense opportunity to attain efficiency and agility for continuity, supply chain resiliency and sustainability through risk aversion, cross functional technical expertize and compliance to international standards requirements. The emergency situations, unexpected challenges may be removed through stream line processes for removal of barriers and developing new robotic solutions including shipments.

It is well understood that complete replacement of existing packaging material may not be feasible in one or two decades but the innovative steps are necessary for attaining the sustainability aspects focused in international sustainable development goals. Smart consumers are looking forward to smart packaging modes to use the concept of amalgamated technologies viz. nano-technology encapsulation for innovative development of new biogenic systems. The real-time monitoring of food technology is to be coupled with commercial and ecological prosperity.

Smart packaging has enormous potential in the food, beverage, pharmaceutical, automotive, and personal care industries, responding to customer preferences for handy packages which ensure product safety and quality. Safe and novel chemical entities with specific characteristics can be artificially inseminated to provide freshness and information backup by changing the color in response to specific changes in pH, gas levels, or temperature. Nowadays, their industrial development is centered on using synthetic polymers since they are cost-effective, operational, light, and versatile but with few limitations.

The interest in replacing these polymers with bioplastics, on the other hand, is an attempt to meet society's demands for environmental stewardship and sustainability. Recent studies have attempted to explore the potential of biopolymer-based food containers to transport, encompass, and limit the spread of these bioactive components. Despite the numerous advantages of intelligent packaging, e-commerce manufacturers are having difficulty meeting the high demand for eco-friendly, sustainable packaging with logistical considerations and new designs. As a result, designers and innovators are required to create environmentally friendly products that can withstand shipping stresses while also being aesthetically pleasing.

Recent research trends prefer the use of technological innovations such as nanotechnology packaging ideas and antimicrobial packaging to aid in the effective integration of active substances and the improvement of designed features. In addition to the packaging, the food safety aspect must also be covered. The quality and the cleanness of the container; the packing of product in the container transportation vehicle with appropriate maintenance of the vehicle; display and handling practices at the distribution point; work health status and attire.

References

[1] J. Jacob-John, C. D'Souza, T. Marjoribanks, S. Singaraju, Synergistic interactions of SDGs in food supply chains: a review of responsible consumption and production, Sustainability 13 (16) (2021) 8809, https://doi.org/10.3390/su13168809.

[2] Innovative and functional shelf-life extending food packaging in a circular economy context. Frontiers Research Topic. (n.d.). Retrieved May 6, 2022, from https://www.frontiersin.org/research-topics/30342/innovative-and-functional-shelf-life-extending-food-packaging-in-a-circular-economy-context.

[3] UN Environment Programme, Sustainable Consumption and Production Policies, UNEP - UN Environment Programme, October 2, 2017. https://www.unep.org/explore-topics/resource-efficiency/what-we-do/sustainable-consumption-and-production-policies.

[4] Smart packaging: Connecting the physical with the digital. (n.d.).https://www.foodengineeringmag.com/articles/99310-smart-packaging-connecting-the-physical-with-the-digital.

[5] I. Ahmed, H. Lin, L. Zou, Z. Li, A.L. Brody, I.M. Qazi, L. Lv, T.R. Pavase, M.U. Khan, S. Khan, L. Sun, An overview of smart packaging technologies for monitoring safety and quality of meat and meat products, Packag. Technol. Sci. 31 (7) (2018) 449—471, https://doi.org/10.1002/pts.2380.

[6] Taking the next steps in the digitization of packaging. Packaging Europe. (n.d.) Retrieved May 6, 2022, from https://packagingeurope.com/comment/taking-the-next-steps-in-the-digitization-of-packaging/7677.article.

[7] Introduction of Radio Frequency Identification (RFID), GeeksforGeeks, July 7, 2020. https://www.geeksforgeeks.org/introduction-of-radio-frequency-identification-rfid/.

[8] Radio Frequency Identification—an overview. ScienceDirect Topics. (n.d.).Retrieved May 6, 2022, from https://www.sciencedirect.com/topics/agricultural-and-biological-sciences/radio-frequency-identification.

[9] A. Iordanskii, Bio-based and biodegradable plastics: from passive barrier to active packaging behavior, Polymers 12 (7) (2020) 1537, https://doi.org/10.3390/polym12071537.

[10] A. TrajkovskaPetkoska, D. Daniloski, N.M. D'Cunha, N. Naumovski, A.T. Broach, Edible packaging: sustainable solutions and novel trends in food packaging, Food Res. Int. 140 (2021) 109981, https://doi.org/10.1016/j.foodres.2020.109981.

[11] P.R. Salgado, L. Di Giorgio, Y.S. Musso, A.N. Mauri, Recent developments in smart food packaging focused on biobased and biodegradable polymers, Front. Sustain. Food Syst. 5 (2021), https://doi.org/10.3389/fsufs.2021.630393.

[12] V. Chaudhary, S. PuniaBangar, N. Thakur, M. Trif, Recent advancements in smart biogenic packaging: reshaping the future of the food packaging industry, Polymers 14 (4) (2022) 829, https://doi.org/10.3390/polym14040829.

[13] H. Ahari, S.P. Soufiani, Smart and active food packaging: insights in novel food packaging, Front. Microbiol. 12 (2021), https://doi.org/10.3389/fmicb.2021.657233.

[14] R. Dobrucka, R. Przekop, New perspectives in active and intelligent food packaging, J. Food Process. Preserv. 43 (11) (2019), https://doi.org/10.1111/jfpp.14194.

[15] K. Kraśniewska, S. Galus, M. Gniewosz, Biopolymers-based materials containing silver nanoparticles as active packaging for food applications—A review, Int. J. Mol.Sci. 21 (3) (2020) 698, https://doi.org/10.3390/ijms21030698.

[16] Wikipedia Contributors, Sustainable Development Goals, Wikipedia; Wikimedia Foundation, February 27, 2019. https://en.wikipedia.org/wiki/Sustainable_Development_Goals.

[17] J.H. Han, in: J.H. Han (Ed.), 6 - Antimicrobial Packaging Systems, Science Direct, Academic Press, 2005. https://www.sciencedirect.com/science/article/pii/B9780123116321500383.

[18] T. Huang, Y. Qian, J. Wei, C. Zhou, Polymeric antimicrobial food packaging and its applications, Polymers 11 (3) (2019) 560, https://doi.org/10.3390/polym11030560.

[19] B. Malhotra, A. Keshwani, H. Kharkwal, Antimicrobial food packaging: potential and pitfalls, Front. Microbiol. 6 (2015), https://doi.org/10.3389/fmicb.2015.00611.

CHAPTER 11

Production of smart packaging from sustainable materials

Adeshina Fadeyibi
Department of Food and Agricultural Engineering, Faculty of Engineering and Technology, Kwara State University, Ilorin, Kwara, Nigeria

1. Introduction

Smart packaging is a technology that can perform a dual function of monitoring transformation in a product or its surrounding [1], and them to promote a desirable physical behavior of the product [2]. In most cases, biosensors or chemical sensors are often used to oversee and secure the food during production, distribution, storage, and consumption [3]. This technology offers an avenue to especially detect and analyze freshness, pathogens, oxygen, pH, carbon-dioxide, time, and temperature of the food [1,2]. There is a variation in the functional properties of the SPMs depending on the nature of the product which can be food, medical products, or any other household products [4]. It also depends on some experimental condition like the temperature, humidity and so on, which can adjust the behavior of the SPMs significantly [5]. This can enhance its ability to track, trace, analyze and control the behavior of the product so that manufacturers and the end-users are informed of its condition before and after processing [6].

Most of the SPMs available in the global food market are produced from nondegradable materials that can pose a major challenge to the industry because they are not sustainable and difficult to recycle or ruse [7–9]. Although government policies on recycling and processing packaging waste have long been in place, practical experience shows how difficult it is to recycle certain kinds of packaging waste [8]. Thus, one of the major concerns of the manufacturers of the SPM is in finding a suitable material that will not add to the already existing environmental problem that are caused using none-degradable polymers [10–12].

The sustainable materials of plant and animal origins have been proposed and used in some cases to produce the SPMs with high biosensing abilities [10,13–15]. Due to the increased consumer demand, which are based on societal developments as well as the nutritional benefits of foods derived from animal and plant materials, and the development of good health and wellness [16], novel ideas are sought to enhance safe handling, and quality maintenance [17,18]. In this context, smart food packaging technology plays major roles in food handling, transporting, and storing [19–22]. It has proved to be effective in reducing postharvest losses of food materials [12,16,23]. This review is aimed at providing an overview of the production processes of the SPMs via the extrusion and injection molding approaches, and the use of sustainable plant and animal products as major ingredients in their production. During the SPM production, Aerotolerant anaerobes, like Cutibacterium acnes, which can protect the system from gaseous interaction, were proposed as a filler to provide extra gaseous and temperature control within the package. This can enhance the performance of the packaging system in food preservation and other applications.

2. Mechanical production of SPMs

2.1 Extrusion technique

The SPM, based on sustainable materials, can be produced via extrusion techniques, as shown in Fig. 11.1 [4]. This normally involves manual mixing of the composite materials externally before introducing into a screw extruder, which further mixes and homogenizes the product to form pellets [24–26]. The pellets are thereafter crushed, and hot water added to form a gelatinous liquid, which is later transformed into thin sheets or film by drying in a mold. Typical procedures were outlined by Vedove et al. [4] and Fadeyibi et al. [27] in their work on the production of the SPM from starchy food products.

2.2 Injection molding technique

The SPMs can be produce using the injection molding techniques (Fig. 11.2) 28,29. This technique is most suitable to produce sophisticated SPMs for commercial applications. Unlike the extrusion technique, which is time wasting and requires the mixing of the constituent materials to get a film, this technology is fast, efficient, and low cost. During

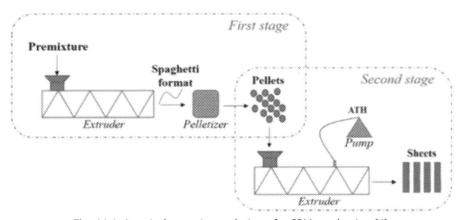

Fig. 11.1 A typical extrusion technique for SPM production [4].

Fig. 11.2 An industrial injection molding machine for production SPMs [36].

the SPM processing, the plastic materials are first heated to form molten substance. It is followed by injection into a mold which has vents to allow air escapes so that bubbles are not formed [30–32]. Once the mold is filled, the workpiece is left to cool for the exact time needed to harden the material. The cooling time is a function of the thickness of the piece and resin type [33–35]. The temperature of the lines during this process is maintained by passing water through the channels.

3. Biochemical production of SPMs
3.1 Production of SPM from sustainable bamboo products

The bamboo is a perennial plant with a high ability for self-regeneration and can be eco-friendly, thus is widely regarded as a sustainable material. It can be incorporated into synthetic materials to enhance strength for food packaging and other useful applications. According to Hai et al. [37], nanocomposite film from a blend of chitin nanofiber and bamboo has been reported to exhibit strong intelligent and active behavior, in terms of its thermal, mechanical, and biodegradable properties, compared to ordinary chitin film. The addition of bamboo leaves into a blend of chitin and zinc oxide nanoparticles generates a smart nanocomposite film which synergistically enhanced the antibacterial activity against *E. coli* and *S. aureus* [38]. It also reduced the UV light transmittance and significantly enhanced the antioxidant activity of the films [38]. The bamboo particles embedded as reinforcement for the cork-bamboo composite were reported to enhance resistance, water absorption and dimensional stability of cork products [39]. This in turn can be used as a smart material for flooring and other structural applications. The application of bamboo in the design of polymer nanocomposite was reported by Santos et al. [40]. The authors reported that bamboo fibers can enhance the surface hydrophilicity, structural and strength of polyethylene material due to the presence of voids in the structure of the composite, as shown in Fig. 11.3 [40]. However, the commercial application of this type of SPM is limited because the bamboo fillers may absorb moisture and swell thereby causing structural distortion and stability of the matrix of the packaging material. Thus, to correct this, this writer suggests the use of biological fillers, such as the Aerotolerant anaerobes, that can suppress this incidence and create a stable system for the smart package.

3.2 Production of SPM from sustainable wool products

The wool is essential type of sustainable material that can influence the performance of many packaging systems. It is used as a blend with other materials to form composite which has enhanced structural and mechanical advantages. The production of SMP from wool sustainable materials has been reported only sparingly in the literature. Typically, a superheated water hydrolysis method has been used to process waste wool fibre-reinforced for active and intelligent design for food packaging. This involves the utilization of a reactor to mix and turn a bulk sample of the wool at variable temperature and pressure for 1 h. This is followed by mixing with a kraft pulp in water suspensions in fractions of dry weight

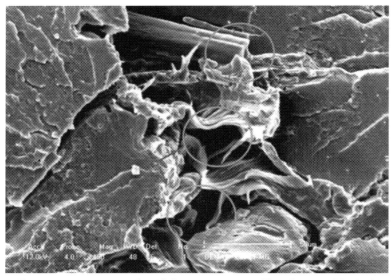

Fig. 11.3 Appearance of void due to presence of bamboo in LDPE [40].

without adding filler or chemical additives [41,42]. Also, wool keratin has been used to prepare other SPMs by dissolving solution of protein in a milk substrate (Fig. 11.4) [43]. This technology has the advantage of improving the ability of the SPMs to control the stability of fresh food products, especially during preservation in temperate conditions. This is because of the high surface absorption ability of the wool-keratin to moisture that may cause cooling and then extends the shelf-life. However, it is rarely applied for food requiring tropical treatments because the cooling effect can encourage microbial growth and encourage deterioration. Thus, the addition of a sustainable nano-wooden material during production into the SPM matrix can help address this problem because its ability to prevent heat transfer, insulate the surface, and prevent subsequent cooling, thereby promoting wider application.

3.3 Production of SPM from sustainable agricultural residues

The use of agricultural residues such as coconut, groundnut, jaropha, sugarcane bagasse, melon shells and other domestic and industrial wastes in the production of SPM is very critical to ensure ecofriendly and sustainable environment. The effectiveness of the production of bioplastics from agricultural residues and its economic viability depend on the type of residues produced in certain areas, treatment costs, biomass availability, transportation costs, raw material expenses and other common industrial expenditure. Since the type of waste generated during each season may vary, the productivity of bioplastics may be affected by seasonal changes, and this can hamper the supply of the raw materials. The production of cellulose acetate biofiber from flax fibers and cotton linters has been reported by Mostafa et al. [44]. This involves the mixture of the refined flax fiber, cotton liter, acetic anhydride, glacial acetic acid as well as sulfuric acid in definite proportions and analytically. A solution of polyethylene glycol was added to ensure the product has sufficient plasticity to withstand external variables such UV light and vibrational loads. A mixture of cellulose

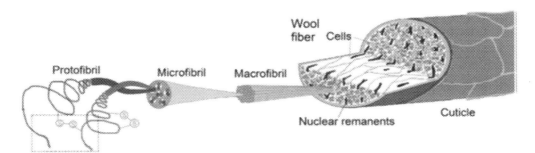

Fig. 11.4 A wool fiber reinforcement for SPM production [43].

cocoa shell incorporated into sugarcane bagasse has been used to formulate excellent bioplastic for smart food packaging [45]. The food waste from diverse sources in an environment can also be chemically transformed to form biodegradable SPMs (Fig. 11.4) [46]. Despite the advantages presented by this technology, the degradation of the bio-wastes can build gas concentration within the package, thereby causing system imbalance that may interfere with the product stability during preservation. This can however be addressed by the inclusion of Aerotolerant anaerobes during production to create an atmosphere that is devoid of excess gases, thereby enhancing the performance of the SPMs.

3.4 Production of SPM from sustainable animal skin product

The inherent properties of the animal skins that can be useful in the design of SPM includes water desorption and absorption abilities, aesthetics and surface pattern, and heat insulation. Although, the production of the SPM based on the animal skin is not very common like the ones from the sustainable plant products, the few reports have demonstrated great potentials for the future needs of the packaging industries. Typically, the production of a skate skin gelatin SPM from the mixture of the stake skin and thyme essential oil has been reported by Lee et al. [47]. The production of another type of SPM from the blend of a skin gelatin brown stripe red and bigeye snappers was also reported by Jongjareonrak et al. [48,49]. In addition, the animal skin substrates are added into the bioplastic majorly to enhance the antimicrobial activities of the resulting SPM. The work of Tkaczewska et al. [50] on the production of SPM by blending a polysaccharide and carpus skin gelatin hydrolysate shows that the skin coatings effectively inhibited microorganism growth and extend the shelf-life of perishable food products, as shown in Fig. 11.5. To the best of knowledge, despite the advantages presented by this class of material, there is no reported research on the commercialization of the SPM derived from the animal skin sources. Also, the production of SPM from the blend of animal skin and other bio-composite ingredients is not yet fully exploited. There is therefore a need for further research on the application of other useful skin sources in the production of the SPM since this can help to improve its inherent properties.

3.5 Production of SPM from sustainable eggshell product

It is a semipermeable membrane which permits gaseous exchange through its pores, and has a thin outer coating called the flower or cuticle that helps keep bacteria and dust from getting in the way. In the food packaging industry, the eggshells have found numerous commercial applications. Here, the eggshells are often transformed into powder with notable particle size in the nano-range and added as fillers to bio-composites to produce thermoplastics for packaging applications. Typically, an eco-friendly SPM with improved functional performance have been developed by combining nano-scale hydroxyapatite synthesized from eggshell and polymer-based protein extracted from soy residues [51]. Also, Kosarsoy [52] developed an ostrich shell and an edible biopolymer made from

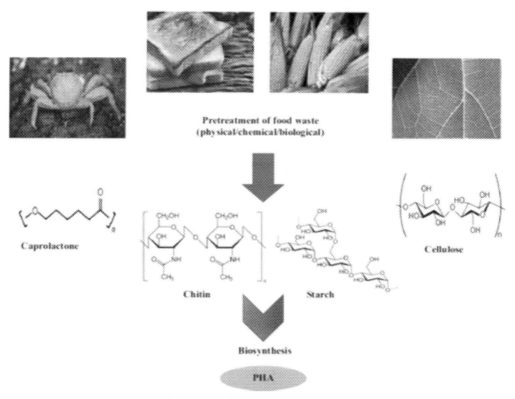

Fig. 11.5 Transformation of some food wastes into bioplastic [46].

nano-levan SPM with excellent mechanical, physical and bioactive properties for food packaging applications. A combined high temperature calcination and hydrothermal synthesis was used to produce SPM from the blending artemia eggshells and zinc oxide nanoparticles for treatment of wastewater in the industry [53]. This was done by calcining the shells high temperature to create a porous organic skeletal material. The structure is then charged with nanoparticles of zinc-oxide to synthesize the target product through a hydrothermal reaction. The material is also used to break down methylene blue, rhodamine B and neutral red under light irradiation (Fig. 11.6) [53]. Moreover, a new strategy to produce an SPM from the mixture of eggshell nanoparticles and starch with characteristics high temperatures and mechanical stabilities has been reported for fire hazard suppression [54]. This involves adding a requisite amount of starch to 30 mL deionized water in a 500 mL reactor and stirred thoroughly. The reaction temperature was controlled using a thermostat and maintained at 80°C. The solution was stirred under nitrogen atmosphere at 25°C for 20 h to synthesize the transplanted copolymer and thereafter synthesized the SPM. The nanomaterials of eggshell incorporated into the package can generate substantial amount of hydroxyapatite, which can collapse the

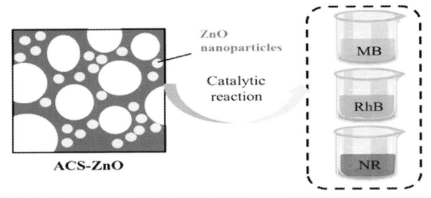

Fig. 11.6 Production of SPM from blend of ACS and ZnO nanoparticles used to degrade MB, RhB and NR material substrates [53].

microsphere into calcium and phosphate ions to cause great instability of the SPM after a long-time use. Accordingly, Piccirillo et al. [55] incorporated titanium dioxide into the hydroxyapatite composite to produce a SPM with improve stability and performance. The hydroxyapatite component of the package did not decompose due to the presence of the oxide. However, the use of biological materials to effect this change has not been widely studied. Hence, there is a need to focus on this in future research.

3.6 Production of SPM from sustainable chicken feathers

The chicken feathers consist mainly of keratin, which is a tough protein like that found in the animal hair, hoofs, and horns, that can enhance the strength and the durability of the SPMs. When the keratin extract is blended into biopolymers, the resulting material, which is characterized with sufficient hydrophilicity, can act intelligently on the food products during packaging improve their quality and durability. Typical is chicken feather protein SPM production containing clove oil was reported by Song et al. [56] for intelligent packaging of smoked salmon fish. This was done by dispersing gelatin solution into a mixture of the chicken feather protein and sorbitol plasticizer. The resulting solution was then heated using a warm 75°C water bath for 30 min, and clove oil and cinnamaldehyde solvent were added to enhance the hydrophilicity of the SPM. Also, Chen et al. [57] reported the production of the SPM from the blend of the feather keratin and plasticizer and stabilized using tris (hydroxymethyl) aminomethane. A bioplastic SPM production by degrading the feather waste using microorganism to obtain the keratin and mixing it with a starch source has been reported by Alshehri et al. [58]. As much as the author knows, the commercial application of the SPMs made from the keratin in the feathers has not been exploited. More research is required to figure out the nature of the macromolecule in the package, and how it influences the surface interaction to effect the change in the food product during packaging and in shelf storage.

4. Prospects and conclusions

The application of sustainable plant materials that influence the environment in the production SPMs for food storage and processing is well known in the literature. Despite this, some bio-wastes used in the SPMs can still decompose, mainly due to overuse, and elevate the concentration of the gases within the package. This can cause a system imbalance that may interfere with the product stability during preservation. In this paper, the inclusion of the Aerotolerant anaerobes during production of the SPMs was proposed to correct this situation. This will create an atmosphere that can tolerate the excess gases, thus enhancing the performance of the package. Other materials, like the wood-powder, which can prevent the transfer of heat, insulate the surface, and prevent subsequent cooling, were also proposed for inclusion during production. However, as much as the author knows, the production and commercialization of the SPMs derived from the blend of the nanoparticles of the sustainable animal products and other bio-composite ingredients have not been fully exploited. It is thus necessary to bridge this gap. Finally, research is needed to understand the nature of the microsphere in the package, and how this influences the surface interaction with the bio-based fillers to control the changes in the food quality during packaging.

References

[1] V. Gomes, A.S. Pires, N. Mateus, V. de Freitas, L. Cruz, Pyranoflavylium-cellulose acetate films and the glycerol effect towards the development of pH-freshness smart label for food packaging, Food Hydrocoll. 127 (2022), https://doi.org/10.1016/J.FOODHYD.2022.107501.

[2] H. Zhang, A. Hou, K. Xie, A. Gao, Smart color-changing paper packaging sensors with pH sensitive chromophores based on azo-anthraquinone reactive dyes, Sens. Actuat. B Chem. 286 (2019) 362–369, https://doi.org/10.1016/J.SNB.2019.01.165.

[3] H. Cheng, H. Xu, D. Julian McClements, L. Chen, A. Jiao, Y. Tian, et al., Recent advances in intelligent food packaging materials: principles, preparation and applications, Food Chem. (2022) 375.

[4] T.M.A.R.D. Vedove, B.C. Maniglia, C.C. Tadini, Production of sustainable smart packaging based on cassava starch and anthocyanin by an extrusion process, J. Food Eng. (2021) 289, https://doi.org/10.1016/J.JFOODENG.2020.110274.

[5] JG de Oliveira Filho, M.R.V. Bertolo, M.Á.V. Rodrigues, C.A. Marangon, G. da C. Silva, F.C.A. Odoni, et al., Curcumin: a multifunctional molecule for the development of smart and active biodegradable polymer-based films, Trends Food Sci. Technol. 118 (2021) 840–849, https://doi.org/10.1016/J.TIFS.2021.11.005.

[6] S. Park, Y. Jeon, T. Han, S. Kim, Y. Gwon, J. Kim, Nanoscale manufacturing as an enabling strategy for the design of smart food packaging systems, Food Packag. Shelf Life 26 (2020), https://doi.org/10.1016/J.FPSL.2020.100570.

[7] D. Kim, S. Thanakkasaranee, K. Lee, K. Sadeghi, J. Seo, Smart packaging with temperature-dependent gas permeability maintains the quality of cherry tomatoes, Food Biosci. 41 (2021), https://doi.org/10.1016/J.FBIO.2021.100997.

[8] P. Madhusudan, N. Chellukuri, N. Shivakumar, Smart packaging of food for the 21st century – a review with futuristic trends, their feasibility and economics, Mater. Today Proc. 5 (2018) 21018–21022, https://doi.org/10.1016/J.MATPR.2018.06.494.

[9] Y. Wu, P. Tang, S. Quan, H. Zhang, K. Wang, J. Liu, Preparation, characterization and application of smart packaging films based on locust bean gum/polyvinyl alcohol blend and betacyanins from

cockscomb (*Celosia cristata* L.) flower, Int. J. Biol. Macromol. 191 (2021) 679–688, https://doi.org/10.1016/J.IJBIOMAC.2021.09.113.

[10] Y. Duan, Y. Liu, H. Han, H. Geng, Y. Liao, T. Han, A dual-channel indicator of fish spoilage based on a D-π-A luminogen serving as a smart label for intelligent food packaging, Spectrochim. Acta A 266 (2022), https://doi.org/10.1016/J.SAA.2021.120433.

[11] M. Cheng, X. Yan, Y. Cui, M. Han, X. Wang, J. Wang, et al., An eco-friendly film of pH-responsive indicators for smart packaging, J. Food Eng. (2022) 321, https://doi.org/10.1016/J.JFOODENG.2022.110943.

[12] B. Kuswandi, Y. Wicaksono, A. Abdullah, L. Yook Heng, M. Ahmad, Smart Packaging: Sensors for Monitoring of Food Quality and Safety, 5, Springer, 2011, pp. 137–146, https://doi.org/10.1007/s11694-011-9120-x.

[13] M.H. Balalzadeh Tafti, M.R. Eshaghi, P. Rajaei, A smart meat packaging to show ciprofloxacin residues based on immunochromatography, Meat Sci. (2021) 181, https://doi.org/10.1016/J.MEATSCI.2021.108605.

[14] M. Alizadeh-Sani, E. Mohammadian, J.W. Rhim, S.M. Jafari, pH-sensitive (halochromic) smart packaging films based on natural food colorants for the monitoring of food quality and safety, Trends Food Sci. Technol. 105 (2020) 93–144, https://doi.org/10.1016/J.TIFS.2020.08.014.

[15] A. Sobhan, K. Muthukumarappan, L. Wei, Biosensors and biopolymer-based nanocomposites for smart food packaging: challenges and opportunities, Food Packag. Shelf Life 30 (2021), https://doi.org/10.1016/J.FPSL.2021.100745.

[16] R. Ahvenainen, E. Hurme, Active and smart packaging for meeting consumer demands for quality and safety, Food Addit. Contamin. 14 (1997) 753–763, https://doi.org/10.1080/02652039709374586.

[17] Y. Yu, Study on transport packages used for food freshness preservation based on thermal analysis, Archiv. Thermodynam. 37 (2016) 121–135, https://doi.org/10.1515/aoter-2016-0031.

[18] E. Poyatos-Racionero, J.V. Ros-Lis, J.L. Vivancos, R. Martínez-Máñez, Recent advances on intelligent packaging as tools to reduce food waste, J. Clean. Product. 172 (2018) 3398–3409, https://doi.org/10.1016/j.jclepro.2017.11.075.

[19] A. Fadeyibi, Z. Osunde, M. Yisa, Optimization of processing parameters of nanocomposite film for fresh sliced okra packaging, J. Appl. Packag. Res. 11 (2019).

[20] A. Fadeyibi, Z. Osunde, M. Yisa, Effects of period and temperature on quality and shelf-life of cucumber and garden-eggs packaged using cassava starch-zinc nanocomposite film, J. Appl. Packag. Res. (2020) 12.

[21] A. Fadeyibi, Z.D. Osunde, M.G. Yisa, A. Okunola, Investigation into properties of starch-based nanocomposite materials for fruits and vegetables packaging- A review, J. Eng. Eng. Technol. 11 (2017) 12–17.

[22] A. Fadeyibi, M.G. Yisa, F.A. Adeniji, K.K. Katibi, K.P. Alabi, K.R. Adebayo, Potentials of zinc and magnetite nanoparticles for contaminated water treatment, Agricult. Rev. (2018), https://doi.org/10.18805/ag.r-113.

[23] K.B. Biji, C.N. Ravishankar, C.O. Mohan, T.K. Srinivasa Gopal, Smart packaging systems for food applications: a review, J. Food Sci. Technol. 52 (2015) 6125–6135, https://doi.org/10.1007/S13197-015-1766-7.

[24] A. Fadeyibi, Z. Osunde, M. Yisa, Effects of glycerol and diameter of holes in breaker plate on performance of screw mixer for nanocomposites, Agricult. Eng. 21 (2017) 15–26.

[25] A. Fadeyibi, Z.D. Osunde, G. Agidi, E.C. Egwim, Design of single screw extruder for homogenizing bulk solids, Agricult. Eng. Int. 18 (2016) 222–231.

[26] A. Fadeyibi, Z.D. Osunde, G. Agidi, E.C. Evans, Mixing index of a starch composite extruder for food packaging application, in: Innamudin (Ed.), Green Polymer Composites Technology, first ed. 1, CRC Press, 2016, pp. 233–258, https://doi.org/10.1201/9781315371184-21.

[27] A. Fadeyibi, Z.D. Osunde, E.C. Egwim, P.A. Idah, Performance evaluation of cassava starch-zinc nanocomposite film for tomatoes packaging, J. Agricult. Eng. 48 (2017) 137–146, https://doi.org/10.4081/JAE.2017.565.

[28] D.P. Wermuth, T.C. Paim, I. Bertaco, C. Zanatelli, L.I.S. Naasani, M. Slaviero, et al., Mechanical properties, in vitro and in vivo biocompatibility analysis of pure iron porous implant produced by

[28] metal injection molding: a new eco-friendly feedstock from natural rubber (*Hevea brasiliensis*), Mater. Sci. Eng. C (2021) 131, https://doi.org/10.1016/J.MSEC.2021.112532.
[29] R. Yavari, H. Khorsand, Numerical and experimental study of injection step, separation, and imbalance filling in low pressure injection molding of ceramic components, J. Eur. Ceram. Soc. 41 (2021) 6915–6924, https://doi.org/10.1016/J.JEURCERAMSOC.2021.07.050.
[30] N. Nagasundaram, R.S. Devi, M.K. Rajkumar, K. Sakthivelrajan, R. Arravind, Experimental investigation of injection moulding using thermoplastic polyurethane, Mater. Today 45 (2021) 2286–2288, https://doi.org/10.1016/J.MATPR.2020.10.264.
[31] H. Huang, M. Sun, X. Wei, E. Sakai, J. Qiu, Effect of interfacial nanostructures on shear strength of Al-PPS joints fabricated via injection moulding method combined with anodising, Surface Coat. Technol. (2021) 428.
[32] R. Siva, S. Sundar Reddy Nemali, S. Kishore Kunchapu, K. Gokul, T. Arun Kumar, Comparison of mechanical properties and water absorption test on injection molding and extrusion - injection molding thermoplastic hemp fiber composite, Mater. Today 47 (2021) 4382–4386, https://doi.org/10.1016/J.MATPR.2021.05.189.
[33] C. Wang, V. Shaayegan, F. Costa, S. Han, C.B. Park, The critical requirement for high-pressure foam injection molding with supercritical fluid, Polymer 238 (2022), https://doi.org/10.1016/J.POLYMER.2021.124388.
[34] V. Speranza, S. Liparoti, R. Pantani, G. Titomanlio, Prediction of morphology development within micro–injection molding samples, Polymer 228 (2021), https://doi.org/10.1016/J.POLYMER.2021.123850.
[35] M.R. Khosravani, S. Nasiri, T. Reinicke, Intelligent knowledge-based system to improve injection molding process, J. Industr. Inform. Integr. 25 (2022), https://doi.org/10.1016/J.JII.2021.100275.
[36] Anonymous, 1, 2018. https://greenindustrylinks.com/metal-injection-molding-technique/. (Accessed 27 September 2022).
[37] L van Hai, E.S. Choi, L. Zhai, P.S. Panicker, J. Kim, Green nanocomposite made with chitin and bamboo nanofibers and its mechanical, thermal and biodegradable properties for food packaging, Int. J. Biol. Macromol. 144 (2020) 491–499, https://doi.org/10.1016/J.IJBIOMAC.2019.12.124.
[38] J. Liu, J. Huang, Z. Hu, G. Li, L. Hu, X. Chen, et al., Chitosan-based films with antioxidant of bamboo leaves and ZnO nanoparticles for application in active food packaging, Int. J. Biol. Macromol. 189 (2021) 363–369, https://doi.org/10.1016/J.IJBIOMAC.2021.08.136.
[39] X. Li, R. Liu, L. Long, B. Liu, J. Xu, Tensile behavior and water absorption of innovative composites from natural cork granules and bamboo particles, Composite Struct. 258 (2021) 113376, https://doi.org/10.1016/J.COMPSTRUCT.2020.113376.
[40] P. Santos Delgado, S. Luiza, B. Lana, E. Ayres, P. Oliveira, S. Patrício, et al., The potential of bamboo in the design of polymer composites, Mater. Res. 15 (2012) 639–644, https://doi.org/10.1590/S1516-14392012005000073.
[41] P. Bhavsar, T. Balan, G. Dalla Fontana, M. Zoccola, A. Patrucco, C. Tonin, Sustainably processed waste wool fiber-reinforced biocomposites for agriculture and packaging applications, Fibres 55 (2021) 1–18, https://doi.org/10.3390/fib9090055.
[42] D.O.S. Ramirez, R.A. Carletto, C. Tonetti, F.T. Giachet, A. Varesano, C. Vineis, Wool keratin film plasticized by citric acid for food packaging, Food Packag Shelf Life 12 (2017) 100–106, https://doi.org/10.1016/J.FPSL.2017.04.004.
[43] B. Fernández-d'Arlas, Tough and functional cross-linked bioplastics from sheep wool keratin, Sci. Rep. 9 (2019) 1–12, https://doi.org/10.1038/s41598-019-51393-5.
[44] N.A. Mostafa, A.A. Farag, H.M. Abo-dief, A.M. Tayeb, Production of biodegradable plastic from agricultural wastes, Arab. J. Chem. 11 (2018) 546–553, https://doi.org/10.1016/J.ARABJC.2015.04.008.
[45] S.N.H.M. Azmin, N.A.B.M. Hayat, M.S.M. Nor, Development and characterization of food packaging bioplastic film from cocoa pod husk cellulose incorporated with sugarcane bagasse fibre, J. Bioresour. Bioprod. 5 (2020) 248–255, https://doi.org/10.1016/J.JOBAB.2020.10.003.
[46] Y.F. Tsang, V. Kumar, P. Samadar, Y. Yang, J. Lee, Y.S. Ok, et al., Production of bioplastic through food waste valorization, Environ. Int. 127 (2019) 625–644, https://doi.org/10.1016/J.ENVINT.2019.03.076.

[47] K.Y. Lee, J.H. Lee, H.J. Yang, K bin Song, Production and characterisation of skate skin gelatin films incorporated with thyme essential oil and their application in chicken tenderloin packaging, Int. J. Food Sci. Technol. 51 (2016) 1465–1472, https://doi.org/10.1111/IJFS.13119.

[48] A. Jongjareonrak, S. Benjakul, W. Visessanguan, T. Prodpran, M. Tanaka, Characterization of edible films from skin gelatin of brownstripe red snapper and bigeye snapper, Food Hydrocoll. 20 (2006) 492–501, https://doi.org/10.1016/J.FOODHYD.2005.04.007.

[49] A. Jongjareonrak, S. Benjakul, W. Visessanguan, M. Tanaka, Skin gelatin from bigeye snapper and brownstripe red snapper: chemical compositions and effect of microbial transglutaminase on gel properties, Food Hydrocoll. 20 (2006) 1216–1222, https://doi.org/10.1016/J.FOODHYD.2006.01.006.

[50] J. Tkaczewska, P. Kulawik, E. Jamróz, P. Guzik, M. Zając, A. Szymkowiak, et al., One- and double-layered furcellaran/carp skin gelatin hydrolysate film system with antioxidant peptide as an innovative packaging for perishable foods products, Food Chem. 351 (2021) 129347, https://doi.org/10.1016/J.FOODCHEM.2021.129347.

[51] M.M. Rahman, A.N. Netravali, B.J. Tiimob, V. Apalangya, V.K. Rangari, Bio-inspired "green" nanocomposite using hydroxyapatite synthesized from eggshell waste and soy protein, J. Appl. Polymer Sci. 133 (2016), https://doi.org/10.1002/APP.43477.

[52] G. Kosarsoy Ağçeli, Development of ostrich eggshell and nano-levan-based edible biopolymer composite films: characterization and bioactivity, Polymer Bull. (2022 2022) 1–15, https://doi.org/10.1007/S00289-021-04069-Y.

[53] C. Qian, J. Yin, J. Zhao, X. Li, S. Wang, Z. Bai, et al., Facile preparation and highly efficient photodegradation performances of self-assembled Artemia eggshell-ZnO nanocomposites for wastewater treatment, Coll. Surf. A 610 (2021) 125752, https://doi.org/10.1016/J.COLSURFA.2020.125752.

[54] D.K. Jena, P.K. Sahoo, New strategies for the construction of eggshell powder reinforced starch based fire hazard suppression biomaterials with tailorable thermal, mechanical and oxygen barrier properties, Int. J. Biol. Macromol. 140 (2019) 496–504, https://doi.org/10.1016/J.IJBIOMAC.2019.08.156.

[55] C. Piccirillo, C.J. Denis, R.C. Pullar, R. Binions, I.P. Parkin, J.A. Darr, et al., Aerosol assisted chemical vapour deposition of hydroxyapatite-embedded titanium dioxide composite thin films, J. Photochem. Photobiol. A 332 (2017) 45–53, https://doi.org/10.1016/J.JPHOTOCHEM.2016.08.010.

[56] N.B. Song, J.H. Lee, M. al Mijan, K bin Song, Development of a chicken feather protein film containing clove oil and its application in smoked salmon packaging, LWT - Food Sci. Technol. 57 (2014) 453–460, https://doi.org/10.1016/J.LWT.2014.02.009.

[57] X. Chen, S. Wu, M. Yi, J. Ge, G. Yin, X. Li, Preparation and physicochemical properties of blend films of feather keratin and poly(vinyl alcohol) compatibilized by tris(hydroxymethyl) aminomethane, Polymers 10 (2018) 1054, https://doi.org/10.3390/POLYM10101054.

[58] W.A. Alshehri, A. Khalel, K. Elbanna, I. Ahmad, Bio-plastic films production from feather waste degradation by keratinolytic bacteria *Bacillus cereus*, J. Pure Appl. Microbiol. (2021) 1–13, https://doi.org/10.22207/JPAM.15.2.17.

CHAPTER 12

Smart packaging for commercial food products

Pinku Chandra Nath[a], Nishithendu Bikash Nandi[b], Shamim Ahmed Khan[b], Biswanath Bhunia[a], Tarun Kanti Bandyopadhyay[c] and Biplab Roy[c]

[a]Department of Bio Engineering, National Institute of Technology Agartala, Jirania, Tripura, India; [b]Department of Chemistry, National Institute of Technology Agartala, Jirania, Tripura, India; [c]Department of Chemical Engineering, National Institute of Technology Agartala, Jirania, Tripura, India

ABBREVIATIONS

BHT Butylated hydroxytoluene
BSE Bovine spongiform encephalopathy
BPE Ball Packaging Europe
EVOH Ethylene vinyl alcohol
IMP Inosine monophosphate
MAP Modified atmosphere packaging
SSOs Specific spoilage organisms

1. Introduction

Environmentally damaging impacts on food products are delayed by traditional food packaging. Today's food packaging has an active purpose in addition to passive functions of confinement, protection, and marketing, allowing it to play a dynamic role in food preservation (during production and packaging) and maintaining quality and food safety along the distribution chain as a result of this useful interaction between packaging and the environment inside, the food is actively protected [1]. Innovative packaging technologies are described using terms such as active, interactive, smart, clever, intelligent, and indications. A new approach to envisioning food packaging has brought together intelligent and active packaging designs [2]. The primary distinction is that whereas intelligent packaging does not directly interact with food, active packaging interacts with the environment around food to extend its shelf life. It takes action, while intelligent packaging detects and communicates information. Smart packaging combines intelligent and active packaging [3].

From a market perspective, active and intelligent packaging in Europe lags well behind international markets, particularly Japan, USA, and Australia [4]. Food packaging regulations in Europe were commonly blamed for this gap. Until now, only Regulation 450/2009 has set down explicit rules for the use and licensing of active and intelligent materials designed to come into contact with meals [5]. The benefits to consumers must be properly stated to stimulate smart packaging development. Marketing and advertising spending must be increased to support significant packaging innovation. For many fast-moving consumer goods, smart packaging might be viewed as a logical step forward in the evolution of packaging innovation [6].

This chapter is largely devoted to the issues facing commercial food product packaging in the future. The food industry is increasingly looking for ways to develop the value and safety of foodstuff, composition, and nutrition, and ways to expand the longevity of food goods and provide customers with more information and ease of use. These considerations must be taken while designing new packaging formats in the future.

2. Smart packaging technologies for food products

2.1 Beverage products

In recent days, the beverage industry developed quickly to adopt different smart technologies by innovative packaging solutions. Thus emphasis is placed on juice and milk-based beverages as they gain popularity, and are ideal to carry out components like nutrients and wellness [7]. Smart packaging is a completely new and growing area for researchers, particularly in food and beverage products [8]. The use of smart technologies is being addressed through the beverage products discussed below.

2.1.1 Gas releasing packaging

"Widget" is an innovative gas-releasing packaging developed by Robertson in the year 2006 for canned and bottled beer products [9]. The widget was created to help build a head on beer by releasing carbon dioxide from the beverage. The Widget is made of a plastic spherical that contains a hollow nitrogen gas canister and has a small hole in the center. The sphere floats on the surface of the beer and it is added to the canned before the seal. Before sealing the can, a tiny amount of liquid N_2 gas is incorporated into the beer which helps to pressurize the can during the entire canning process [10]. As the pressure gradually increases inside the can, nitrogen is compressed. Upon opening the can, the internal pressure decreases, and the pressurized gas propels the beer out across the small opening, causing CO_2 bubbles to develop. However, a smaller bubble formed because of the presence of dissolved nitrogen formed creaminess of the head as smaller bubbles need high interior pressure to maintain the superior surface tension.

2.1.2 Nutrient releasing packaging

In order to release nutrients to the beverage products, fresh can have been paid developed by the ball packaging group. In collaboration with Degussa FreshTech Beverages LLC, Ball Packaging Europe (BPE) reported Fresh Can Wedge in which dry sensitive ingredients such as vitamins, probiotic additives, vitamins, and trace elements are dispensed into a canned beverage. The additives are stored in a canned bottle in a dry state and mixed with the beverage as soon as the canned is opened. In the sports food sector, a ready-to-drink creatine product was introduced by Atlantic Multipower Germany. Adding 4.6 g creatine citrate to Cranberry flavored Crea Max enhances the performance by engaging intensive muscle workouts. When the can is opened, the creatine is blended with the beverage inside of the container. For the sports people, the advantage is that previously creatine is available only in powder or tablet form, not in the liquid state. Portola Company designed a fusion cap in such a way that vitamins/flavors are added to the beverage bottles by twisting the cap.

2.1.3 Flavor releasing packaging

To preserve the value of the items, the packaging device is designed in such a way that it releases flavor to the beverage products at the time of consumption. Companies attract consumers by incorporating aromas into the polymer material. US Company tested the effect of flavor and aroma released for nutrient and water bottle packaging technology. However, Markarian developed milk-based, chocolate-flavored, drinks with smart packaging polyethylene materials having chocolate fragrance. Li Ting et al. (2020) [11] engaged in developing mess free milk flavoring solution for children by smart technology.

2.1.4 Enzyme release packaging

Till now, several antimicrobials like bacteriocins, fungicides, silver substitute zeolite, and inorganic gases have been utilized in food packaging applications. However, the employment of bioactive compounds such as enzymes and proteins that are frequently incorporated into the packaging system has been widely studied by the researcher. Lopez-Rubio et al. (2004) [12] facilitated the smart packaging process by incorporating enzymes into the polymeric materials. A work studied by Haghighi-Manesh and Azizi [13], discovered that the bitterness of citrus juices could've been greatly alleviated by using cellulose acetate sheets, which were produced by the enzyme naringinase. In milk products, immobilized enzyme viz. cholesterol reductase incorporated into packaging structure which allows the reduction of cholesterol helps to transport the package to consumer in time.

2.2 Bakery based products

Mold spoilage is a major problem in bakeries and a product's shelf life mostly depends on the growth of the mold [14]. However, fat rancidity is a common problem that arises in dried bakery products, especially breakfast cereals. Fat rancidity occurs mainly due to oxidation of fats/oils in presence of air or light or bacterial infection results in unpleasant odor and taste [15]. So the emphasis is placed on the employment of different strategies which delay the spoiling of bakery products. Generally, the two ways have been utilized to delay the spoilage viz. oxygen elimination inside the package (Oxygen scavenger) and introduction of chemicals that inhibit the mold growth (Ethanol emitters) [16].

2.2.1 Oxygen scavengers

The most common ways of removing oxygen from oxygen sensitive foods were used of MAP and vacuum packaging [17]. However, these techniques have drawbacks of the presence of residual gas and/or penetration of oxygen through the polymeric packaging materials. However in recent days, to eliminate oxygen concentration, O_2 scavengers are used as a chemically reducing agent. The reducing agents generally removed oxygen concentration by combining with oxygen within the package [18]. An oxygen scavenger absorbs residual oxygen present inside the packaging material that helps to minimize the spoiling of foods [19,20]. The substance to acts as an oxygen scavenger fulfills some basic criteria as follows:

- The oxygen scavenging substances should have a multilayer structure as it is sensitive to external oxygen.
- The agent should not interact with the processing as well as physical properties of polymeric material.
- Preventing the release of by-products that degrade food nutrients.
- It must be stable in presence of air.

The scavenging films are achieved by dispersing or blending the reducing substances in polymeric material in a smart polymer technology [12]. Usually utilized O_2 scavengers are based on iron. Iron powder introduced to low density polyethylene show an excellent O_2 absorber [21]. However, in some cases, nanocomposites have been utilized which claimed to activate the reaction at the iron containing films in addition to stabilizing the scavengers [22]. Unsaturated fatty acids also serve the purpose of an O_2 absorber inside the package [23]. While the incorporation of fatty acids into the polymer matrix, the catalyst is required to activate the reaction by ultraviolet radiation, which is the functional barrier between the scavenging layer and food products helps to stop the production of undesired products [18,24]. In the case of breakfast cereals, synthetic antioxidant like Butylated hydroxytoluene (BHT) has been incorporated into flexible thermoform able plastic materials [24]. However, because of drawbacks to consumption of BHT include some health related issues; some natural antioxidants are being explored. Till now, at present numerous naturally occurring anti-oxidants have been applied as O_2 scavengers. Among them, lecithin, organic acids, and rosemary extract are the few mentioned natural occurring compounds showing antioxidant properties [25]. In addition to the antioxidant property, the incorporation of Vitamin E and C increases the nutritional characteristics of food products.

2.2.2 Ethanol emitters

To enhance the longevity of food products, ethanol has been widely used as a mold growth inhibitor as it has the property of fungi static. If ethanol is directly dispersed onto bakery products, it will depart from the target mold growth in addition to strong flavor [26]. Thus, ethanol has been incorporated into the polymer by smart technology owing to better release of compounds into the surface of matrix resulting in better inhibition of mold growth. To trap the molecules and decrease the volatility, cyclodextrin complexes are encapsulated as during solution casting or extrusion processes, volatilization of molecules occurs [27]. In a study conducted by the researcher, efficiency was improved by addition of sodium lauryl sulfates, which allows the reduction of dextrin used while preparing microcapsules. In some other technology, EVOH copolymers have been combined with polymer to increase the efficiency during lamination processes [28].

2.3 Fruits and vegetable products

Smart or intelligent packaging can detect, sense, record, trace, communicate, and use science logic to enhance longevity, improve protection, develop value, give identification, and advise regarding potential concerns [29]. TTIs, gas indicators, biosensors, and radio-frequency tags are examples of smart packaging technologies (figure). Smart packaging may contain physical shock alerts as well as intelligent labeling. A diabetic customer may be alerted about the product's sugar content, and a patient may be reminded to take prescribed medications on time [30]. In some cases, the appropriate or harmful temperature of a beer bottle (which should be chilled) or the temperatures at which to cook pre-made soups can already be shown using thermochromic inks [31].

Temperature is by far the most essential factor in slowing down the rotting of fruits and vegetables. Generally, each 10°C raise in temperature doubles or triples the rate of respiration [32]. An equation of the Arrhenius low type can usually be used to represent the change in the rate of change. 40–105 kJ/mol is the amount of energy it takes to start a chemical reaction [33]. This is dependent on the surrounding air composition. However, some foods, like bananas, lemons, and mangoes, can be damaged by low temperatures. Most products can't be fixed when the temperature drops below −1°C. Because these products need to be kept at the right temperature at all times during transportation and storage, temperature control and monitoring are very important. The amount of gas in the environment of the product also has a high impact on how quickly the product can respirate [34]. MAP can help the value and longevity of the product a lot. The rate at which the body breathes slows down as the amount of oxygen in the air drops. Carbon dioxide usually has the opposite effect; however, this depends on the product, maturation, concentration range, and exposure period [35]. The figure shows the range of O_2 and CO_2 concentrations that can help fruits and vegetables stay fresh longer. To keep a product as fresh as possible, it requires a certain gas concentration ratio.

2.4 Milk-based products

The main sources of essential nutrients are milk and dairy foods. About 70% of the World's population, consumption of dairy products, and suffer from gastrointestinal discomfort due to low levels of intestinal enzyme lactase [36]. In addition, the cholesterol content of lactose, the main constituent of carbohydrates in milk is also high. Therefore, the use of enzymes in smart packaging can help to solve the problems. However, beyond these, the other factors including the presence of oxygen and growth of microorganisms also need to control.

2.4.1 Reduced content of cholesterol and lactose

Some enzymes are being considered in food packaging applications though initially these were employed in production lines. Enzyme lactase immobilized with the surfaces of low density polythene attached by covalent interaction retained 10% of the enzyme activity [37]. It reduced the lactose content in milk during storage by converting it into glucose and galactose. Similarly, enzyme cholesterol reductase employed on the surfaces of polymer reduced the cholesterol content by splitting into coprostanol and coprosterol [38]. However, in both cases the products formed are poorly digested and easily excreted from the body. The main objective of using this kind of product is to produce a high-value product without changing the production process.

2.4.2 Oxygen scavenger (for Yoghurt)

Food Science Australia, CSIRO synthesized a reducible organic compound viz. substituted anthraquinone serve as oxygen scavenger additive for Yoghurt that dispersed into a polymer in a packaging film. Before packaging the substituted anthraquinone is subjected to UV light. It has been revealed that combined with an excessive barrier polymer in yogurt bowls reduces the oxygen concentration [39]. In the case of other smart technologies in Diary products including the development of anti-microbial films for cheese using nisin or lacticin, improved inhibitor property in both mold and lactic acid bacteria on attachment with polymers.

2.5 Fish and seafood products

According to recent data, busy consumers increasingly prefer ready-to-eat fresh chilled or frozen fish items. Today's hurried lifestyles have resulted in a rise in the necessity for home consumption products. Customers are increasingly preferring superior semi-processed frozen pre-packaged solutions, in addition to making economic, social, and lifestyle choices [40]. Undoubtedly, this trend is being fueled by improved awareness of the nutritional aspects of seafood (rich in protein, reduced fat, and high in omega-3) [41]. Seafood, and especially fish, has the advantage of not being harmed by recent health problems like BSE, avian flu, and foot-and-mouth. These crises may have increased fish consumption in some industries. New smart packaging technologies are being developed to minimize the threat to the well-being of the consumers and the credibility of the suppliers [42]. The quantity of microorganisms on the skin and gill surfaces grows steadily after death and throughout storage. Depending on the environment and product composition, specific spoilage organisms (SSOs) will outgrow others, releasing a variety of breakdown and metabolic chemicals [43]. Table 12.1 lists SSOs and typical fish spoilage chemicals.

Table 12.1 Specific spoilage organisms and typical spoilage compounds in groups of fresh and lightly preserved sea foods.

Fish product	Specific spoilage organisms	Number at rejection[a]	Typical volatile compounds	Ref.
Fresh fish, chilled, modified atmosphere package	*P. phosphoreum*[b] Lactic acid[c]	$>10^6$ $>10^7$	Hydrogen sulfide, Acetic acid, Trimethylamine, Ketones & alcohols	[44]
Fresh fish (less than 10° C)	*Vibrionaceae, Aeromonas* sp.	10^6-10^7	Trimethylamine, volatile sulfides	[45]
Sugar-salted herring	Halophilic, anaerobes[d] osmotolerant yeast	10^6-10^7 10^3-10^4	Indole, hydrogen sulfide	[46]
Fresh refrigerated products stored in atmosphere	*Shewanella putrefaciens*[b] *Pseudomonas* sp.[c]	10^7-10^8 10^7-10^8	Trimethylamine, hydrogen sulfide, hypoxanthine, ammonia	[47]
Cooked fish	*Clostridium* sp.	$>10^7$	Strong fecal, sulfydryl odors	[48]

[a]Numbers in cfu/g.
[b]Typical of marine, temperature water fish.
[c]Typical of fish from warm and fresh water.
[d]Not identified.

2.6 Meat and poultry products

Food preservation packaging for meat ensures that the appearance, aroma, and flavor of the meat remain acceptable while also stabilizing the food composition and preventing microbiological decomposition. Fresh and cooked meats and meat products are included in the scope of muscle food packaging systems and technologies. Packaging can be carried out on trays lined with O_2 permeable membrane or even in a stationary gaseous modified atmospheric packaging (MAP) atmosphere containing 75–85% O_2 and 30–40% CO_2 [49]. With 65%–75% nitrogen and 25%–35% CO_2, fresh poultry items can be repackaged in overwrap or MAP type packaging [50].

Essential elements affecting the color stability of cured, cooked, and packaged meat products include the proportion of residual O_2, volume fraction of product to headspace, and oxygen transportation charge of the packing substances [51]. Other important factors to consider include storage temperature (temperature range), light intensity (light intensity range), and product composition (composition range). During the cooking process, Nitrosylmyoglobin, which is generated by a reaction between myoglobin and nitrite, denatures to form nitrosylmyochrome, which gives cooked cured ham its distinctive pink color [52]. Nitrosylmyochrome is formed through a reaction between myoglobin and nitrite [53]. Because even oxygen depletion can induce nitrosylmyochrome to be transformed to denatured metmyoglobin, exposure to light in aggregate with oxygen is important for the color stability of cooked cured ham [54,55]. Due to the oxidization of nitrosylmyochrome, which results in the formation of denatured metmyoglobin, the beef surface might develop an unattractive gray hue. Observations have been made in the commercial setting that the discoloration of pre-packaged, cooked, preserved ham is associated with oxygen depletion phenomena, which can be addressed by the use of oxygen scavengers [2]. As well as this, when it comes to clean red meat, the employment of oxygen scavengers alongside a CO_2/nitrogen gasoline blend can appreciably prolong the shelf-lifestyles of sparkling pork. Oxygen scavengers have been among the most well-known instances of smart packaging gadgets which might be used to defend oxygen-touchy meat-primarily based objects from deterioration [56].

2.6.1 Oxygen scavengers

The oxygen scavenger can be integrated into the packaging itself as an alternative to sachets. As a result, adverse reactions from customers are lessened, and improved outputs may lead to a financial advantage. It also reduces the possibility of sachets rupturing accidentally and their contents being accidentally consumed [54]. In the United States, the Cryovac Div. of Sealed Air Corporation created the polymer-based O_2 scavenging film Cryovac® 0S2000™. It takes 4—10 days for this UV light-activated O_2 scavenging film (Fig. 12.1) to lower headspace oxygen levels from 1% to ppm, making it comparable to oxygen sachets in terms of effectiveness. In addition to dried and smoked meats, processed meats can also benefit from the OS2000™ scavenging films [57]. The OSPTM system is another commercially successful example of meat processing (Chevron Philips Chemical Company, USA). The oxygen scavenging process in OSPTM systems is activated by mixing ethylene methacrylate and cyclohexene methacrylate with a catalyst or photoinitiator. Most commercial oxygen scavengers work by oxidizing iron:

$$Fe \longrightarrow Fe^{2+} + 2e^-$$
$$1/2\ O_2 + H_2O + 2e- \longrightarrow 2OH^-$$
$$Fe_2^+ + 2OH^- \longrightarrow Fe(OH)_2$$
$$Fe(OH)_2 + 1/4\ O_2 + 1/2\ H_2O \longrightarrow Fe(OH)_3$$

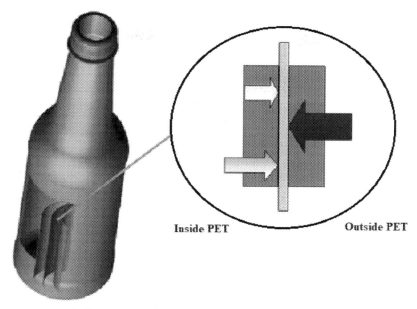

Fig. 12.1 Light-activated oxygen scavenging films Cryovac® OS Films.

3. Consumer advantages and comfort factors of smart packaging

Until now, packaging has completed a fantastic job of minimizing waste and handing over gadgets in appropriate condition, but we stay in an unexpectedly converting global in which the existing device is not appropriate. When it comes to many consumer products, smart packaging can be seen as a logical next step in the evolution of package design. Indeed, shelf space is limited, and competition is changing away from media and toward the moment of purchase, rendering packaging increasingly essential than it's ever been. According to the magazine over 50 readers, 71% have been injured when attempting to open food packaging, and 98% believe there is just "too much unnecessary packing" in the food industry today. According to these new criteria, the current packaging is subpar. Smart packaging may be able to meet at most a few of these future packaging ambitions through the user interface and consumer experience [58]. To support customers in their daily lives, new types of beneficial functionality will be introduced to aid promote efficient and effective product usage, storage, and disposal.

3.1 Assessing the consumer's value proposition

Consumer advantages must be crucial for a successful new consumer product introduction for smart packaging targeted at the consumer. This is a prevalent issue when introducing new technologies into packaging, as well as in other areas. Therefore, Gao et al. (2011) [59] came up with an interesting way to evaluate the benefits of new products and services by mapping usefulness to various stages of customer experience and then comparing them. To boost client productivity, a product or service must be easy to purchase or utilize. A new smart packaging concept may demonstrate how it generates a different benefit proposition from existing products by placing it in one of the three most crucial stages for packaging: purchase, use, and disposal [60]. As shown in Fig. 12.2, smart packaging may considerably improve the buyer experience by enhancing product interaction during use and disposal.

3.2 Improving convenience in product use

Paint containers have long been made of metal and opened and closed with a screwdriver. With this innovative packaging, the days of requiring a screwdriver to open metallic paint cans are finally over [61]. We started with a survey of the public, which revealed that 75% of paint purchases are made by women. It was developed to reduce paint spilling from round metal paint cans by incorporating a twist-off top and a pour-spout. Because of its square design and rounded corners, the container was ideal for pallet presentations and distribution. The square design and integral handle made them easy to carry, display, and stack. The brand discovered unmet consumer needs in this industry through smart structural design. Due to the lightweight plastic container, printing timber and metal become easier, cleaner, and higher pleasurable. That it's ergonomically made

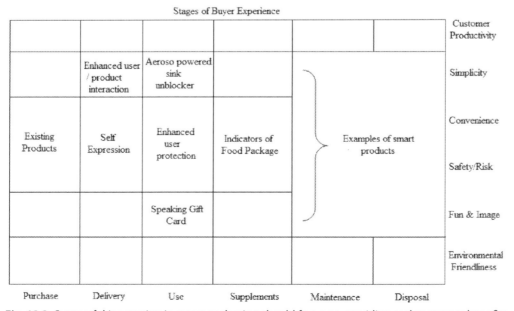

Fig. 12.2 Successful innovation in smart packaging should focus on providing real consumer benefits at points of purchase, use and disposal.

to accommodate the hand makes painting a breeze. Whenever a free arm is necessary, the container rim acts as a stable resting place for the brush, eliminating the need for screwdrivers. Excess paint is returned to the container, eliminating undesired drips down the side of the can.

4. Conclusion

Most people would agree that smart packaging extends beyond the use of basic packaging materials and standard printed characteristics like alphanumeric, graphics, and simple barcodes, but there is no official definition for it. The main difference is that, whereas intelligent packaging does not directly interact with food, active packaging does so through interacting with the environment around it to improve the shelf life of the food. While intelligent packaging detects and distributes information, it takes action. The term "smart packaging" refers to packaging that is both intelligent and active. The food industry is increasingly seeking ways to improve the quality, safety, composition, and nutrition of food products, as well as ways to extend product shelf life and provide consumers with more information and convenience of use. These factors must be taken into account while developing new packaging formats in the future. This chapter focuses mostly on the difficulties that may face commercial food product packaging in the future.

References

[1] S. Gottardo, A. Mech, J. Drbohlavová, A. Małyska, S. Bøwadt, J.R. Sintes, et al., Towards safe and sustainable innovation in nanotechnology: state-of-play for smart nanomaterials, NanoImpact 21 (2021) 100297.
[2] J.P. Kerry, New packaging technologies, materials and formats for fast-moving consumer products, Innov. Food Packag. (2014) 549–584.
[3] E.A. Helmy, Nano-biotechnology breakthrough and food-packing industry-A Review, Microb. Biosyst. 1 (1) (2016) 50–69.
[4] B.P. Day, Active packaging of food, in: Smart Packaging Technologies for Fast Moving Consumer Goods 1, 2008.
[5] E.F.S. Authority, Administrative Guidance for the Preparation of Applications on Substances to Be Used in Active and Intelligent Materials and Articles Intended to Come into Contact with Food, Wiley Online Library, 2021, pp. 2397–8325. Report No.
[6] V. Amenta, K. Aschberger, M. Arena, H. Bouwmeester, F.B. Moniz, P. Brandhoff, et al., Regulatory aspects of nanotechnology in the agri/feed/food sector in EU and non-EU countries, Regulat. Toxicol. Pharmacol. 73 (1) (2015) 463–476.
[7] D. Sun-Waterhouse, The development of fruit-based functional foods targeting the health and wellness market: a review, Int. J. Food Sci. Technol. 46 (5) (2011) 899–920.
[8] R. Dobrucka, R. Cierpiszewski, Active and intelligent packaging food-research and development-a review, Pol. J. Food Nutr. Sci. 64 (1) (2014).
[9] M.J. Kirwan, R. Coles, Food and Beverage Packaging Technology, John Wiley & Sons, 2011.
[10] H.A. Abdulmumeen, A.N. Risikat, A.R. Sururah, Food: its preservatives, additives and applications, Int. J. Chem. Biochem. Sci. 1 (2012) (2012) 36–47.
[11] T. Li, K. Lloyd, J. Birch, X. Wu, M. Mirosa, X. Liao, A quantitative survey of consumer perceptions of smart food packaging in China, Food Sci. Nutr. 8 (8) (2020) 3977–3988.

[12] A. Lopez-Rubio, E. Almenar, P. Hernandez-Muñoz, J.M. Lagarón, R. Catalá, R. Gavara, Overview of active polymer-based packaging technologies for food applications, Food Rev. Int. 20 (4) (2004) 357−387.
[13] S. Haghighi-Manesh, M.H. Azizi, Active packaging systems with emphasis on its applications in dairy products, J. Food Process. Eng. 40 (5) (2017) e12542.
[14] M. Garcia, M. Copetti, Alternative methods for mould spoilage control in bread and bakery products, Int. Food Res. J. 26 (3) (2019) 737−749.
[15] M.S. Rahman, Handbook of Food Preservation, CRC press, 2007.
[16] C. Mohan, C. Ravishankar, Active and Intelligent Packaging Systems-Application in Seafood, 2019.
[17] D. Narasimha Rao, N. Sachindra, Modified atmosphere and vacuum packaging of meat and poultry products, Food Rev. Int. 18 (4) (2002) 263−293.
[18] L. Vermeiren, L. Heirlings, F. Devlieghere, J. Debevere, Oxygen, ethylene and other scavengers, Novel Food Packag. Tech. (2003) 22−49.
[19] S.A. Cichello, Oxygen absorbers in food preservation: a review, J. Food Sci.Technol. 52 (4) (2015) 1889−1895.
[20] A. Mills, Oxygen indicators and intelligent inks for packaging food, Chem. Soc. Rev. 34 (12) (2005) 1003−1011.
[21] K.K. Gaikwad, S. Singh, Y.S. Lee, Oxygen scavenging films in food packaging, Environ. Chem. Lett. 16 (2) (2018) 523−538.
[22] A. Llorens, E. Lloret, P.A. Picouet, R. Trbojevich, A. Fernandez, Metallic-based micro and nanocomposites in food contact materials and active food packaging, Trends Food Sci. Technol. 24 (1) (2012) 19−29.
[23] S. Mexis, M. Kontominas, Effect of oxygen absorber, nitrogen flushing, packaging material oxygen transmission rate and storage conditions on quality retention of raw whole unpeeled almond kernels (*Prunus dulcis*), LWT Food Sci. Technol. 43 (1) (2010) 1−11.
[24] R.S. Cruz, G.P. Camilloto, A.C. dos Santos Pires, Oxygen scavengers: an approach on food preservation, Struct. Funct. Food Eng. 2 (2012) 21−42.
[25] M. Brewer, Natural antioxidants: sources, compounds, mechanisms of action, and potential applications, Comprehens. Rev. Food Sci. Food Safety 10 (4) (2011) 221−247.
[26] F. Baghi, A. Gharsallaoui, E. Dumas, S. Ghnimi, Advancements in biodegradable active films for food packaging: effects of nano/microcapsule incorporation, Foods 11 (5) (2022) 760.
[27] B. Bhandari, B. D'Arcy, G. Young, Flavour retention during high temperature short time extrusion cooking process: a review, Int. J. Food Sci. Technol. 36 (5) (2001) 453−461.
[28] D. Cava, C. Sammon, J. Lagaron, Sorption-induced release of antimicrobial isopropanol in EVOH copolymers as determined by ATR-FTIR spectroscopy, J. Appl. Polym. Sci. 103 (5) (2007) 3431−3437.
[29] K.L. Yam, P.T. Takhistov, J. Miltz, Intelligent packaging: concepts and applications, J. Food Sci. 70 (1) (2005) R1−R10.
[30] J. Avorn, Powerful Medicines: The Benefits, Risks, and Costs of Prescription Drugs, Vintage, 2008.
[31] R. Chelliah, S. Wei, E.B.-M. Daliri, M. Rubab, F. Elahi, S.-J. Yeon, et al., Development of nanosensors based intelligent packaging systems: food quality and medicine, Nanomaterials 11 (6) (2021) 1515.
[32] M.E. Saltveit, Respiratory metabolism, in: The Commercial Storage of Fruits, Vegetables, and Florist and Nursery Stocks 68, 2004.
[33] R.K. Mishra, K. Mohanty, Pyrolysis kinetics and thermal behavior of waste sawdust biomass using thermogravimetric analysis, Bioresour. Technol. 251 (2018) 63−74.
[34] S.A. Zahra, W.C. Bogner, Technology strategy and software new ventures' performance: exploring the moderating effect of the competitive environment, J. Business Ventur. 15 (2) (2000) 135−173.
[35] A.A. Kader, D. Zagory, E.L. Kerbel, C.Y. Wang, Modified atmosphere packaging of fruits and vegetables, Crit. Rev. Food Sci. Nutr. 28 (1) (1989) 1−30.
[36] N. Silanikove, G. Leitner, U. Merin, The interrelationships between lactose intolerance and the modern dairy industry: global perspectives in evolutional and historical backgrounds, Nutrients 7 (9) (2015) 7312−7331.

[37] J. Goddard, J. Talbert, J. Hotchkiss, Covalent attachment of lactase to low-density polyethylene films, J. Food Sci. 72 (1) (2007) E036—E41.
[38] D. Givens, MILK Symposium review: the importance of milk and dairy foods in the diets of infants, adolescents, pregnant women, adults, and the elderly, J. Dairy Sci. 103 (11) (2020) 9681—9699.
[39] C.W. Miller, M.H. Nguyen, M. Rooney, K. Kailasapathy, The control of dissolved oxygen content in probiotic yoghurts by alternative packaging materials, Packag. Technol. Sci. 16 (2) (2003) 61—67.
[40] E. Alves Da Silva Oliveira, Food Packaging Trends for the Refrigerator, 2015. Master's thesis.
[41] M. Venegas-Calerón, O. Sayanova, J.A. Napier, An alternative to fish oils: metabolic engineering of oil-seed crops to produce omega-3 long chain polyunsaturated fatty acids, Progr. Lipid Res. 49 (2) (2010) 108—119.
[42] S. Matindoust, M. Baghaei-Nejad, M.H.S. Abadi, Z. Zou, L.-R. Zheng, Food quality and safety monitoring using gas sensor array in intelligent packaging, Sens. Rev. (2016).
[43] M. Sivertsvik, W.K. Jeksrud, J.T. Rosnes, A review of modified atmosphere packaging of fish and fishery products—significance of microbial growth, activities and safety, Int. J. Food Sci. Technol. 37 (2) (2002) 107—127.
[44] M. Sivertsvik, J.T. Rosnes, H. Bergslien, Modified atmosphere packaging, in: Minimal Processing Technologies in the Food Industry, 2002, pp. 61—86.
[45] V. Lougovois, V. Kyrana, Freshness quality and spoilage of chill-stored fish, Food Pol. Control Res. 1 (2005) 35—86.
[46] L. Gram, Microbiological spoilage of fish and seafood products, in: Compendium of the Microbiological Spoilage of Foods and Beverages, Springer, 2009, pp. 87—119.
[47] L. Galaviz-Silva, G. Goméz-Anduro, Z.J. Molina-Garza, F. Ascencio-Valle, Food safety Issues and the microbiology of fish and shellfish, Microbiol. Safe Foods (2009) 227—273.
[48] S. Kose, G.M. Hall, Sustainability of fermented fish products, in: Fish Processing: Sustainability and New Opportunities, 2010, pp. 138—166.
[49] N. Hutchings, B. Smyth, E. Cunningham, C. Mangwandi, Development of a mathematical model to predict the growth of *Pseudomonas* spp. in, and film permeability requirements of, high oxygen modified atmosphere packaging for red meat, J. Food Eng. 289 (2021) 110251.
[50] A.L. Brody, A Perspective on MAP Products in North America and Western Europe, Technomic Publishing Company, Inc., Lancaster, PA, 1995.
[51] S. Yildirim, B. Röcker, M.K. Pettersen, J. Nilsen-Nygaard, Z. Ayhan, R. Rutkaite, et al., Active packaging applications for food, Comprehen. Rev. Food Sci. Food Safety 17 (1) (2018) 165—199.
[52] H. Walsh, J. Kerry, Packaging of ready-to-serve and retail-ready meat, poultry and seafood products, Adv. Meat Poultr. Seafood Packag. (2012) 406—436.
[53] J.K. Møller, J.S. Jensen, M.B. Olsen, L.H. Skibsted, G. Bertelsen, Effect of residual oxygen on colour stability during chill storage of sliced, pasteurised ham packaged in modified atmosphere, Meat Sci. 54 (4) (2000) 399—405.
[54] J. Kerry, M. O'grady, S. Hogan, Past, current and potential utilisation of active and intelligent packaging systems for meat and muscle-based products: a review, Meat Sci. 74 (1) (2006) 113—130.
[55] G. Nellis, Investigation of Practical Hurdle Technologies for Preventing Photooxidation of Cured Lunch Meats in Prepackaged Sandwiches, 2015.
[56] B. Kuswandi, Active and intelligent packaging, safety, and quality controls, in: Fresh-Cut Fruits and Vegetables, 2020, pp. 243—294.
[57] S. Hogan, J. Kerry, Smart packaging of meat and poultry products, in: Smart Packaging Technologies for Fast Moving Consumer Goods, 2008, pp. 33—54.
[58] R. Holdway, D. Walker, M. Hilton, Eco-design and successful packaging, Des. Manag. J. 13 (4) (2002) 45—53.
[59] J. Gao, Y. Yao, V.C. Zhu, L. Sun, L. Lin, Service-oriented manufacturing: a new product pattern and manufacturing paradigm, J. Intel. Manuf. 22 (3) (2011) 435—446.
[60] N.P. Mahalik, A.N. Nambiar, Trends in food packaging and manufacturing systems and technology, Trends Food Sci. Technol. 21 (3) (2010) 117—128.
[61] E. Shove, L. Araujo, Consumption, materiality, and markets, Reconnect. Market. Markets (2010) 13—28.

CHAPTER 13

Smart applications for fish and seafood packaging systems

Oya Irmak Sahin[a], Furkan Turker Saricaoglu[b], Ayse Neslihan Dundar[b] and Adnan Fatih Dagdelen[b]

[a]Department of Chemical Engineering, Faculty of Engineering, Yalova University, Yalova, Turkey; [b]Department of Food Engineering, Faculty of Engineering and Natural Science, Bursa Technical University, Bursa, Turkey

1. Introduction

Smart packaging aims to indicate, monitor, and trace food quality and safety by giving information through various signs. Therefore, play an especially important role in food supply chains by facilitating the flow of both material and information" [1]. Innovation activities in food packaging have been in the direction of the development of smart packaging since the 2000s [2]. "Smart packaging" has been defined by some researchers, as a whole term that describes both "active packaging" and "intelligent packaging" [3,4], however, some researchers have specified the "smart" term also as "intelligent" for the packaging systems.

Considering these diverse definitions by authors, we can define "smart packaging" as; packaging with the capabilities to carry out smart functions that can detect, record, monitor, or briefly trail the product, and communicate with customers during the food shelf life. Auxiliary structures of these smart technologies are barcodes, RFID tags, indicators, and biosensors used in smart packaging because of their potential to sense, monitor, and indicate [5]. These systems can be classified as; *sensors* - which aim to inform the level of spoilage based on the quantity of analytes, *indicators* - which aim to provide information about the food quality, and *data carriers* or *tracing devices*-which purpose is food storage, distribution, and traceability, and presented in Fig. 13.1.

The new concepts of packaging focused on food safety and food trailing system, have been rapidly growing in the past two decades. There has been extensive research on intelligent materials monitoring the packaged food conditions and/or the surrounding environment of food without interacting with food [2,6–10].

1.1 Smart technologies for fish and seafood

Fish and seafood are extremely sensitive foods whose freshness and quality deteriorate rapidly after death. Maintaining the freshness of seafood and preventing spoilage are sophisticated and many factors affect these processes. For many years, the goal of the research of fish and seafood technology has been focused on developing trustworthy

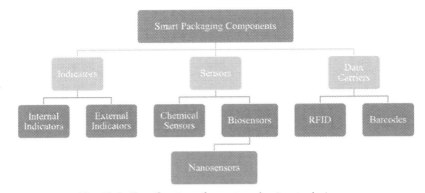

Fig. 13.1 Classification of smart packaging techniques.

freshness measuring and quality determining methods [4]. In the food industry physical, chemical, microbiological, and sensorial methods are used to monitor fish and seafood products' quality. Currently, the perception of the freshness of fish and seafood is largely determined by features such as appearance (eyes, skin, and gills), texture, smell, and color. The Quality Index Method (QIM), which is widely used in the quality assessment of fresh fish in Europe, uses these basic parameters without varying according to the aquaculture type. According to the prolonged, expensive and labour-skilled QIM analysis, there has been a profound need for fast and cheap methods that can be easily adapted to any level of the food supply chain [11]. The invention of smart packaging is the one possible solution to the issue of being informed about the quality of fish and seafood. The software programs designed and used for QIM are continually being developed and adapted to these "smart technologies".

As is well known, post-mortem changes in fish and seafood make consumption impossible. For this reason, it is imperative that aquaculture products are kept in the cold chain, which will slow down these processes from the time they are caught until they are transferred to the consumer. Post-mortem changes are mostly chemical changes, which are mainly formed autolytic and bacteriologically by the degradation products; volatile bases such as ammonia, methyl amines, biogenic amines, and volatile acids. Based on monitoring these degradation products, smart systems can be assigned as valuable quality monitoring for fish and seafood products.

2. Sensors

In order to measure the freshness of seafood products, which are very important for both producers and consumers, it is constantly being tried to develop faster and higher standards quality control methods that will respond to the demands of the sector. In this respect, sensor technology is very convenient to be used for quality measurements, as it provides the expected speed and high standard and is open to development. While biosensors are a promising monitoring quality system it can be concluded that this technology is still in its infancy for food technology and industry.

Sensor, according to Patel & Beveridge [12] is a device or system, an interconnection network, with control and processing electronics, based on a receptor and a transducer principle. Physical and chemical data were converted to signals by receptors that were received from the source. These signals can be electrical, thermal, chemical or optical [13]. In order, for the sensors, to perform their measurements, they must be in a continuous communication environment with the source. Sensors transmit the oxygen, carbon dioxide, water vapor, and ethanol amounts that occur as a result of various physical and chemical reactions occurring in the food in the package to the reader with the receiver and transducer parts they have and thus ensure the food quality and safety.

There are many sensors with different operating principles and these sensors can be classified according to their intended use. Various sensors: chemical sensors, biosensors, gas sensors, printed electronics, and electronic nose, are used in smart packaging technology to measure the degree of microbiological deterioration, oxidative rancidity, and other changes caused by temperature [14] or to determine the amount of some chemical substances mixed with foods [15].

2.1 Chemical sensors

An electrochemical sensor for exposure to ammonia, tetramethylammonium hydroxide, ethylamine, cadaverine, and putrescine incorporate with trifluoroacetylazobenzene dye, was developed by Lin et al. [16]. Morsy et al. [17] also developed another electrochemical sensor for the detection of trimethylamine, dimethylamine, cadaverine, and putrescine with a sensitivity of 1 ppm. Both sensors were found to be successful in monitoring the spoilage based on volatile biogenic amines.

Electrochemical sensors also can be named opto-chemical sensors, are used for quality control by detecting compounds such as carbon dioxide or amines released as a result of spoilage of foods due to microbial contamination or other effects. Opto-chemical sensors, which detect volatile amines, hydrogen sulfide and carbon dioxide, are classified into three categories: fluorescence-based systems sensitive to pH changes, absorption-based systems sensitive to color changes, and fluorometric-based. In recent years, sensors have been developed for monitoring "colourimetric changes". pH-sensitive sensors can be improved for the detection of volatile amines in fish, meat and poultry. A colourimetric sensor study was carried out with a methyl red/cellulose membrane to observe fish spoilage volatile compounds [18]. Ocular oxygen sensors based on the theory of luminescence of analytes were expressed by Papkovsky et al. [14].

The favorite of recent years, colourimetric sensors were produced as a "sensor film" made of food-derived materials; pectin and red cabbage and are used to monitor gaseous amines that occur in fish and seafood microbial spoilage [19]. Freshness of shrimps was evaluated by Hashim et al. [20] in which agar and sugarcane wax films were incorporated with butterfly pea flower anthocyanins for their pH sensing potential. Remarkably, films were displayed complete protection against UV–Vis light and valuable reduction in visible light.

2.2 Biosensors

Biosensors have been featured throughout the last two decades. The main difference between chemical sensors and biosensors is the recognition layer that biosensors are the devices monitoring the enzyme-catalyzed reaction either by the gases or small ions. A schematic diagram of a biosensor is shown in Fig. 13.2. Biosensors based on the detection of biogenic amines according to microbial growth and amino acid decarboxylation were created [21]. This biosensor technology has been applied for fish and seafood products the monitoring the microbial spoilage degradation products concentrations. A biosensor based on colourimetric changes of enzymatic reactions of hypoxanthine and trimethylamine oxide of fish products was developed [22]. Early studies of analytical and chemical biosensors are summarized in Table 13.1.

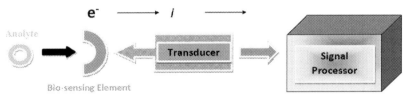

Fig. 13.2 Schematic diagram of biosensor systems.

Table 13.1 Analytical and chemical biosensors [23,24].

Analyte	System	Product	Detection range	Reference
Xanthine	Amperometric enzymatic	Crucian carp	0.03–21.2 µM	[25]
Xanthine	Amperometric enzymatic	Channa striatus	0.4–2.4 nM	[26]
Xanthine	Amperometric enzymatic	Labeo fish	0.5–500 µM	[27]
Xanthine	Amperometric enzymatic	Hake	0.05–12 µM	[28]
Xanthine	Glassy carbon electrode	Salmon	0.001–50 µM	[29]
Histamine	Amperometric enzymatic	Carp, Prussian carp, Wels catfish, European perch	0.1–300 µM	[30]
Histamine	Carbon electrode	Spiked tuna, Mackerel	—	[31]

Table 13.1 Analytical and chemical biosensors [23,24].—cont'd

Analyte	System	Product	Detection range	Reference
Hypoxanthine	Fluorescent	Fish, Shrimp, Squid	8–250 µM	[32]
Volatile ammines	Ammonium ion-selective electrode	Cod	1–250 ppm	[33]
Volatile ammines	Organic gas sensor	Tilapia, Beltfish, Mackerel	0.1–1 ppm	[34]
Hypoxanthine Xanthine Uric acid	Graphite electrode	Barracuda, Lady fish, Mackerel, blue catfish, Channel cat fis	6–30 µM 8–36 µM 3–21 µM	[35]
Hypoxanthine Xanthine Uric acid	Pyrolitic graphite electrode	Tuna, Hake, Myleus paku, Silverside	6–30 µM 8–36 µM 3–21 µM	[36]

Researchers conclude a high enzymatic reaction selective, simple and cost-effective H biosensor based on the etching of gold nanorods which has a vivid color change during the reaction of Hypoxanthine (Hx) and dissolved oxygen to form of Xanthine oxide (XOD). A pioneering research group, Watanabe's group, has evaluated these enzyme sensors since the 1980s. One of these researches is the specific enzyme-based sensor for detection of Hx using an XOD-immobilized oxygen probe. During the reaction due to oxygen consumption decreasing the current of the oxygen probe and the probe could be used more than 100 assays [37]. Another sensor system based on a previous study was also studied for fish, seabass, mackerel, yellowfish and etc. with a good correlation [38,39].

An optical biosensor was improved with a chromo-ionic dye based on the enzymatic pH changes which is related to creatine concentration [40]. The optical biosensor showed a high reflectance (16–48 nM) to increased creatine concentration. As biosensors response time is a prominent issue, Fazal et al. [40] reported a 7 min response time biosensor for the detection of amine-based compounds.

Another detection and monitoring capability of biosensors is microbial growth, especially pathogenic detection based on antimicrobial peptides were known [4]. Unfortunately, its time-consuming processes and specialized procedures were limiting theirs *in situ* detection potential with limited applications of fish and seafood industry.

Nowadays, sensor studies are based on usage in combination with different technologies. One of the important examples of enzyme-based ultrasensitive electrochemical biosensors, for XOD immobilization, was created by Yazdanparast et al. [29]. Poly amino acids were used as a biosensing platform on GCE surface and the freshness of fish meat by determination of xanthine was carried out successfully with high sensitivity of 0.35 nM of the detection limit for xanthine. Sensor applications are being developed due to their non-portable properties which limited their applications in industry. Currently, a portable all-in-one enzyme-based biosensor for fish freshness monitoring has been evaluated [41]. The highlight of this biosensor is the reagentless HX detection provided by ceria nanoparticles.

3. Indicators

Indicators provide visual, qualitative or semi-quantitative information about food through color change or dye diffusion on the packaging. Indicators used in smart packaging, described below, are leakage, freshness, and temperature-time indicators [3].

3.1 Freshness indicators

Freshness indicators supply information about biochemical or microbiological changes. They are sensitive to pH, volatile nitrogen compounds, hydrogen sulfide and microbial metabolites. Salgado et al. [4] summarized these indicators as, monitoring the certain biochemicals that are considered quality indicators based on color change. As these quality indicators of fish and seafood products, profoundly volatile amines, biogenic amines, and organic acids, concentrations change generally an observation of a change in color of indicators within the packaging system is monitored. Additionally, freshness indicators can be supply information about the remaining shelf life of fishery products [42].

Freshness indicators have been considered during the last two decades. Volatile amines had been the main subject of these indicators and FreshTag® (COX Technologies, Belmont, NC, USA) which is known and applied in the industry is secured by patents [43,44]. The basic schematics of the freshness indicator were presented in Fig. 13.3. The color change of the indicator actualized gradually.

Chun et al. [45] investigated bromocresol green as a pH-sensitive dye for monitoring the microbial spoilage of Mackerel fillets based on the volatile amines production with RFID (radio frequency identification) technology. After 48 h of storage, the color indicator turned from yellow to blue due to the total volatile amine compounds formed. Using anthocyanin is best known for the color change ability of indicators. Extracts of anthocyanin obtained from sweet purple potatoes were integrated into the carboxy methyl cellulose/starch composite and the pH change, especially due to the release of ammonia, was monitored as the color changed from red to blue-green [46]. As mentioned before anthocyanins were the most frequent color indicator and the supply

Fig. 13.3 Color change distribution of indicators.

of anthocyanins can be made from different sources. In a study with shrimps, *Echium amoenum* anthocyanins were immobilized in bacterial cellulose and used as an indicator sensitive to pH change for a naked eye qualification based on the color change [47].

Based on the fish's volatile amines, another pH-sensitive indicator was prepared with curcumin (CR) and anthocyanins (ANT) with a ratio of 2:8 (v:v). They were successfully immobilized into starch-polyvinyl alcohol-glycerol films and Bighead carp (*Hypophthalmichthys nobilis*) was used for application. The colourimetric film which is sensitive to volatile ammonia was found to be proper for the real-time monitoring of fish freshness [48].

Moradi et al. [49] developed and characterized a pH-sensing indicator, for rainbow trout and common carp fillet, to detect microbial spoilage during refrigerator storage. This intelligent black carrot anthocyanins (CA) embedded bacterial nanocellulose (BC) membrane-based pH-sensing indicator, were found to distinguishably identify the spoilage that can be easily recognized by the naked eye.

3.2 Leakage indicators

Leakage indicators are systems to recognize the presence or absence of gases and provide information about leakages by a color change [50]. They can be in the form of tablets, labels, or prints, or they can be formulated by coating a polymer film. Generally, leak indicators used as oxygen (O_2) and carbon dioxide (CO_2) indicators, were used incorporation with modified atmosphere packaging (MAP). They are based on color reactions indicating the presence of a particular amount of O_2 and CO_2 in the headspace of the package. Examples of patented and commercially available leak indicators are VitalOn, Samso-Checker, Ageless Eye, FreshPax [51,52], Tell-Tab™, O_2Sense™, Novas Insignia Technologies and Shelf Life Guard [53].

3.3 Temperature-time indicators (TTI)

The temperature-time indicators are small measurement tools that provide a visual indication of the temperature history of the food along the distribution chain with irreversible reactions [54,55]. The basis of this indicator is the change of color depending on the pH decrease caused by the breakdown of lipids in foods as a result of enzymatic reactions. In systems where the indicators are used, the control of the entire distribution chain can be carried out effectively. These types of indicators can be placed on food products, parcels or containers. These indicators are divided into 4 groups according to their working principles and being patented in the market:
1. Diffusion based indicators,
2. Enzymatic time and temperature indicators,
3. Time temperature indicators,
4. Time-temperature indicators based on polymers.

Advantages of TTIs compared to other systems are their low-cost and is easily placed in the packaging material. Another possibility offered by TTI is new strategies such as "delivering the product on time or with the shortest shelf life" instead of "first in first out or last in last out". Since the 1980s, TTI systems were studied and proposed with limited commercial applications [42]. The limitation of TTI is the necessity of combined usage with other indicators or sensors to assess the freshness and predict the shelf life of fisheries such as chilled boque [56], seabream [57], tuna [58], turbot [59], grouper [60], and cod [61]. Currently, research focuses on the commercial application of TTI in fishery products (Table 13.2) [62]. The CheckPoint® TTI, which is a polymer based TTI, has been reported as a primary preventive measure tool for vacuum and modified atmosphere-packed fish and fishery products. Polymer based TTIs are activated by temperature and give a warning with the color change on the label as a result of the polymerization of diacetylene monomers. Other TTIs are named as diffusion based (e.g. 3 M Monitor Mark®) and enzymatic based indicators. Effective TTI applications were proposed to be combined with radiofrequency identification (RFID) technology for the accurate traceability of fish and seafood industry [63].

Table 13.2 Commercial TTIs available in fish and seafood industry.

Product name	Applications
3M™ MonitorMark® (www.3m.com)	Meat
	Beverage
	Bakery
Fresh-Check® Temperature Intelligence™ (www.fresh-check.com)	Fresh foods
Insignia Deli Intelligent Labels™ (www.insigniatechnologies.com)	Chilled foods
OnVu™ (www.packworld.com)	Meat
	Fish
	Dairy
WarmMark® (www.deltatrak.com)	Transport, storage, processing
Cold chain iToken™ (www.deltatrak.com)	Supply Chain
TempDot® (www.deltatrak.com)	Seafood
	Meat
Freshtag/Check point® (www.vitsab.com)	Meat
	Fish
	Dairy
TOPCRYO (www.cryolog.com)	Cold chain
Traceo® (www.cryolog.com)	Chilled foods
eO® (www.cryolog.com	Cold chain
Keep-it® (www.keep-it.com)	Fish and other fresh foods

Adapted from E. Drago, R. Campardelli, M. Pettinato, P. Perego, Innovations in smart packaging concepts for food: an extensive review, Foods 9 (2020) 1628.

4. Blockchain systems

Blockchain is a common system used for seafood supply chain management and contains information from the moment of fish capture to the retailer or consumer handling. For this blockchain system, some unique tags can be used, RFID (Radio Frequency Identification), NFC (Near Field Communication) and QR code, to give information and track the products [64]. Ahamed et al. [64] also describe a blockchain system that can be applied to the seafood supply system (Fig. 13.4).

4.1 RFID systems

RFID is a system that uses radio waves to track the food, in areas of increasing food safety, such as supply chain management, food traceability and recalls. In the systems applied, there are smart tags containing RFID on the pallets coming to the warehouse. Through the RFID antennas to be placed at the entrances and exits of the warehouse, all

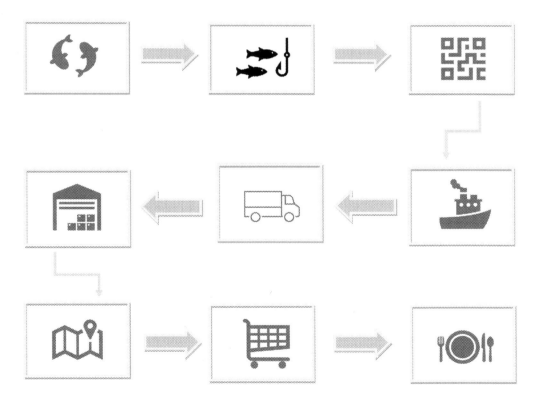

Blockchain: Monitoring Seafood Supply Chain

Fig. 13.4 Blockchain Seafood supply system.

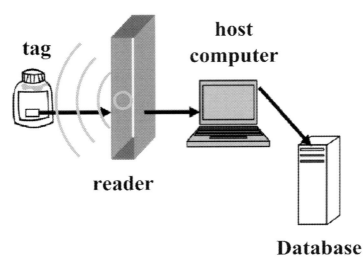

Fig. 13.5 Scheme of RFID working principle.

Table 13.3 Commercial RFID technologies in fish and seafood industry.

Product name	Applications
Easy2Log® (www.environmental-expert.com)	Seafood Meat
CS8304 (www.convergence.com.hk)	Cold chain
TempTRIP (www.temptrip.com)	Cold chain

Adopted from E. Drago, R. Campardelli, M. Pettinato, P. Perego, Innovations in smart packaging concepts for food: an extensive review, Foods 9 (2020) 1628.

information is automatically transferred to the system, together with the reading of the pallets in the warehouse. In this way, it is ensured that the products are on the right shelves and in sufficient quantities, and shipment and placement processes can be carried out on time. With these benefits provided by RFID systems, a great deal of savings can be achieved from costs, labor and time [65].

An application has been made to the changing environmental factors such as light, temperature and humidity in the fish supply chain with RFID smart tags. Working principle of RFID systems is schematized in Fig. 13.5. With the help of RFID tags placed in polystyrene fish boxes, the application of traceability was carried out by measuring the temperature and humidity inside the box from a distance of 10 cm before the boxes were opened [66]. Commercial RFID technologies are presented in Table 13.3.

5. Electronic sensing systems

Artificial odor and taste sensor technology first emerged in 1982 with the invention of the gas multi-sensor array. Innovations in aroma sensor technologies have enabled the development of devices that can measure and characterize volatile aroma compounds. Developed as a simulation of human olfactory system, these devices are known as "electronic noses (e-nose)" and "electronic tongue (e-tongue)" allow the identification and classification of aroma mixtures and can offer reproducible measurements [23,67].

The classification of the components that form the basis of the structure and odor characteristics forms the basis of the e-nose. Quantitative and quality of the volatile components that emerge during the storage of the fish, and the selection of the sensors to be used in the measurement of quality monitoring is important. Sensor types generally used in electronic nose systems are conductive metal-oxide derivatives, polymers, oligomers, and quartz crystal microbalance sensors. When the odor reaches these sensors, they change the electrical conductivity of the active substances of the sensors that cause a changing voltage value. The change in voltage then is measured, and an idea about the intensity and characteristics of the odor can be obtained.

The mechanism established to measure the freshness of fish meat consists of e-nose sensor arrays and data collection card (Data Acquisition System-DAQ or ADC) box, dry air system, a compartment in which fish is placed and usually made of Teflon, mass flow controller (MFC) and signal consists of a computer system for processing [68]. The schematic drawing of the experimental setup of the related study is shown in Fig. 13.6.

Olafsdottir et al. [69], used a series of gas sensor prototypes called FishNose, for the observation of vacuum and modified atmosphere packaged cold-smoked salmon quality

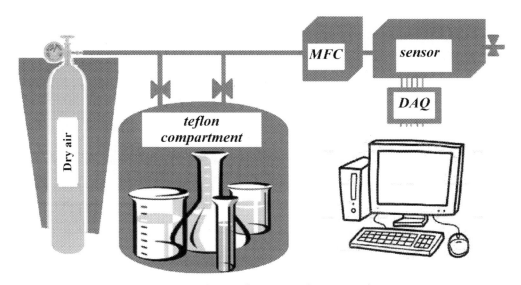

Fig. 13.6 Scheme of e-nose working principle.

changes. Sensorial (sweet/sour, rancid and undesirable) and microbiological qualities were determined and classified according to FishNose results. They reported that the data obtained from the gas sensors and the sensory analysis values are compatible with each other, and therefore the system can be used as a fast and reliable quality control tool in measuring the freshness of products.

Haugen et al. [70] investigated the direct applicability of Fish-Nose for smoked salmon. For this purpose, in the selection of gas sensors, they chose to detect the changes in the volatile components that occur during the deterioration. Researchers reported that FishNose showed positive results in measuring quality changes and could also predict quality-related properties such as microbial load, off-odours, and sweetness/sourness.

The concentrations of the fish odor chemicals are directly incorporated with the degree of degradation. Among long-chain alcohols, carbonyls, bromophenols, amines, sulfur compounds and N-cyclic compounds, amines are considered typical biomarkers for the determination of freshness [71].

E-tongue refers to the identification of taste, which early versions were designed based on the biosensor principle. E-tongue systems can be electrochemical, optical and enzymatic. The e-tongues based on taste patterns are commonly electrochemical and designed as a sensor array with potentiometric, voltametric, amperometric, impedimetric, and conductimetric [72]. Recent applications of e-tongue for fish and seafood were summarized by Zaukuu et al. [67]. Commonly potentiometric type e-tongue was used for tilapia fillets to determine the changes in volatile compounds [73], grass carp to validate the sensory analysis [74], sea bream [75] and crucian carp [76] to determine and predict the microbiological shelf life. Additionally, Nakano et al. [77] used e-tongue with the corporation of e-eye to characterize the fish sauce.

E-tongue and e-nose, as electronic sensor devices, have been used not only for fish and seafood but also for meat and poultry. Besides all these research and validated applications, a single measurement to measure changes during fish spoilage may not be sufficient for an accurate assessment of quality. As both devices have the highlights of being non-waste, cost-effective and highly sensitive, it is profound to overcome the shortcomings. One way of tackling this situation is to develop instruments that measure a variety of attributes for better estimation of quality or freshness, rather than simply measuring a single attribute.

6. Conclusions and potential trends

According to consumer demands and developments in science and technology, emerging technologies have enabled in packaging systems. These innovative technologies called "smart technology" provide a real-time monitoring information chain for various food products in the terms of from farm to fork. As known, fish and seafood products are the most perishable foods that need profound attention throughout the entire food supply chain.

For this purpose, the above-mentioned smart packaging techniques should be adapted to the "farm to fork" approach and should be of the quality and mixture to monitor all stages of fish and seafood and to provide consumers with detailed information about the product they buy. So far, sensors, biosensors and indicators are the most preferred mechanisms in food applications to improve and monitor the quality of the product during shelf-life. Moreover, portable and inexpensive E-sensor systems can be developed for future studies based on machine learning.

Considering how sensitive fish and seafood are, the importance of determining their quality and consumability with fast techniques throughout the supply chain is also understood. Future trends are driving research into non-invasive food quality assessment through compatible devices that can report the status of packaged food products either through labels placed inside the package or via spectroscopic techniques that can remotely query the package contents. It can be concluded that using modified techniques instead of a single technique will be more accurate and increase the efficiency of monitoring and indicating food quality.

References

[1] K. Huff, Active and Intelligent Packaging: Innovations for the Future, Department of Food Science & Technology, Virginia Polytechnic Institute and State University, Blacksburg, Va, 2008, pp. 1–13.

[2] M. Vanderroost, P. Ragaert, F. Devlieghere, B. De Meulenaer, Intelligent food packaging: the next generation, Trends Food Sci. Technol. 39 (2014) 47–62.

[3] J.P. Kerry, 20 - Application of smart packaging systems for conventionally packaged muscle-based food products, in: J.P. Kerry (Ed.), Advances in Meat, Poultry and Seafood Packaging, Woodhead Publishing, 2012, pp. 522–564, https://doi.org/10.1533/9780857095718.4.522.

[4] P.R. Salgado, L. Di Giorgio, Y.S. Musso, A.N. Mauri, Recent developments in smart food packaging focused on biobased and biodegradable polymers, Front. Sustain. Food Syst. 5 (2021). https://www.frontiersin.org/article/10.3389/fsufs.2021.630393 (Accessed 25 March 2022).

[5] H. Yousefi, H.-M. Su, S.M. Imani, K. Alkhaldi, C.D.M. Filipe, T.F. Didar, Intelligent food packaging: a review of smart sensing technologies for monitoring food quality, ACS Sens. 4 (2019) 808–821, https://doi.org/10.1021/acssensors.9b00440.

[6] E.L. Tan, W.N. Ng, R. Shao, B.D. Pereles, K.G. Ong, A. Wireless, Passive sensor for quantifying packaged food quality, Sensors (Basel). 7 (2007) 1747–1756, https://doi.org/10.3390/s7091747.

[7] S. Köse, Evaluation of seafood safety health hazards for traditional fish products: preventive measures and monitoring issues, TrJFAS 10 (2010) 139–160.

[8] A. Pacquit, J. Frisby, D. Diamond, K.T. Lau, A. Farrell, B. Quilty, D. Diamond, Development of a smart packaging for the monitoring of fish spoilage, Food Chem. 102 (2007) 466–470, https://doi.org/10.1016/j.foodchem.2006.05.052.

[9] D. Hofmann, P.-G. Dittrich, C. Gärtner, R. Klemm, Multi-hybrid instrumentations with smartphones and smartpads for innovative in-field and POC diagnostics, in: Microfluidics, BioMEMS, and Medical Microsystems XI, SPIE, 2013, pp. 15–22, https://doi.org/10.1117/12.2005885.

[10] S. Grassi, E. Casiraghi, C. Alamprese, Handheld NIR device: a non-targeted approach to assess authenticity of fish fillets and patties, Food Chem. 243 (2018) 382–388, https://doi.org/10.1016/j.foodchem.2017.09.145.

[11] S. Pons-Sánchez-Cascado, M.C. Vidal-Carou, M.L. Nunes, M.T. Veciana-Nogués, Sensory analysis to assess the freshness of Mediterranean anchovies (*Engraulis encrasicholus*) stored in ice, Food Contr. 17 (2006) 564–569, https://doi.org/10.1016/j.foodcont.2005.02.016.

[12] P.D. Patel, C. Beveridge, 12 - In-line sensors for food process monitoring and control, in: I.E. Tothill (Ed.), Rapid and On-Line Instrumentation for Food Quality Assurance, Woodhead Publishing, 2003, pp. 215–239, https://doi.org/10.1533/9781855737105.2.215.

[13] M. Ghaani, C.A. Cozzolino, G. Castelli, S. Farris, An overview of the intelligent packaging technologies in the food sector, Trends Food Sci. Technol. 51 (2016) 1–11, https://doi.org/10.1016/j.tifs.2016.02.008.

[14] D.B. Papkovsky, M.A. Smiddy, N.Y. Papkovskaia, J.P. Kerry, Nondestructive measurement of oxygen in modified atmosphere packaged hams using a phase-fluorimetric sensor system, J. Food Sci. 67 (2002) 3164–3169, https://doi.org/10.1111/j.1365-2621.2002.tb08877.x.

[15] M.V. Traffano-Schiffo, M. Castro-Giraldez, V. Herrero, R.J. Colom, P.J. Fito, Development of a non-destructive detection system of deep pectoral myopathy in poultry by dielectric spectroscopy, J. Food Eng. 237 (2018) 137–145, https://doi.org/10.1016/j.jfoodeng.2018.05.023.

[16] J.-F. Lin, J. Kukkola, T. Sipola, D. Raut, A. Samikannu, J.-P. Mikkola, M. Mohl, G. Toth, W.-F. Su, T. Laurila, K. Kordas, Trifluoroacetylazobenzene for optical and electrochemical detection of amines, J. Mater. Chem. A. 3 (2015) 4687–4694, https://doi.org/10.1039/C4TA05358C.

[17] M.K. Morsy, K. Zór, N. Kostesha, T.S. Alstrøm, A. Heiskanen, H. El-Tanahi, A. Sharoba, D. Papkovsky, J. Larsen, H. Khalaf, M.H. Jakobsen, J. Emnéus, Development and validation of a colorimetric sensor array for fish spoilage monitoring, Food Contr. 60 (2016) 346–352, https://doi.org/10.1016/j.foodcont.2015.07.038.

[18] B. Kuswandi, Jayus, T.S. Larasati, A. Abdullah, L.Y. Heng, Real-time monitoring of shrimp spoilage using on-package sticker sensor based on natural dye of curcumin, Food Anal. Methods. 5 (2012) 881–889, https://doi.org/10.1007/s12161-011-9326-x.

[19] I. Dudnyk, E.-R. Janeček, J. Vaucher-Joset, F. Stellacci, Edible sensors for meat and seafood freshness, Sens. Actuat. B 259 (2018) 1108–1112, https://doi.org/10.1016/j.snb.2017.12.057.

[20] S.B.H. Hashim, H. Elrasheid Tahir, L. Liu, J. Zhang, X. Zhai, A. Ali Mahdi, F. Nureldin Awad, M.M. Hassan, Z. Xiaobo, S. Jiyong, Intelligent colorimetric pH sensoring packaging films based on sugarcane wax/agar integrated with butterfly pea flower extract for optical tracking of shrimp freshness, Food Chem. 373 (2022) 131514, https://doi.org/10.1016/j.foodchem.2021.131514.

[21] K. Pospiskova, I. Safarik, M. Sebela, G. Kuncova, Magnetic particles–based biosensor for biogenic amines using an optical oxygen sensor as a transducer, Microchim Acta 180 (2013) 311–318, https://doi.org/10.1007/s00604-012-0932-0.

[22] Z. Chen, Y. Lin, X. Ma, L. Guo, B. Qiu, G. Chen, Z. Lin, Multicolor biosensor for fish freshness assessment with the naked eye, Sens. Actuat. B 252 (2017) 201–208, https://doi.org/10.1016/j.snb.2017.06.007.

[23] L. Franceschelli, A. Berardinelli, S. Dabbou, L. Ragni, M. Tartagni, Sensing technology for fish freshness and safety: a review, Sensors 21 (2021) 1373, https://doi.org/10.3390/s21041373.

[24] A.T. Lawal, S.B. Adeloju, Progress and recent advances in fabrication and utilization of hypoxanthine biosensors for meat and fish quality assessment: a review, Talanta 100 (2012) 217–228, https://doi.org/10.1016/j.talanta.2012.07.085.

[25] X. Tang, Y. Liu, H. Hou, T. You, A nonenzymatic sensor for xanthine based on electrospun carbonn nanofibers modified electrode, Talanta 83 (2011) 1410–1414, https://doi.org/10.1016/j.talanta.2010.11.019.

[26] K. Thandavan, S. Gandhi, S. Sethuraman, J.B.B. Rayappan, U.M. Krishnan, Development of electrochemical biosensor with nano-interface for xanthine sensing – a novel approach for fish freshness estimation, Food Chem. 139 (2013) 963–969, https://doi.org/10.1016/j.foodchem.2013.02.008.

[27] J. Narang, N. Malhotra, C. Singhal, C.S. Pundir, Evaluation of freshness of fishes using MWCNT/TiO$_2$ nanobiocomposites based biosensor, Food Anal. Methods. 10 (2017) 522–528, https://doi.org/10.1007/s12161-016-0594-3.

[28] B. Borisova, A. Sánchez, S. Jiménez-Falcao, M. Martín, P. Salazar, C. Parrado, J.M. Pingarrón, R. Villalonga, Reduced graphene oxide-carboxymethylcellulose layered with platinum nanoparticles/PAMAM dendrimer/magnetic nanoparticles hybrids. Application to the preparation of enzyme electrochemical biosensors, Sens. Actuat. B 232 (2016) 84–90, https://doi.org/10.1016/j.snb.2016.02.106.

[29] S. Yazdanparast, A. Benvidi, S. Abbasi, M. Rezaeinasab, Enzyme-based ultrasensitive electrochemical biosensor using poly (l-aspartic acid)/MWCNT bio-nanocomposite for xanthine detection: a meat freshness marker, Microchem. J. 149 (2019) 104000, https://doi.org/10.1016/j.microc.2019.104000.

[30] I.M. Apetrei, C. Apetrei, Amperometric biosensor based on diamine oxidase/platinum nanoparticles/graphene/chitosan modified screen-printed carbon electrode for histamine detection, Sensors 16 (2016) 422, https://doi.org/10.3390/s16040422.

[31] R. Torre, E. Costa-Rama, H.P.A. Nouws, C. Delerue-Matos, Diamine oxidase-modified screen-printed electrode for the redox-mediated determination of histamine, J. Anal. Sci. Technol. 11 (2020) 5, https://doi.org/10.1186/s40543-020-0203-3.

[32] J. Chen, Y. Lu, F. Yan, Y. Wu, D. Huang, Z. Weng, A fluorescent biosensor based on catalytic activity of platinum nanoparticles for freshness evaluation of aquatic products, Food Chem. 310 (2020) 125922, https://doi.org/10.1016/j.foodchem.2019.125922.

[33] J.K. Heising, M. Dekker, P.V. Bartels, M.A.J.S. van Boekel, A non-destructive ammonium detection method as indicator for freshness for packed fish: application on cod, J. Food Eng. 110 (2012) 254–261, https://doi.org/10.1016/j.jfoodeng.2011.05.008.

[34] L.-Y. Chang, M.-Y. Chuang, H.-W. Zan, H.-F. Meng, C.-J. Lu, P.-H. Yeh, J.-N. Chen, One-minute fish freshness evaluation by testing the volatile amine gas with an ultrasensitive porous-electrode-capped organic gas sensor system, ACS Sens 2 (2017) 531–539, https://doi.org/10.1021/acssensors.6b00829.

[35] N. Vishnu, M. Gandhi, D. Rajagopal, A.S. Kumar, Pencil graphite as an elegant electrochemical sensor for separation-free and simultaneous sensing of hypoxanthine, xanthine and uric acid in fish samples, Anal. Methods. 9 (2017) 2265–2274, https://doi.org/10.1039/C7AY00445A.

[36] G.D. Pierini, S.N. Robledo, M.A. Zon, M.S. Di Nezio, A.M. Granero, H. Fernández, Development of an electroanalytical method to control quality in fish samples based on an edge plane pyrolytic graphite electrode. Simultaneous determination of hypoxanthine, xanthine and uric acid, Microchem. J. 138 (2018) 58–64, https://doi.org/10.1016/j.microc.2017.12.025.

[37] E. Watanabe, K. Ando, I. Karube, H. Matsuoka, S. Suzuki, Determination of hypoxanthine in fish meat with an enzyme sensor, J. Food Sci. 48 (1983) 496–500, https://doi.org/10.1111/j.1365-2621.1983.tb10775.x.

[38] E. Watanabe, K. Toyama, I. Karube, H. Matsuoka, S. Suzuki, Determination of inosine-5-monophosphate in fish tissue with an enzyme sensor, J. Food Sci. 49 (1984) 114–116, https://doi.org/10.1111/j.1365-2621.1984.tb13684.x.

[39] E. Watanabe, K. Toyama, I. Karube, H. Matsuoka, S. Suzuki, Enzyme sensor for hypoxanthine and inosine determination in edible fish, Appl. Microbiol. Biotechnol. 19 (1984) 18–22, https://doi.org/10.1007/BF00252811.

[40] F.F. Fazial, L.L. Tan, S.I. Zubairi, Bienzymatic creatine biosensor based on reflectance measurement for real-time monitoring of fish freshness, Sens. Actuat. B 269 (2018) 36–45, https://doi.org/10.1016/j.snb.2018.04.141.

[41] F. Mustafa, A. Othman, S. Andreescu, Cerium oxide-based hypoxanthine biosensor for Fish spoilage monitoring, Sens. Actuat. B 332 (2021) 129435, https://doi.org/10.1016/j.snb.2021.129435.

[42] T.N. Tsironi, P.S. Taoukis, Current practice and innovations in fish packaging, J. Aquat. Food Product Technol. 27 (2018) 1024–1047, https://doi.org/10.1080/10498850.2018.1532479.

[43] J. Williams, K. Myers, M. Owens, M. Bonne, Food Quality Indicator, Google Patents, 2006.

[44] J. Williams, K. Myers, Compositions for Detecting Food Spoilage and Related Methods, US20050153452A1, 2005, https://patents.google.com/patent/US20050153452A1/en (Accessed 27 March 2022).

[45] H.-N. Chun, B. Kim, H.-S. Shin, Evaluation of a freshness indicator for quality of fish products during storage, Food Sci. Biotechnol. 23 (2014) 1719–1725, https://doi.org/10.1007/s10068-014-0235-9.

[46] G. Jiang, X. Hou, X. Zeng, C. Zhang, H. Wu, G. Shen, S. Li, Q. Luo, M. Li, X. Liu, A. Chen, Z. Wang, Z. Zhang, Preparation and characterization of indicator films from carboxymethyl-cellulose/starch and purple sweet potato (*Ipomoea batatas* (L.) lam) anthocyanins for monitoring fish freshness, Int. J. Biol. Macromol. 143 (2020) 359–372, https://doi.org/10.1016/j.ijbiomac.2019.12.024.

[47] S. Mohammadalinejhad, H. Almasi, M. Moradi, Immobilization of *Echium amoenum* anthocyanins into bacterial cellulose film: a novel colorimetric pH indicator for freshness/spoilage monitoring of shrimp, Food Control 113 (2020) 107169, https://doi.org/10.1016/j.foodcont.2020.107169.

[48] H. Chen, M. Zhang, B. Bhandari, C. Yang, Novel pH-sensitive films containing curcumin and anthocyanins to monitor fish freshness, Food Hydrocoll. 100 (2020) 105438, https://doi.org/10.1016/j.foodhyd.2019.105438.

[49] M. Moradi, H. Tajik, H. Almasi, M. Forough, P. Ezati, A novel pH-sensing indicator based on bacterial cellulose nanofibers and black carrot anthocyanins for monitoring fish freshness, Carbohydrate Polym. 222 (2019) 115030, https://doi.org/10.1016/j.carbpol.2019.115030.

[50] J.K. Heising, M. Dekker, P.V. Bartels, M.A.J.S. (Tiny) Van Boekel, Monitoring the quality of perishable foods: opportunities for intelligent packaging, Crit. Rev. Food Sci. Nutr. 54 (2014) 645–654, https://doi.org/10.1080/10408398.2011.600477.

[51] J.P. Kerry, M.N. O'Grady, S.A. Hogan, Past, current and potential utilisation of active and intelligent packaging systems for meat and muscle-based products: a review, Meat Sci. 74 (2006) 113–130, https://doi.org/10.1016/j.meatsci.2006.04.024.

[52] S.A. Hogan, J.P. Kerry, Smart Packaging of Meat and Poultry Products, Smart Packaging Technologies for Fast Moving Consumer Goods, 2008, pp. 33–54.

[53] E. Drago, R. Campardelli, M. Pettinato, P. Perego, Innovations in smart packaging concepts for food: an extensive review, Foods 9 (2020) 1628.

[54] M. Serdaroğlu, Ç. Purma, Su Ürünlerinde Kalitenin saptanmasında Kullanılan hızlı teknikler, Ege J. Fish. Aquat. Sci. 23 (2006).

[55] Ç. Purma, M. Serdaroğlu, Akıllı ambalajlama sistemlerinin gıda sanayinde kullanımı, Türkiye. 9 (2006) 24–26.

[56] P.S. Taoukis, K. Koutsoumanis, G.J.E. Nychas, Use of time–temperature integrators and predictive modelling for shelf-life control of chilled fish under dynamic storage conditions, Int. J. Food Microbiol. 53 (1999) 21–31.

[57] M.C. Giannakourou, K. Koutsoumanis, G.J.E. Nychas, P.S. Taoukis, Field evaluation of the application of time temperature integrators for monitoring fish quality in the chill chain, Int. J. Food Microbiol. 102 (2005) 323–336, https://doi.org/10.1016/j.ijfoodmicro.2004.11.037.

[58] T. Tsironi, E. Gogou, E. Velliou, P.S. Taoukis, Application and validation of the TTI based chill chain management system SMAS (Safety Monitoring and Assurance System) on shelf-life optimization of vacuumpacked chilled tuna, Int. J. Food Microbiol. 128 (2008) 108–115, https://doi.org/10.1016/j.ijfoodmicro.2008.07.025.

[59] M. Nuin, B. Alfaro, Z. Cruz, N. Argarate, S. George, Y. Le Marc, J. Olley, C. Pin, Modelling spoilage of fresh turbot and evaluation of a time–temperature integrator (TTI) label under fluctuating temperature, Int. J. Food Microbiol. 127 (2008) 193–199, https://doi.org/10.1016/j.ijfoodmicro.2008.04.010.

[60] H.-I. Hsiao, J.-N. Chang, Developing a microbial time–temperature indicator to monitor total volatile basic nitrogen change in chilled vacuum-packed grouper fillets, J. Food Process. Preserv. 41 (2017) e13158, https://doi.org/10.1111/jfpp.13158.

[61] N.T.T. Mai, M. Gudjónsdóttir, H.L. Lauzon, K. Sveinsdóttir, E. Martinsdóttir, H. Audorff, W. Reichstein, D. Haarer, S.G. Bogason, S. Arason, Continuous quality and shelf life monitoring of retail-packed fresh cod loins in comparison with conventional methods, Food Control 22 (2011) 1000–1007, https://doi.org/10.1016/j.foodcont.2010.12.010.

[62] D. Wu, M. Zhang, H. Chen, B. Bhandari, Freshness monitoring technology of fish products in intelligent packaging, Crit. Rev. Food Sci. Nutr. 61 (2021) 1279–1292, https://doi.org/10.1080/10408398.2020.1757615.

[63] P.S. Taoukis, 14 - Commercialization of time-temperature integrators for foods, in: C.J. Doona, K. Kustin, F.E. Feeherry (Eds.), Case Studies in Novel Food Processing Technologies, Woodhead Publishing, 2010, pp. 351–366, https://doi.org/10.1533/9780857090713.3.351.

[64] N.N. Ahamed, P. Karthikeyan, S.P. Anandaraj, R. Vignesh, Sea food supply chain management using blockchain, in: 2020 6th International Conference on Advanced Computing and Communication Systems (ICACCS), 2020, pp. 473–476, https://doi.org/10.1109/ICACCS48705.2020.9074473.

[65] F. Bibi, C. Guillaume, N. Gontard, B. Sorli, A review: RFID technology having sensing aptitudes for food industry and their contribution to tracking and monitoring of food products, Trends Food Sci. Technol. 62 (2017) 91–103, https://doi.org/10.1016/j.tifs.2017.01.013.

[66] E. Abad, F. Palacio, M. Nuin, A.G. de Zárate, A. Juarros, J.M. Gómez, S. Marco, RFID smart tag for traceability and cold chain monitoring of foods: demonstration in an intercontinental fresh fish logistic chain, J. Food Eng. 93 (2009) 394–399, https://doi.org/10.1016/j.jfoodeng.2009.02.004.

[67] J.L.Z. Zaukuu, G. Bazar, Z. Gillay, Z. Kovacs, Emerging trends of advanced sensor based instruments for meat, poultry and fish quality— a review, Crit. Rev. Food Sci. Nutr. 60 (2020) 3443–3460, https://doi.org/10.1080/10408398.2019.1691972.

[68] I. Ahmed, H. Lin, L. Zou, Z. Li, A.L. Brody, I.M. Qazi, L. Lv, T.R. Pavase, M.U. Khan, S. Khan, L. Sun, An overview of smart packaging technologies for monitoring safety and quality of meat and meat products, Packag. Technol. Sci. 31 (2018) 449–471, https://doi.org/10.1002/pts.2380.

[69] G. Olafsdottir, E. Chanie, F. Westad, R. Jonsdottir, C.R. Thalmann, S. Bazzo, S. Labreche, P. Marcq, F. Lundby, J.E. Haugen, Prediction of microbial and sensory quality of cold smoked Atlantic salmon (*Salmo salar*) by electronic nose, J. Food Sci. 70 (2005) S563–S574.

[70] J.E. Haugen, E. Chanie, F. Westad, R. Jonsdottir, S. Bazzo, S. Labreche, P. Marcq, F. Lundby, G. Olafsdottir, Rapid control of smoked Atlantic salmon (*Salmo salar*) quality by electronic nose: correlation with classical evaluation methods, Sens. Actuat. B 116 (2006) 72–77, https://doi.org/10.1016/j.snb.2005.12.064.

[71] A. Alimelli, G. Pennazza, M. Santonico, R. Paolesse, D. Filippini, A. D'Amico, I. Lundström, C. Di Natale, Fish freshness detection by a computer screen photoassisted based gas sensor array, Analytica Chim. Acta 582 (2007) 320–328, https://doi.org/10.1016/j.aca.2006.09.046.

[72] A. Mimendia, J.M. Gutiérrez, L. Leija, P.R. Hernández, L. Favari, R. Muñoz, M. del Valle, A review of the use of the potentiometric electronic tongue in the monitoring of environmental systems, Environ. Model. Soft. 25 (2010) 1023–1030, https://doi.org/10.1016/j.envsoft.2009.12.003.

[73] C. Shi, X. Yang, S. Han, B. Fan, Z. Zhao, X. Wu, J. Qian, Nondestructive prediction of Tilapia fillet freshness during storage at different temperatures by integrating an electronic nose and tongue with radial basis function neural networks, Food Bioprocess. Technol. 11 (2018) 1840–1852, https://doi.org/10.1007/s11947-018-2148-8.

[74] P. Pattarapon, M. Zhang, B. Bhandari, Z. Gao, Effect of vacuum storage on the freshness of grass carp (*Ctenopharyngodon idella*) fillet based on normal and electronic sensory measurement, J. Food Process. Preserv. 42 (2018) e13418, https://doi.org/10.1111/jfpp.13418.

[75] L. Gil, E. Garcia-Breijo, J.M. Barat, R. Martinez-Manez, J. Soto, F. Baena, J. Ibanez, E. Llobet, J. Brezmes, Analysis of fish freshness by using metallic potentiometric electrodes, in: 2007 IEEE International Symposium on Industrial Electronics, IEEE, 2007, pp. 1485–1490.

[76] F. Han, X. Huang, E. Teye, H. Gu, Quantitative analysis of fish microbiological quality using electronic tongue coupled with nonlinear pattern recognition algorithms, J. Food Safety 35 (2015) 336–344, https://doi.org/10.1111/jfs.12180.

[77] M. Nakano, Y. Sagane, R. Koizumi, Y. Nakazawa, M. Yamazaki, K. Ikehama, K. Yoshida, T. Watanabe, K. Takano, H. Sato, Clustering of commercial fish sauce products based on an e-panel technique, Data Brief 16 (2018) 515–520.

CHAPTER 14

Smart packaging for medicinal food supplements

Vipul Prajapati and Salona Roy
Department of Pharmaceutics, SSR College of Pharmacy, Affiliated to Savitribai Pule Pune University, Silvassa, Union Territory of Dadra Nagar Haveli and Daman Diu, India

1. Introduction

The primary goal of any package for the product is to prevent the product from any deterioration due to outside exposure and use. Additionally, product packaging might be a marketing technique for reaching out to the consumers. It offers the range of sizes and forms, and as an interface, it enables users ease and elegance. The most important roles of the product packaging include protection, communication, easiness of use, and confinement. Packing generally serves the functions such as to restrict the product from spilling or leaking, as well as from being contaminated. Packing also gives an important information about medicinal products, such as its nutritional value and preparation directions, making things easier for clients, to provide shipping and handling containment [1,2].

In light of evolving consumer demands, complex products, and most recently, legislative initiatives aimed at promoting a circular economy, and reduction in manufacturing and production of carbon footprints. Traditional packaging is now no longer acceptable. Innovation in the types of package with increased usability is also necessary to meet the client's expectations. Conventional and smart packaging also provide legal protection for many goods such as the foods with or without preservatives, the products that adhere to more stringent regulatory requirements, and the medicinal products that permits continuous monitoring. Smart packaging, in the context of globalization, not only aids in the acceptance of stricter international and domestic food safety regulations, but it also guards against the threat of food-borne biological weapons [3,4].

The words "smart packaging" and "active packaging," both are used interchangeably, have all been mentioned in various literatures. Food, drinks, pharmaceuticals, cosmetics, and a range of other perishable things are all packaged in these smart/active systems. There is a distinction to be made between an intelligent, smart, and active packaging in a strict sense. There is need of further advancement in the packaging system using special and novel ingredients in packaging materials and newer technologies to improve the product characteristics like its quality and shelf life [3,5].

2. Medicinal food products and its types

According to the definition of food safety and standards, special medical food is the food specially prepared for a particular diet or food intended for the management of a patient's diet. They are therefore only used for either on medical advice or foods requiring separate/partial feeding of people with limited, impaired ability to absorb, digest, and metabolize conventional foods or certain nutrients or metabolites. Medical foods may be divided into two categories namely dietary supplements and medicinal items. The difference between both the terms is mentioned in a Table 14.1 [6,7] and Fig. 14.1 indicates the step of identification of a medicinal product.

3. Concept of traditional and smart packaging

Traditionally, any types of packaging for pharmaceutical items are used to enhance quality, safety, stability of product and to make product's essential information available to the customers for safe use. Instead of standard packaging, active packaging might be utilized as a first step. This innovative method of food and medicinal supplements package was created in response to the need of consumer's preference and industry trends. In order to preserve food additives and to improve their shelf life, active packaging technology incorporates the components inside the package that may absorb or receive chemicals from the surrounding environment [11,12].

A full packaging system that monitors and responds to change in a product or its surroundings is known as smart packaging. Chemical or biological sensors are employed in the smart packaging system regarding monitoring the quality and safety of foods, medicinal supplements and to aware the consumer if any changes are observed. Similar to the technology stated before, smart packaging employs various sensors to track the quality and safety of food with the goal of identifying and analyzing pathogens, leaks, pH levels, time, and temperature. Specific smart packaging technologies have different functionality depending on the products being wrapped. Likewise, the precise state to be observed, communicated, or modified varies. Pictorial diagram of the smart packaging for medicinal food supplements on the basis of active and intelligent packaging is depicted in a Fig. 14.2. With aim to notify the product's manufacturer, merchant, or consumer of the product's current form, smart packaging enables the asset monitoring of a product through time as well as the analysis and management of the environment within and outside the box [13].

4. Necessity of the smart packaging

The fundamental objective of packing is to keep a product safe from degradation due to environmental exposure and utilization. Furthermore, the product packaging may be

Table 14.1 The difference between medicinal products and food supplements as per EU and MHRA.

Sr. No.	Food supplements	Medicinal products	References
1.	Food supplements are foods that are meant to augment a healthy person's regular diet.	Medical items have the ability to prevent, treat, or even cure human diseases and are designed for use in or on human bodies.	[6]
2.	For any kind of violation or biasness of the law, the general food law is only applicable.	For any kind of violation or biasness of the law, the pharmaceutical legislation is applicable.	[7]
3.	There is no need for preliminary regulatory verification of effectiveness or safety. The product's safety is the responsibility of the business owner. The federal states are the competent authorities in charge of ensuring that products comply with food law rules.	During the approval process, manufacturers must submit clinical research demonstrating the medication's effectiveness and safety.	[7]
4.	The amount on the package may differ by up to 50% from the actual quantity of the product.	The amounts on the label must not differ by more than 5% from the actual dose of the active components.	[6,7]
5.	Vitamins, minerals, and other compounds usually have no precise requirements or maximum values.	Within the framework of the authorization procedure, all ingredient dosages are evaluated and defined.	[6,7]
6.	The goal of the law is to provide excellent consumer protection while promoting consumer choice when it comes to adding vitamins, minerals, and other ingredients to the diet. The directive was also approved to promote commerce and remove barriers to the application of the free movement of goods concept.	The three main goals of the law were to create a well-functioning pharmaceutical industry internal market, provide a high degree of public health protection for EU residents, and streamline and standardize the drug approval process.	[8]
7.	The advertisement criteria for dietary food are prescribed food information to consumers regulation (FIC) of MHRA.	The advertisement for medicinal product is prescribed broadly in regulation 7 of MHRA to highlight qualities of medicinal product.	[9]

Continued

Table 14.1 The difference between medicinal products and food supplements as per EU and MHRA.—cont'd

Sr. No.	Food supplements	Medicinal products	References
8.	Food supplements must not be promoted with claims that they are MHRA approved on the labeling.	Medicinal products must be promoted with claims that they are MHRA approved on the labeling.	[9]
9.	Green tea extract classified as food supplements under EU guidelines.	Green tea extracts classified as medicinal products under USA guidelines.	[10]
10.	Food supplements can be regarded specifically as iron supplements or herbal supplements on the product label.	Medicinal product cannot be regarded from specific terms on product label.	[10]

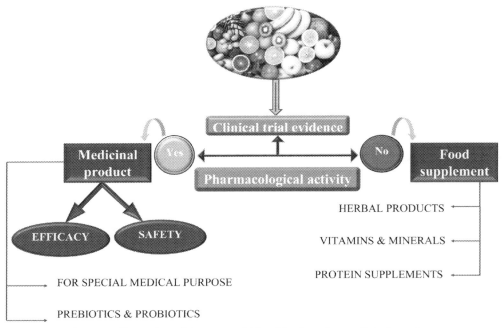

Fig. 14.1 Illustration of the steps of identification of a medicinal product.

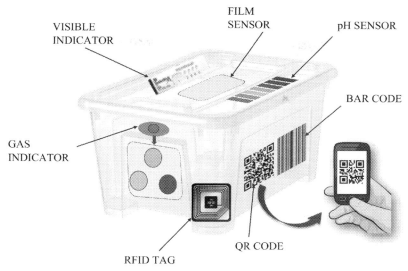

Fig. 14.2 Pictorial presentation of smart packaging.

utilized as a marketing tool to interact or communicate with buyers. It is available in a variety of materials of different size, shapes, and it gives consumers with both simplicity of use and convenience as a user interface. The major roles of packaging of items are protection, communication, convenience and confinement. For illustration, medicinal food packaging aims at preventing the commodity from rupturing or busting, as well as to guard it from pathogens; to connect vital information of the enclosed food product, like nutritive value and intake data; and to provide longevity, such as extending the medicinal product's shelf life for long-term use.

Traditional packaging system is no longer acceptable in light of rising customer expectations, a rise in number of complex products, and currently, national and international legislation aimed at promoting resource efficiency and lowering the carbon footprints of produced goods. To meet a range of new client demands, innovative packaging with improved usability is also necessary. These are just a few examples: providing odour-free food, products that satisfy tougher regulatory criteria, and packaging that enables for start to finish tracking and hence, protects against liability. Furthermore, smart packaging aids in the acceptance of food safety rules that are more stringent on a national and international level, as well as protecting against the threat of biochemical warfare in the food supply in the context of globalization [13].

5. Smart packages

Active/Smart package uses the unique characteristics of composites or incorporating specialized ingredients into them to determine improvements in safety, quality, information gathering and provision of information on food, medicines or nutraceutical issues. Smart packages, on the other hand, use intelligent means such as detection devices, sensors and tracking devices to spread information on pharmaceutical and food standards and to make choices on the decisions to improve overall safety, quality, data collection, and information on products' issues. The surface of the packing film may be coated with the active chemical or its multilayer construction may acquire the coating. Additionally, labels, bottle caps, and miniature packets all contain the active component. Phytochemical constituents, chemical compounds, enzymes, antibacterial and antifungal chemicals, and other active substances come in a variety of forms and sizes. Fig. 14.3 depicts the full classification of active, intelligent types packaging as the smart packaging system for medicinal food supplements [14].

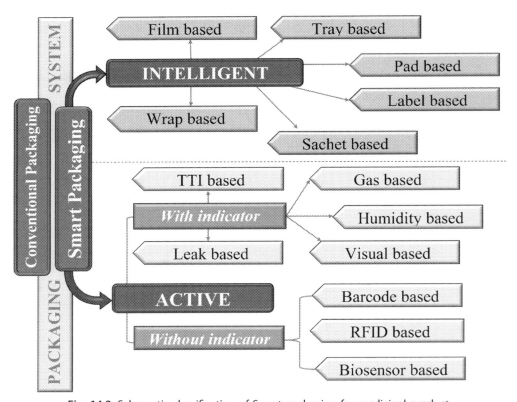

Fig. 14.3 Schematic classification of Smart packaging for medicinal product.

5.1 Active packaging

Over the last few decades, there has been a significant technical advancement in food packaging to meet customer desires for better natural preservation methods and techniques to manage packing and food warehousing to ensure food safety. It is merely a phenomenon of the current era and may be characterized as packaging that modifies the properties of the container while balancing product quality, enhancing shelf-life, and improving safety or data related to it. The objective of active packaging is to have a purpose rather than just serving as a non-active mediator between the inner product and the outer environment. The main objectives are like increasing product quality and acceptance by using potential food-container interactions. Active packaging (AP) for foods is a multifaceted notion that encompasses a wide variety of options for extending shelf life and making food preparation and consumption easier [15].

AP is described as a packaging system that intentionally integrates components which have potential either releasing or absorbing chemicals in to the food package food or the environment close to the food packet, according to the European Regulations (EC). Using AP instead of adding active ingredients such as antibacterial and antioxidants directly to the product can reduce the amount of these chemicals required. The use of AP in fresh foods lowers the levels of active chemicals, reduces local activity and particle transport from the film to the food and creates unwanted industrial activity that can introduce bacteria into the product. Common elements of AP systems include scavengers like ethylene, oxygen, taste and odor absorbers or releasers, antibacterial agents, and antioxidants [13,16–18].

5.1.1 Classification of scavengers, emitters for active packaging

Absorbents remove disagreeable constituents from food product or the environment, such as water, carbon dioxide, and scents. Emitters also add substances to pack product or extra space, like antibacterial substances, carbon dioxide, antioxidants, tastes, ethylene, or ethanol. The detailed data of commercial scavengers and emitters are mentioned in a Table 14.2 [16].

5.1.1.1 Oxygen scavengers

Oxygen may be present in addition to the packing material's oxygen permeability, in free air in the food and in packaging compounds, due to minor leaks caused by inadequate drainage and closure. Oxygen can also change organoleptic qualities while also promote the bacterial development. Oxidation of iron powder, ascorbic acid as well as enzymes and other mechanisms are involved in oxygen scavenging. Normally, these components are packed in a sachet. In dairy and fermented goods, oxygen scavenging is an efficient approach to limit the growth of erobic bacteria and molds. For this, oxygen concentrations in the headspace must be less than 0.1% v/v. Instead of oxygen scavengers, CO_2 and N_2

Table 14.2 Commercial products available in the market for the oxygen, carbon dioxide, ethylene and ethanol scavengers.

Trade name	Scavengers mechanism	Packaging form	Manufacturer	Country	References
Oxygen scavenging commercials					
Ageless	On the basis of iron	Sachets & labels	Mitsubishi gas chemical company limited	Japan	[19]
Seagul	Iron based		Nippon soda co. Ltd.		[19]
ATCO	On the basis of iron	Labels	EMCO packaging systems	UK	[19]
Shelf plus O_2	PET co-polyester	Plastic type film	Ciba speciality chemical	Switzerland	[19]
Freshilizer	On the basis of iron	Sachets	Toppan printing company limited	Japan	[20]
Vitalon			Toagosei chem. Industry company limited		[20]
Sanso-cut			Finetec company limited		[21]
ActiTUF	On the basis of iron	Polyester bottles	M&G	Italy	[21]
Oxyguard	On the basis of iron	Plastic trays	Toyo seikan kaisha limited	Japan	[22]
Oxyeater		Sachets and labels	Ueno seiyaku company limited		[22]
FreshMax FreshPax & fresh pack	On the basis of iron	Labels	Multi-sorb technologies, Inc.	USA	[23]
—	Benzyl Acrylate	Plastic film	Chevron chemicals		[23]
ZerO$_2$	Photosensitive Dye	Plastic films	Food science	Australia	[23]
Pure seal	Ascorbate/Salts of Metal	Bottle crowns	W.R., grace company limited	USA	[24]
Carbon dioxide scavenging commercials					
CO$_2$ pads	Sodium Bicarbonate and Citric Acid	Pad	CO$_2$ technologies	USA	[25]
Super-fresh		Sachet			[25]
Ethylene scavenging commercials					
Evert-fresh	Activated Zeolite	Plastic type film	Evert fresh corporation	USA	[17]
BO film	Crysburite Ceramic	Plastic type film	Odja shoji company limited	Japan	[17]
Halofresh	Active Carbon	Paper/board	Honshu paper limited		
	Tetrazine Derivatives	Plastic film	Food science	Australia	[19]
	Potassium Permanganate	Sachets/blankets	Air repair products, Inc Extenda life systems	USA	[19]

Bio-Kleen	Titanium Dioxide Catalyst	—	Kes irrigations systems	[19]
	Potassium Permanganate	Sachets/blankets	Ethylene control, Inc.	[24]
Neupalon	Activated Carbon	Sachet	Sekisui jushi limited	Japan	[24]
PEAKfresh	Activated Clays/Zeolites	Plastic type film	Peak fresh products limited	Australia	[24]
Bio-fresh			Grofit plastics	Israel	[24]
Sendo-mate	Activated Carbon	Sachet	Mitsubishi gas chemical company limited	Japan	[26]
Orega	Activated Clays/Zeolites	Plastic type film	Cho Yang Heung San company limited	South Korea	[27]

Ethanol scavenging commercials

Ageless	Ethanol	Encapsulated	Mitsubishi gas chemical company limited	Japan	[28]

have been used for the same purpose but they have not proven to be completely effective because the anaerobic environment is not solely responsible for preventing the growth of psychrotropic pathogens like *Listeria monocytogenes* and anaerobic bacteria like *Clostridium botulinium*. The O_2 scavengers can be incorporated into the packaging framework itself as a substitute to sachets. This decreases the negative consumer reactions while also potentially offering a financial gain through enhanced output. It also eliminates the possibility of the sachets unintentionally exploding and the contents being mistakenly ingested.

Scavenging oxygen is beneficial for items that are labile to oxygen and light. The significant merit of AP on top of Modified Atmosphere Packaging (MAP) is the significant reduction in the capital investment requirement. In certain circumstances, all that is required is the closure of the apparatus containing the sachet of the oxygen absorber. This is especially important for small and mediocre sized medicinal food businesses, as packing equipment is sometimes the most expensive component. Chemically synthesized, Ultraviolet light activated oxygen scavengers that may be tailored to fit into deeper components of packaging technologies. The closing lining is directly used in oxygen scavenging technology.

Oxygen scavengers have opened up entirely new perspectives and possibilities when it comes to maintaining nutrient levels and extending shelf life. Detailed information on how the O_2 scavenger works in a variety of situations is awaited before developing the optimal, safe and cost-effective package. This is especially true for recently marketed oxygen gas absorbent films, labels, sheets, sachets, and trays [29].

5.1.1.2 Ethylene scavengers

The hormone ethylene accelerates maturity, causes blooming, encourages quick softening, enhances photosynthetic breakdown, also reduces the shelf life of unprocessed, fresh fruits and vegetables. Some absorbents contain potassium permanganate ($KMnO_4$), but they are toxic and are not directly incorporated into foods. It is commonly used in sachets within the product containers. Diffusion through the sachet material is not a limiting issue because it is very permeable to ethylene. Potassium permanganate converts hydrocarbons to acetate and alcohol. On substrates including alumina, silica gel, vermiculate, and activated carbon, like potassium permanganate-based scavengers normally contain 4% to 6% of potassium permanganate.

The capacity and effectiveness of ethylene scavengers to scavenge ethylene is largely reliant on the absorber plate area and potassium permanganate reagent concentration. Another ethylene scavenger is activated carbon loaded with inorganic compounds of the bromine type. The combination of carbon and bromine is mixed into a pouch or corrugated box that contains high-quality pharmaceuticals. Other approaches to ethylene removal rely on the ability of uniformly suspended minerals such as zeolites, directly activated carbon, and pumice to remove ethylene. These inorganic substances can be found in a wide variety of items, including the polyethylene bags used to wrap organic fruits and food-stuffs.

While the inserted minerals can absorb ethylene, they can also alter the film's fragility, making it more permeable to ethylene, CO_2, and oxygen than if it were formed entirely of pure polyethylene. These effects, which are not dependent on ethylene absorption, can increase expiration life and reduce ethylene concentrations in the headspace [30].

5.1.1.3 Ethanol scavengers
Ethanol is commonly employed in medical and pharmaceutical packaging showing that it possesses vapor phase inhibitor properties. Intermediate moisture foods (IMFs) and fermented items are all protected against microbial deterioration. It also slows down the oxidative and staling processes. In such a sachet, fresh product ethanol is contained in a fine, inert powder. To regulate the rate of ethanol mist discharge, the susceptibility of the sachet may be adjusted. The use of an ethanol sachet has been shown to suppress yeast growth and lengthen the product's lifespan.

Ethanol vapor emission has been shown to be effective against eleven mold species, including fungal species as well as several bacterial species which include common gut bacteria and putrid bacillus species. Various experiments undertaken for the expiration life extension of a fermented products vulnerable to secondary deterioration by yeast, revealed ethanol vapor to be effective. Ethanol vapor is emitted into the container airspace when a sachet is placed adjacent to food. Ethanol vapor is released in concentrations ranging in few percent in v/v concentration, solidifies on the product surface, and functions as a bacterial inhibitor or antibacterial agent [15].

5.1.1.4 Odor or taint scavengers
A few packing materials have been utilized commercially to eliminate unwanted flavor or scent components of the product. The use of cellulose triacetate in fruit juice such as in case of orange juice packaging might help to eliminate bitter taste components like limonin that accumulate during storage and/or pasteurization. Additional bitter component found throughout most tropical fruit juices is naringin. Another possible use is the elimination of amines produced by the degradation of protein in non-vegetarian medicinal products. Strongly basic molecules are formed, making them sensitive to solvent like citric acid and vitamin C. However, there was little evidence to substantiate the usefulness of the concept or the products safety. During the production and preparation of some packaged goods or products, aromas created by the growth of mercaptanes and hydrogen sulfide are absorbed by various commercial odor scavenger sachets [30].

5.1.1.5 Moisture scavengers
The amount of too much humidity in the food products promotes bacterial development as well as smudging of the polythene bag. It also softens dry, crunchy objects and moistens hygroscopic items. The need for a vapor monitoring packing material can keep the relative humidity (RH) within a product at a controlled level, avoiding condensation and, as a result, limiting mold growth in the storage environment. Excess moisture in packaging

can be hazardous to moisture-sensitive foods. On the other hand, even a significant loss of water from food can lead to dehydration of the product. Humidity control chemicals regulate, remove snowmelt water from frozen produce, prevent condensation from fresh foods and limit rancidity, all of these prevent the growth of microorganisms.

For packaged foods, substances like silica gels, natural clay minerals, and hydrated lime are employed; however, for high-moisture meals, internal humidity controls are used. The most popular forms of desiccants are internal permeable sachets or permeable hydrogels plastic cartridges containing desiccants. A certain amount of water may be produced or absorbed by solid foods, which may cause it to be preserved during packing or to leak within the container. If not removed, it may form a condensate, leading in degradation and low commercial potential. Moisture issues can appear in many different of circumstances [15].

5.1.1.6 Carbon dioxide (CO$_2$) absorbers & emitters

CO_2 as a molecule has a bacteriostatic impact on some microbes, prolonging the lag phase and slowing development during the bacterial growth cycle. The features and size of the bacterial population at starting point, storage conditions like temperature, etc. and oxygen content all have an impact on this bacteriostatic effect. All species of bacteria are not equally inhibited. Microbial activity is known to be suppressed by carbon dioxide. Microbial growth on surfaces is inhibited by relatively high CO_2 levels generally 60% to 80% which extends the shelf life. Most plastic films have a three to four times greater greenhouse gas permeability than that of molecular oxygen, thus it must be created on a continuous basis and keep the proper concentration within the package. On the other hand, high CO_2 levels may cause product flavor changes and the emergence of undesirable mesophilic agents have been developed in fruits. As a result, only a few uses, such the packaging of dairy, fruit, and vegetable-like products, are appropriate for a carbon dioxide production. The combination of O_2 scavenger and CO_2 generator might be considered in tandem to avert packaging failure owing to oxygen gas absorption.

Incorporating a CO_2 producing mechanism within a film or adding it as a sachet is a complimentary technique to O_2 management. This approach is often employed in MAP and regulated environmental food storage, where a higher CO_2 level is necessary to slow down various undesired metabolic processes. CO_2 is three to five times more permeable through plastic sheets than O_2. As a result, for some applications, a generator is required. Commercially available CO_2 emitters are the reaction of bicarbonate, an acid with moisture to form CO_2. The principal quality deteriorative reaction is the loss of fragrance compounds as a result of age of the product. To guarantee that the product has the right scent, the CO_2 created in this process must be eliminated. Calcium hydroxide (slacked time) is the most often utilized CO_2 scavenger and is found in a variety of formulations. Controlling both O_2 and CO_2 levels, on the other side, may not have a positive impact on the metabolic activity of fruits and vegetables as a medicinal product [15,29,30].

5.1.1.7 Preservative releasers

In the antimicrobial AP approach, antimicrobial chemicals are adsorbed using polymerization frameworks or incorporated in conventional plastics before film casting. Stabilizers that slowly diffuse onto the surface of food after being packaged or preservatives that bond firmly and do not transfer to food are two varieties of this technique. Food packaging materials are assumed to possess antibacterial qualities if organic chemicals such as antibacterial peptides, antibiotics, enzymes, proteins, chelating agents, parabens, and metals are present or added to them.

Certain metals adsorbents such as silver and zinc adsorbents are two of the most often utilized antimicrobial packaging materials. The microporous surface of the film releases zinc and silver ions when it comes into touch with food, which upsets the biochemistry of microbial cells. The development of spoilage bacteria was significantly suppressed by lysozyme-containing whey protein films. To prevent microbial development in food, only a few chemical antimicrobial treatments are utilized commercially due to negative impact on the food items in the medicinal product. Many of the compounds, such as sodium propionate, have been used for decades in the packaging of food items with no evidence of danger in humans [15].

5.1.2 Types of antimicrobial active packaging

Antimicrobial packaging comes in a variety of shapes and sizes which includes the packets that remains to the package's headspace by direct addition of antimicrobial agents, volatile functionality into polymers. Ion or covalent linkages are employed to immobilize the antimicrobials to the polymer surfaces after coating or adsorbing them with the antimicrobials. Fig. 14.4 indicates the illustration of various forms of active packaging of a medicinal product.

5.1.2.1 Addition of packets in headspace of package

5.1.2.1.1 Film-based technique Packaging may have a direct impact on medicinal product microbiology in ways other than modifying the environment. Surface treatment of items with antimicrobial chemicals, such as spraying or dusting, is common method to achieve AP. Antimicrobial agents might be included precisely into packaging sheets that might remain in contact of the food's exterior. There are two types of antimicrobial films: (1) effective against top growth without causing the active agents to migrate to the food and (2) without an antimicrobial agent that shifts to the food's surface. Silver can be utilized as an antibacterial agent in rare cases. The zeolite's role involves allowing the silver ions to be dispersed gradually into the object.

According to studies, adding conventionally produced fungicidal chemicals to cellulose-based edible films has been demonstrated to be beneficial. To reduce sorbic acid and potassium sorbate emissions, cellulose derivative- and fatty acid-based films were used. Fruit and vegetable coatings appear to be the best use for these films. In

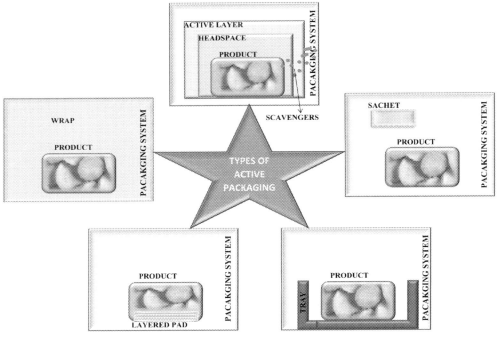

Fig. 14.4 Depiction of various forms of active packaging system.

high humidity environments, cellulose sheets that cannot be heat sealed are ineffective barriers. Anhydrides were utilized to prevent polarity variations from causing incompatibility between the polymer and the antibacterial agent. When dry, anhydrides are typically thermally stable, but they are hydrolyzed in watery environments, such as liquid products. Hydrolysis results in the creation of a free acid, which then migrates from the polymer's surface to the food, where the free acids can act as antimycotics. This is an example of transitioned packaging, in which the active ingredient is retained in the film until it is exposed to food. The activity is started by the moisture in the food product.

Antimicrobial enzymes thought to be linked with the inner surface of food contact films. Microbial poisons would be produced by these enzymes. There are several such enzymes, including glucose oxidase, which produce hydrogen peroxide inside the product. There are enzymes that produce hydrogen peroxide, such as glucose oxidase. Incorporating radiation emitting compounds into antimicrobial coatings is another alternative to prevent product degradation. This is expected to be efficient against microbes while avoiding the dangers of higher-energy radiation. However, there is little concrete proof of the effectiveness of this technology in the scientific literature [31].

5.1.2.1.2 Tray-based technique The tray-based active package was first established as a thermoformed tray structure in inner and outer cardboard boxes that was widely used in MAP. Multilayer MAP system with specific structural support and film interlining between upper and lower warps that can be customized for different purposes. Many other MAP tray compatible systems have also been created. Color issues such as the mottled appearance of the product surface plagued early packaging advances due to inadequate temperature condition during storage, inadequate sealing process and force from the packages kept on top. These issues have been fixed primarily by improving the package layout. The products contained in it are said to have the four weeks shelf-life at 32°F using this process. The item was placed on an opaque tray, covered with a permissible film, and the tray flange was sealed with heat. A high barrier dome cover was then sealed onto the tray following evacuation and flushing with 70% nitrogen and 30% carbon dioxide. For MAP applications, the business has developed an in-line, highly protected version of the conventional extended polystyrene tray that may be coated with conventional barrier top-web materials [32].

5.1.2.1.3 Pad-based technique The food industry has long used absorbent pads to avoid this hazard and maintain sensory characteristics of packed foods for customer acceptance. Typically, the absorbent pad is placed below the trays or canisters to catch liquids that have seeped from food during storage. Absorbent pads have been included in bags in the case of displaying food. Absorbent pads are normally made up of three layers of absorbent material: an upper, lower, and intermediary layer, which is also known as an inner lining. In other instances, the top and bottom layers are made of coated cellulose fibers or impermeable polymers, preventing direct contact between the absorbent material and the food product. To enhance capillary absorption of discharged liquids, perforations are provided on the surface of either the top or bottom layers, or at least one of them.

Classical soaking pads have long been used for preservative, but their wrapping techniques have significant limitations, such as the pad's absorbent capacity being constrained by the bulk of the smart package of medicinal food supplements, limiting the pad from collecting all of the liquid released by the meal. The perforation of the absorbent pad's outer layers is another limitation. To avoid film breaking and, as a result, the release of the absorbent material, these holes should be spaced appropriately. The practicality of bacteriophage incorporation in this system has been established by recent developments in the absorbent pads for packaging applications [33].

5.1.2.1.4 Label-based technique Traditional food labels include information about the item's nature, origin, production date, predicted expiration date, materials, and nutritional composition, allowing consumers to make informed decisions. This can be supplemented by data from sensors that detect influences on food quality, maturation, or suitability in intelligent food packaging materials. Label-metabolite interactions are understood in the same way that dyes change color when they finally came into contact with active compounds generated by food breakdown. The label has an inner semipermeable layer to restrict metabolite diffusion, and exhibits a change in color to signal the presence of metabolites with a protective outside layer. The semi-permeable layer is essential to the colorimetric tag's good functioning because it permits metabolites to pass through the label while restricting dye migration from the colorimetric layer into the food product [33].

5.1.2.1.5 Sachet-based technique Antimicrobial packaging in the form of loose sachets or fastened to the interior of a container has shown to be the most successful commercial use. AP approaches such as oxygen absorbers or emitters, ethanol vapor generators, and moisture absorbers have all proven popular. In order to avoid oxidation and water condensation, the absorbers for both oxygen and water are frequently employed in medicinal food supplements and dairy packaging. Despite the fact that oxygen absorbers are not intended for antibacterial purposes, the reduction of oxygen will inhibit the development of erobic bacteria, especially molds. Dairy and poultry items, coffee and dry food all have oxygen absorbers in sachets. Additionally accessible carbon dioxide and oxygen-absorbing sachets are most frequently discovered in roasted or ground coffee containers. To improve the shelf life of high moisture fermented items, certain sachets can release ethanol as an antibacterial agent. Liquid foods cannot be prepared using sachets. They can't be utilized in a flexible film box because the film will stick to the sachet and restrict it from doing its action. Consumers may inadvertently swallow sachets, which may explain for their limited economic success [33–35].

5.1.2.2 Inclusion of various antimicrobial agents directly into polymers

The introduction of various bioactive compounds such as antibacterial agents into polymers has facilitated the delivery of drugs and pesticides, household products, tissues, implantable devices and other biological devices. Numerous microorganisms have been tested against GRAS (Generally Regarded as Safe), non-GRAS, and endogenous antibacterial agents. It includes *L. monocytogenes*, and *Escherichia coli* as the pathogenic bacteria. The spoilage organisms degrade like mold, and is currently used in paper, thermoplastic elastomers, and composite materials. The most often used antibacterial polymer addition for food applications is silver substituted zeolites. In zeolites, silver ions substitute sodium ions in order to control wide range of bacteria and molds. Low-density polyethylene, polypropylene, nylon, and olefin styrene all include these modified zeolites in concentrations of 1%–3%. Silver ions infiltrate into microbial cells, disrupting their enzymatic activity.

The macromolecular substance has the potential to contain other substances in addition to the antibacterial agent. Among other things, there are antimicrobial enzymes like lactoperoxidase, peptides like magainin, defensin, natural and antioxidant phenols, antibiotics, and metals like copper. The goal of using antimicrobials in packaging is to reduce spoilage and contamination of food by surface growth, a major cause of spoilage and contamination. This procedure can cut down on the amount of antimicrobial agent that would otherwise be absorbed into the product's bulk after being applied on the meal. It could be better to let the antimicrobial ingredient slowly escape from the packing film on the product's exterior rather than dipping and spraying.

Antimicrobial agents can be incorporated into polymers during smelting or by mixing solvents. Thermomechanical polymeric treatments such as plastic injection can be used to produce heat-stable antimicrobials. For example, silver-substituted zeolite can withstand extreme heat. As a result, they are co-extruded with other polymers to form a thin layer. Heat-sensitive antimicrobial agents such as enzymes and volatile compounds could be better incorporated into polymers through solvent formulation. Surface properties and diffusion kinetics become important when non-volatile antimicrobial packaging materials come into contact with food surfaces for antimicrobial chemicals to diffuse to the surface. Multilayer membranes are utilized to produce correct controlled release on food surfaces, with the innermost layer regulating the bioactive compound's diffusion rate, the mixed layer containing the active component, and the barrier layer preventing the agent from escaping the package.

5.1.2.3 Coating antimicrobial on polymer

Antimicrobials are often utilized after the material has been created or added to cast films since they can withstand lower temperatures than those used in polymer synthesis. Molded edible coatings have been used as antimicrobial carriers as well as coatings for packaging and food. Proteins have a higher adsorption capability due to their amphiphilic nature. Nisin adsorption on polyethylene, ethylene-vinylacetate, polypropylene, polyamide, polyethyleneterepthalate, acrylics, and polyvinylchloride are only a few examples. Antimicrobial adsorption can be improved by manipulating the solvents or polymer architectures. The sodium hydroxide-treated and acetone-swelled polyethylene-*co*-metacrylic acid film had a greater capacity to absorb and diffuse benzoic and sorbic acids than the untreated membranes. Molds were likewise inhibited the most by sodium hydroxide treated films. The reason for this is that the greater polarity of sodium hydroxide treated sheets increased antibacterial absorption. Polyamide resins and other binders have also been employed to improve the compatibility of polyolefin surfaces with bacteriocins [36].

5.1.2.4 Surface fixation of antimicrobials to polymers

Antibacterial and polymer functional groups are both required for various forms of surface modification. Functional group antibacterial agents include polypeptides, enzymes, polyamines and aromatic compounds. The usage of spacer molecules for the linkage between the polymer surface and bioactive agent can serve as an additional form of attachment in addition to the use of functional antibacterial agents and polymer carriers. These spacers provide for enough mobility for the drug's active component to interact with microbes on the food's surface. Antibacterial packaging includes dextran, polyethylene glycol, ethylenediamine, polyethyleneimine, etc., which have low toxicity and are widely used in foods. Surface fixation can lead to reduced antibacterial activity, which must be addressed. Structural changes and denaturation with solvents can reduce the activity of proteins and peptides per unit area. Antibacterial agents are ionically bonded to the polymer and are gradually released into food. If the antimicrobial agent is covalently bonded to the polymer, diffusion into the product would be less of a problem, unless circumstances inside the product cause processes like hydrolysis. This can happen if a high acid meal is cooked for a brief period of time. However, integrating substrates into the system and controlling undesired products from reactions are substantial hurdles [34].

5.1.2.5 Antimicrobial attributing polymer

Some polymers are naturally antibacterial, they are used in films and coatings. Cationic polymers that interact with the cell membrane's negative charge and leach intracellular components, including chitosan and polylysine, aid in cell adhesion. Chitosan has been used to wrap natural food stuffs, and it appears to protect them against fungal degradation. Although the antifungal effects of chitosan are said to have an antibacterial effect, it's also possible that the chitosan makes food resistant to both germs and the nutrients in the food. Additionally, organic acids and species have been transported using antimicrobial chitosan films. With respect to increase the shelf life of food goods, bactericidal acrylic polymers that were created by the co-polymerization of acrylic free amine co-monomer have been proposed. Polymers with biguanide substituents also have antimicrobial action. Due to the interaction of ions and other ions with the surfaces, adding nutrients may be able to control harm to cell membrane, limit bacterial recovery, and reduce cell adherence to the surface [34].

5.2 Intelligent packaging

In order to gather and give details on the quality of the packed product throughout transit and preservation, packaging methods or smart sensors are frequently employed to assess the state of packaged consumables like food. Communication-based smart packaging solutions can help maintain food quality, increase shelf life, and improve food safety in general. It is capable of capturing and transferring specific kinds of data in addition to cognitive tasks like sensing, detecting, and tracking. As an outcome, time & temperature indicators, degree of freshness and maturity indicators, gas detectors, and radio frequency identification (RFI) systems, etc. are all included in smart packaging solutions. To convey the necessary information, basic functionality can be created and implemented using indicators and sensors. Indicators offer information on changes that have been seen in a product or its surroundings, such as variations in temperature or pH levels. In order to detect, record, and send data regarding the biological activities and reactions taking place within the container, such as oxygen levels and freshness, biosensors are frequently utilized in food packaging [37–41].

5.2.1 Supported with indicator

Indicators are those devices that reads and shows the information on changes in the different properties as a result of interactions with chemicals such as gases, humidity and microorganisms whether these substances are present or not. These devices lack a receptor and a transducer, which sensors do. They are primarily based on visible changes. There are four types of primary devices of the same category such as the humidity indicators, the gas indicators, the time-temperature indicators (TTIs), leak as well as visual indicator devices.

5.2.1.1 Time temperature indicator (TTI)-based technique

These indications are markings on each box that are applied from the exterior to provide information about the package temperature history. By tracking the temperature of a product, TTIs can be used as a visual or electronic indicator of its freshness. Irreversible temperature fluctuations generated by chemical, electrical, or even nanoparticle changes create the indication. These indicators are divided into three categories as critical temperature indicators that give limited product information and shown when a reference temperature has been surpassed its usual limit. TTIs with a partial history that provide ordered temperature and time comparative variations in comparison to a reference element such as temperature and indicators with a complete history showing all temperature changes during the storage duration.

The operational notion of these indicators is color change against the different photochemical, enzymatic, polymer, microbiological or diffusion-based processes. In the case of an indicator on enzymatic activity, the combination of pH dye, lipolytic enzyme and lipid substrate tends to cause free fatty acid reactions, lowering the pH and causing a visual shift. The active solid-state polymer's quick color change at high temperatures and delayed color change at low temperatures are used to create a polymer-based indication. Due to this kind, it is based on changes in the active polymer's optical properties with respect to fluctuating temperature changes where no further activation is necessary prior to use. Microbial-based TTIs rely on higher growth rates of microorganisms implanted into the label of the product due to maximum temperature exposure. The use of photogenic substances in photochromic type processes leads in a reaction based on temperature-based color changes across time. In the indicator based on diffusion, the colored lipids produces time and temperature based diffusion over an indication track. They were more cost competitive than other intelligent systems due to their inexpensive cost, small size, diffusivity, and ease of use. However, freezing can harm and may also result in sunken, pitted regions or even interior browning in some vegetables consequently, the difficulty in establishing whether the packed product was exposed to subminimum temperatures is a limitation of TTIs. Freezing damage, for example, can result in sunken, pitted portions or even interior browning if carrots are frozen at temperatures below $1.8°C$. As a result, more complex technologies, including the use of radio frequency to quantitatively and effectively monitor moisture, temperature, and gas concentration, may be necessary depending on the packed items.

Many TTIs have been commercialized and are commonly used to monitor perishable goods. A solid-state polymerization procedure is used to create a self-adhesive substance based on polymer where the chemical indication in the form of labels. When the package is faced to varying temperatures over time, the active central polymer changes the color of the inner circle from red to black, offering a visual indication of packed food quality. As much as the temperature raises to which the product was contacted to, the faster the color changed was observed [42–46].

5.2.1.2 Gas indicator-based technique

Gas as well as humidity indicators are made to use to check the composition of gases and material inside the food container. The gas indicators-based techniques are printed labels, films, or tablets incorporated into packaging to offer information on any sort of alteration in the gas's mixture within the headspace of packed items and other directly eatable packaged products. Eatable such as vegetable respiration, leaks, or the presence of bacteria induce changes in the gas composition, which cause the indicator color to vary. The most prevalent indicators in the food industries are the water vapors, $O_{2(g)}$ and $CO_{2(g)}$. The bulk of these indicators rely on a gas-indicator binding relationship or redox reactions that change the dyes used to generate visual indications, such as methylene blue.

There are only a few gas indicators intended to exhibit any change in the gas level in the compartment of packaged vegetables. Using MAP technology, the gas incorporation ratio is adjusted during the entire packaging process to increase the shelf life of the vegetables and medicinal food supplements. When the oxygen level or concentration within the package exceeds a certain limit, an oxygen indicator changes color from pink to purple. This kind does not require activation. As a result, it must be stored in oxygen-absorbing packs until it can be packaged. When the oxygen concentration within the package falls below a certain threshold, the color changes from pink to blue. Carbon dioxide travels through the permeable membrane and reacts chemically with pH-responsive dyes like cresol red, creating a color shift. As the carbon dioxide level rises, the color changes from red to yellow. Moisture in the packaging has no effect on this reversible signal.

It was created that the humidity cards with visible color indicators of relative humidity for moisture indicators. The cobalt chloride solution in these cards changes color from light blue to purplish pink when the humidity level rises. A film made up of carbon-coated copper nanoparticles distributed over a glass substrate that expands under low humidity due to increased space between the nanoparticles (NPs) is another fascinating example of humidity monitoring. When humidity levels rise, water vapor condenses on the films of a strip, causing destructive interference at perpendicular incidence angles. The color shifts from orange to blue, pink, and eventually green as the water vapor saturation rises. This metallic color film has a reversible color shift, making it possible to reuse and recycle such indicators. By placing it within the packaging as a film or labels along with the impregnation of nanoparticles, the humidity percentage may be measured in real time. The humidity percentage may be seen when compared to a pre-set color reference. This strip can detect other gases as well, such as ethanol vapor [43,47].

5.2.1.3 Humidity indicator-based technique

Humidity sensors are often made with electronic hygrometers, which monitor the property such as capacitive and resistive that change due to change in humidity. It has been recently utilized that wheat gluten can be used to measure relative humidity in packaged foods using a tiny layer of wheat gluten protein as an IDC device's relative humidity monitoring polymer. The digital mid-capacitor is mostly used to investigate the physical characteristics of wheat gluten and assess any potential interactions with water molecules. It was discovered that the permittivity and dielectric loss grew along with the relative humidity, showing that the system's permittivity and dielectric loss had become sensitive to relative humidity. The viability of using this gluten to monitor the relative humidity or RH in the space of a packaged meal was then demonstrated by measuring the relationship between dielectric permittivity and electrical loss through relative humidity. For improved product monitoring and assessment, the suggested relative humidity sensor may be employed in combination with an RFID system. Inorganic salts are used in most colorimetric moisture indicators. Cobalt chloride ($CoCl_2$) is commonly used in the manufacture of such indicators, although it is not suitable for basic applications requiring irreversible and toxicity is a serious problem to be solved and decide. Although various metal-related chloride-based moisture indicator devices have been developed, $CoCl_2$ is the most cost-effective and safest. When it comes to trash removal, these sensors also take care of contamination. These factors have increased the emphasis of research on cobalt-free moisture indicators for use in product packaging systems. The developed indicator has the capacity to detecting relative humidity (RH) in between 5% and 100%, which makes it appropriate for active packaging applications. The detected indication still contains low toxic parts; however, they are safer than the cobalt (II) chloride-based sensors [48].

5.2.1.4 Visual indicator-based technique

Chemical diffusion, polymerization, and enzymatic reactions are the basic mechanisms of action for visual indicators that change color in response to heat exposure. These items are used to defend against temperatures that are too hot or too cold during transit and storage. For the producer, these indications are a sign of quality since they ensure that the offered product to the user is in the best possible state or condition. The visual detectors among all the detection techniques are found to be the simplest technique due to the ease of interpretation about the status and quality of the medicinal product eligible to be consumed [49].

5.2.1.5 Leak indicator-based technique

The leakage outside the container due to content activity, the gas composition in the medicinal food package changes, altering the nature of the contained items and the surrounding environment. Some indicators such as $O_{2(g)}$ and $CO_{2(g)}$ are used to check product quality and evaluate the performance of oxygen absorbents as a seal or leak indication. Chemical or enzymatic processes cause the color change of many gaseous based indicators which indicate, for instance, the color change of signal, that there is heavy amount of oxygen molecules in a sealed item container. The primary issue with these indicators is that they must be stored in non-oxygenated or anaerobic conditions because they get decay quickly when exposed to air.

The two-component leakage indicator that both detects leaking and absorbs leftover oxygen has been developed and patented. This oxygen indicator was created with the goal of detecting airway leaks in the designed modified atmosphere packages. Utilizing CO_2 indicators to track the altered CO_2 level has been successfully accomplished in MAP devices. Due to the bacteriostatic properties of carbon dioxide, it is commonly employed in MAP, along with some inert gas such as nitrogen. The modified environment for non-respiring goods generally has a low oxygen level which accounts for about 0%–2% and a high carbon dioxide level which accounts for about 20%–80%. The MAP conditions to induce respiration must be tailored to each fresh fruit and vegetable species. A leak in a high CO_2 pack results in a lower CO_2 and a higher O_2 level. Leak indicators are used to monitor these changes [14].

5.2.2 Supported without indicator
5.2.2.1 Barcode-based technique

Because of their simplicity of use and economic, barcodes are one of the most often used data analysis techniques. Product related information is provided in the form of lines of varied lengths separated by spaces, with numbers written beneath. The barcode known as the UPC (Universal Product Code) is now extensively utilized. As each barcode has a distinct UPC, barcodes are frequently selected to track the location of each cargo in supply-oriented chains. By removing or modifying the barcode to render it unreadable, this type of data carrier may also be used to assess the freshness of packed vegetables.

Barcode-tagged antibodies have been employed as biosensors in a variety of ways. For example, the Product Sentinel System is a barcode that can detect germs and determine whether a food is tainted. This method employs two barcodes: one for product data that can be read by a barcode reading device and another one for a contamination based coded symbol that is generally not readable in the event of safe food. The preformed antibodies are labeled and made into a membrane that is attached to a barcode portion. When limited contamination is determined, one bar code will not be legible, while the other will be readable. The reaction involves antigens binding to antibodies on the barcoding system, causing lines to appear, while antigens dissociating from the matrix in the presence of bacterial metabolites, causing the lines to disappear. This approach employs tagged antibodies on the barcode's membrane to identify certain antigens. Another form of this method turns a virus adhered to an antibody-containing membrane into red ink, which is unreadable by scanners. As a result, this barcode will become unreadable, indicating that the goods within the pack is no longer edible.

5.2.2.2 RFID tags-based technique

Radio-frequency identification (RFID) techniques are those tools that make the usage of electro-magnetic fields to keep track or sense the variety of measures and may communicate the results using a device called reader. This reader generates the radio-frequency waves to read the data through an antenna. Since RFID tags are only equipped with antennae, other sensors and indications are commonly used to detect freshness. In order to optimize the usefulness of this sensing technology, several sensing device and indicators have been connected to these tags.

RFID tags have been categorized into three groups. The electromagnetic induction created by placing the reader close to the tag activates passive tags without batteries. This kind has a long shelf-life; however, the reading distance is short. The tags that are semi-passive utilizes a battery-based component to power the chip but the signal transmission must be powered by the reader. Active tags, which are self-contained and have their own battery, have been more common as part of smart packaging systems in recent years, these tags are highly durable and have a longer shelf life. As RFID tags can accommodate or store more data and enable real-time monitoring for many sorts of data such as temperature, humidity, and product information, they are more efficient than barcode technology. In the monitoring of packed vegetables, the high frequency or HF range is often utilized. However, there are other parameters to consider when implementing this technology in smart packaging, such as cost, which must be kept to a minimum for commercial use. The lifespan of these types of active tags should match the lifespan of the items under consideration or being monitored by using batteries with an acceptable life span. Furthermore, signal breakage due to water absorption in human body or presence of moisture in packages could result in difficult-to-detect low-level microwave signals.

The idea was based on measuring the proportion of respiration gases produced in the package, such as $CO_{2(g)}$ and $O_{2(g)}$. Since it is previously stated, ratio of these two gases determines the freshness of vegetables and fruits. The respiration quotient (RQ) is defined as the $CO_{2(g)}$ to $O_{2(g)}$ ratio per unit time to calculate the rate respiration. The freshness of the packaged fruits, medicinal food supplements and vegetables decreases if RQ exceeds one unit. The LED or light emitting diode color changes from blue indicating the fresh content to red indicating the old content to show freshness and not fresh. However, further scientific study is needed to find less bulky, low-cost circuits so that such approaches may be utilized to assess freshness without significantly affecting or interfering the end product expenses related to packaging.

Another smart sensor of RFID label used temperature, humidity, and nitrogen sensors. This strategy may be used to keep track of supply networks. To monitor true temperature differences, these labels can be put in the storage area or on the display units. One of these smart sensor tags might be placed to boxed vegetable. As a consequence, additional data on the freshness of items and even the rate of disease progression may be retrieved. A screen-printed RFID antenna label is then linked to the multimodal printed electronics platform, from which data may be read using an RFID reading devices. Data from sensors, in case of the measurement mode, is generally measured one at a time, then scripted to memory and read by an external reading device. Due to the considerable dependence of the absolute sensitivity of the ambient and humidity of their sensor layers on temperature, the Resistance Temperature Detector (RTD) on RFID tag provides temperature data and also acts as a temperature compensation element for two other sensors such as humidity and ammonia. This RFID label has the advantage of avoiding the problem of temperature and humidity cross-sensitivity that has been identified in prior studies utilizing various humidity and resistance temperature type sensors. The sole reason for this is the protective top layer of the low water vapor permeability of RTD, which prevents immediate contact of water vapor and the RTD. This system is designed to be powered by a battery with a limit of 150 mA per hour. Due to the measurements are gathered every 60 min for 30 s, it can last up to 57 days [50–55].

5.2.2.3 Biosensors-based technique

It is a scientific sensor that detects, evaluates, records, and converts a biological input or response into measurable signals such as an electric current or a fluorescent signal. Transducers are components of biosensors that transform the target response into a measurable signal, while biosensors are components that recognize the targeted analyte. Cells, DNA, antibodies, enzymes, biomimetic, and phage are all examples of bio-receptors, whereas the transducer might be electrical, optical, chemical, magnetic, or micromechanical. The first two bio-receptor types are often employed in the food packaging sector and have shown promise in detecting contamination in vegetable packing. Antibody-based biosensors will be the focus of this section.

Antibody-based bioreceptors biosensors work on the principle of lock and key model as an antigen fit. Generally, the antigens are proteins that bind to a range of pathogens and are found on the surface of diseases. Antibodies have two distinct regions: a variable area that changes structure depending on the antigen and a constant area with a fixed structure. Antibodies are typically used as biosensors on the detector's surface, allowing them to detect pathogens by attaching to antigens and identifying molecular structure. Antibody labeling is the sequence of connecting antibodies to various labels like fluorophores, isotopes and enzymes, etc. for the suitability of detection. Antigen-specific antibody couplings are converted to digital signals that electrochemical, magnetic, optical and mass-based transducers can read as specified in Table 14.3. The transducer's sensitivity as

Table 14.3 List of the various componential requirements with respective type of biosensors.

Components of biosensors [56–61,62,63]			
Bioreceptor	Transducer	Electrochemical recognition	Analyzer
Enzymatic bioreceptor	Electrochemical recognition	Amperometric type	Optical analyser
Antigen-antibody bioreceptor	Optical recognition	Potentiometric type	Electrical analyser
Aptamer bioreceptor	Electronic recognition	Voltametric type	Electronic analyser
Whole cell bioreceptor	Thermal recognition	Conductometric type	Thermal analyser
Nano bioreceptor	Gravimetric recognition	Impedometric type	Magnetic analyser
	Acoustic recognition		Mechanical analyser

well as the quality of the antibodies used, define the sensitivity of these sensors. This section provides a rapid review of many types of antibody-based biosensors [56–59].

5.2.2.3.1 Optical biosensors Optical systems in biosensors have got a lot of attention recently. The sensing technique utilize light-based interactions to quantify or measure biological responses. Among one of the most used technologies for identifying the detectability of foodborne pathogens is surface Plasmon resonance (SPR). Electrons within a metallic surface get energy transfer when a monochromatic light of near-infrared or visible area passes via a prism into an object surface. Photodiodes would detect this opto-electronic phenomena, which is based on Plasmon resonance. Electrons will be released as a result of the photon energy imparted to them, causing electromagnetic waves which is surface Plasmon's to resonance and absorb light, resulting in the lowest reflectivity at this precise angle. The mass of immobilized antibodies determines the specific angle, which is a function of internal property i.e., refractive index. The bimolecular interaction will affect the refractive index of the reflected beam. As a consequence, evaluating this modification will qualitatively discover the existence of these analyses without the need for enrichment or culture. The high sensitivity and selectivity of this SPR technology, as well as the fact that it can be incorporated into tiny circuits, are all benefits. In contrast, prisms are not always suited for sensors which are based on chip or flat like object. Furthermore, there is a limitation to the number of concurrent readings that may be experimented. Multiple channel sensors, which allow for parallel measurements, were proposed as a solution to these issues. SPR imaging material such as biochips were used to analyze changes in refractive index to exclude unwanted signals from noise.

In grating-coupled SPR imaging, which is low-cost and has a broad dynamic range, optical diffraction gratings are employed for couplings and angle measurements. An SPR-based protein chip was used to detect *E. coli*, Salmonella, and other pathogens. This suggests that such a technique has a lot of promise in the packaging industry. By including this chip in the box, it can be tested using a light source and SPR spectroscopy to detect the presence of infections. Antigens binding to antibodies immobilized on gold or Au substrates are also enhanced using G protein or guanine nucleotide binding protein. However, before detection, these approaches need culture to increase the pathogen's biomass. Furthermore, because food contact with nanoparticles is unavoidable, they are better suited to a separate stage of the supply chain, such as pre-packaging. Other methods of bacterial identification as an assessment marker of freshness include surface plasmon resonance, ultrasound imaging, and polymerase chain reaction, etc. However, these systems need costly equipment, which makes them unsuitable for inspecting packaged items in supermarkets [37,60,61,64–68].

5.2.2.3.2 Electrochemical biosensors In electrochemical biosensors, antibodies act as binding elements, and electrochemical transducers function as transducers. The electrodes will convert the antigen-antibody binding process into an electrical signal. This biosensor category has two advantages: low cost and the flexibility to interact with other bio sensing modalities. However, it has a lesser sensitivity than optical biosensors. Amperometry, potentiometry, impedimetry and conductimetry are the output forms of this category. When a voltage is given to electroactive species, oxidation or reduction takes place, and current flows. The current generated is equal to the concentration of the analyte. This method of measuring has a higher sensitivity than potentiometry. The voltage difference between the two electrodes with near negligible current is recorded in a potentiometric biosensor as the potential rises in response to the biorecognition process. A conductive polymer converts the analyte into an electrical signal in a conductimetric biosensor. In this case, adding nanoparticles to the solution can increase sensitivity by boosting conductivity. The last kind is impedimetric, which uses microbial metabolites to reduce impedance and increase conductance and capacitance [23,69].

5.2.2.3.3 Commercial biosensors A visual biosensor created by some manufacturer was based on antigen or antigen-specific antibodies in the form of antibodies sandwiched with a certain thickness were used in this biosensor immunoassay to detect infections. Capture antibodies and detection antibodies are the two types of antibodies employed. Capture antibodies are trapped in a tiny layer of polymer plastic sheets such polyethylene and then imprinted as layer in a nutrient gel. The detector antibodies which change color when they connect to antigenic sites of antigen which are free to migrate across the layer of nutrient agar gel are identified using colorimetric enzymes or brilliant chromophore dyes. The feeding gel will encourage infections that have been penetrated to proliferate quickly while the detector ligands will attach to pathogen antigens. The detector antibodies bind to pathogen antigens and migrate or move toward the trapped ligand due to the affinity or interaction, resulting in a particular pattern form with concentrated chromophore. The reaction time for various patterns is few minutes, and it might take up to several hours for a pronounced dark color shift. The ability of this biosensor to detect and identify many bacterial species simultaneously including *E. coli*, Salmonella and Listeria species as well as its stability and adaptation to both warm and cold conditions and a one-year shelf life, are among its benefits. Minimal quantities of bacteria that may responsible for illness are not bounded due to sensitivity limitations [62,70,71].

E. coli was used as the model-based system for the detection of pathogenic bacteria dubbed emulsion of Janus droplets as a new method. Janus particle-based technique has the potential to be employed as disease detectors in vegetable packing. Janus particles are utilized in biomedicine and extremely selective sensors, among other sensing applications. These nanoparticles' distinct hemispheres have diverse physical and biological properties. These particles can be utilized as an *E. coli* detector since their surface nature is similar to that of live cells. The droplets are made with carbohydrate-based surface-active chemicals, and pathogen recognition via carbohydrate—lectin interactions change their optical properties to pass light through it. Half of the hemispheres will attach to the virus, clubbing of the droplets that depending on pathogen concentration, produce light scattering. Quantitative measurement using a fast response code based digital reading system on such surfaces using a simple smartphone app while introducing the droplets into the transparent cavity or performing a quantitative measurement. By digital processing techniques. Advantages include the speed and simplicity of the application, as it is detectable with the naked eye or using a mobile phone. The detection method has not yet been commercialized. The chance of embedding in the packaging is greatly increased if these droplets are formed into a film or a thin film used as part of the package for disease detection [72—75].

Traditional biosensors can be replaced by NP-based microbiological sensors. In microorganisms such as *E. coli*, silver nanoparticles (AgNPs) as well as gold nanoparticles (AuNPs) type particulates have been found interactive with pathogenic proteins and inhibit DNA replication. The advantages of using NPs in sensors over biological sensors include their unique optical and electrical characteristics, which increase sensitivity, reaction time, and selectivity. Immunomagnetic separation-based (IMS-based) detection is a method for detecting infections that employs NPs as a magnet. Magnetic nanoparticles such as ferric oxide may attach to certain antibodies, which can be used to detect infections. Pathogen identification will be possible thanks to visible optical and electrical changes after attachment as well as pathogen separation using an external electric field. The IMS operational principles are depicted in this diagram. However, because of concerns about migration to food included within the packaging, the use of these nanoparticles is now being researched. Safety and dangers must be included in order to preserve public health. Hence, the validation procedure for such sensors should be extensively explored [70].

5.2.2.3.4 Gas sensors Gas sensors handle quantitative monitoring of gas concentrations, especially oxygen and carbon dioxide. This is because the gas sensor has a great impact on the disassembly and overall quality of the package. Optical-based gas sensors for carbon dioxide detection in food packaging are a promising technology. When CO_2-sensitive dyes are utilized, commonly used kinds rely on fluorescence, whereas pH sensitive dyes rely on colorimetric change.

An electro-optical dry solid CO_2 sensor employs fluorescent dyes whose brightness varies in response to carbon dioxide levels. A pH sensitive fluorescent dye namely 1-hydroxypyrene-3,6,8-trisulfonate (HPTS) is incorporated in the film. The fluorescence of the dye reduces with increasing the concentration of CO_2. The reaction time of this type of dye is below 2 min and the recovery time is below 40 min. HPTS plastic film CO_2 sensors are suitable for a large number of applications in aqueous and acidic solutions, having high sensitivity, and shelf-life of 6 months or more when stored in the dark. Finding acceptable fluorescent compounds for food applications is a challenge, and improved sensitivity is required to cover the 0%—100% measurement range. The type of dry CO_2 sensor was studied in ready-to-eat packaged items using the Förster Resonance Energy Transfer (FRET). In the FRET process, an energy is transferring from the fluorophore group (referred as a donor) to the chromophore group (referred as an acceptor). The dye, Pt-porphyrin, having characteristic of phosphorescent donor absorbs energy produced by the pH indicator chromophore such as the naphthol phthalein (NP) acceptor in this experiment. As carbon dioxide concentrations grow, the interaction of the two complexes will form films with varying optical responses. Carbon dioxide content has an impact on FRET. When there is no CO_2, it is at its highest, and as CO_2 levels rise, it decreases. When CO_2 levels are between 0% and 10%, the energy transfer for this sensor ranges from blue to colourless, making it suitable for modified environment vegetable packing with the same CO_2 levels. The sensitivity of sensor against several days and even at high temperature made it suitable for medicinal product packing. This sensor has been tested for safety and has been found to be free of residual ink that could be transferred to packaged foods. This is a big plus for commercial applications. As a result, the sensor may be used to simultaneously monitor CO_2 and O_2 concentrations [50,76—78].

The 1-hydroxypyrene-3,6,8-trisulfonate (HPTS) as a pH indicator was adsorbed on a polar-based silica matrix to form a membrane and a solgel-based photocarbon dioxide sensor. The operation of this sensor depends on CO_2 quenching the brightness and causing it to emit light. The dual chromophore reference (DLR) approach is then used to translate the fluorescence intensity lifetime and measure it in the phase domain. In this approach, the fluorescent dye binds to an inert standard chromophore, such as a ruthenium complex. It lasts a very long time and absorbs light from blue to green with great efficiency. The fluorescence intensity signal of the HPTS dye has a short lifetime, so the phase angle of the phase domain is zero, but the reference chromophore has a phase shift. The sum of the measured phases is different because the sum of the measured signals includes the superposition of both signals. As the CO_2 concentration changes, the amplitude of the fluorescence intensity signals from both chromophores fluctuates. The phase change of the total recorded signal, which depends on the amount of CO_2 in the sample gas, correlates with this amplitude shift. The resolution of such types of sensors is greater than 1%, below 0.8% detection limit, and below 0.6% a cross-sensitivity to O_2. It has a seven-month shelf life, which makes it perfect for vegetable packing. This sensor, however, suffers from oxygen sensitivity and temperature sensitivity since the generally used standard luminophore such as ruthenium complex is influenced by presence of oxygen molecules. As a result, to get accurate gas concentration measurements, a correction technique is necessary [45,46,79].

5.2.2.3.5 Humidity sensors The amount of water in packed vegetables has a big influence on their freshness because it creates an excellent habitat for mold and bacteria to develop. Since humidity sensors may be produced as thin films or system of membranes, they can be used as part of packaged product monitoring, especially if they can be integrated with RFID tags. Based on the sensor materials employed as polymers, semiconductor ceramics, and hybrid polymers or ceramic materials, related thin films may categorize into three main groups. On the basis of the type of conversion, they are further divided into capacitive and resistant. Ceramics, although, have higher tensile qualities and physical durability than polymers, polymers are a more appealing and cost-effective alternative for humidity monitoring in packed vegetables with sufficient sensitivity. Detectable changes in electrical properties are required for thin porous films to sense water absorption. The interactions of water molecules alter humidity-sensitivity qualities through a number of prominent processes. At low relative humidity, electronic conduction dominates water molecule interaction. At high relative humidity, ionic conduction dominates, whereas both contribute at medium relative humidity [80,63].

6. Patented products of smart packaging

The patents for various smart packaging devices or technologies have been granted for many years for food and eatable things. The list of few patents which can be utilized in the medicinal product packaging sector as it was used in food packaging is enlisted in a Table 14.4.

7. Future challenges and scope of smart packaging

Smart packaging is a powerful strategy for moving away from traditional quotes of use by or best before quality evaluation models for packaged vegetables and toward more dynamic approaches. There are, however, a number of challenges and considerations to

Table 14.4 List of few patented products of smart package.

Patent no.	Title	Filing date	Assignee	Inventor(s)	Reference(s)
EP 1743282 B1	Activating a data tag using load/orientation/user control	January 19, 2005	Callahan Cellular L.L.C. Wilmington, DE 19808 (US)	Lindsay, Jeffrey, Velazquez, Heriberto, Flores Neenah, Chen, Fung-Jou, Wagner, Eric, Francis Sherwood, WI 54169 (US)	[81]
AU 2016325872 B2	Antioxidant active food packaging	September 23, 2016	International consolidated business group private limited	Withers, Philip	[82]
WO 01/04548 A1	Self-heating flexible packages	January 08, 2001	TDA research Inc., US.	Bell, William, Dippo, James, L., US	[83]
WO 2014/126654 A1	System, method for product track & trace utilizing unique item-specific URLs & quick-response (QR) codes	January 06, 2014	Agileqr, Inc., US	Sayers, Foster, Joseph Ho, Albert Kim, Charles, Hyung	[84]
US 8337923 B2	Encapsulated antimicrobial Material	August 06, 2004	DuPont nutrition Biosciences APS, copenhagen K (DK)	Bob coyne, Lenexa, john Faragher, richfield, sébastien Hansen, richar. Ingram, Torben Isak, Linda Valerie, kathryn. Louize Tse, københaven S (DK)	[85]
US 7,205,016 B2	Packages and methods for processing food products	April 07, 2003	SafeFresh technologies, LLC, Mercer Island, WA (US)	Anthony J.M. Garwood, Mercer Island, WA (US)	[86]
EP 2 047 413 B1	Packaging systems integrated with	August 03, 2007	Binforma group limited liability	Mingerink, Corey, Benjamin Appleton,	[87]

Table 14.4 List of few patented products of smart package.—cont'd

Patent no.	Title	Filing date	Assignee	Inventor(s)	Reference(s)
	disposable RFID devices		company Dover, DE 19904 (US	Madsen, Gary, Fabian Greenville, O'shea, Michael, Donald Neenah, Wisconsin 54956 (US)	
US 10,610,128 B2	Pharma—informatics system	October 20, 2017	Proteus digital health, Inc, Redwood City, CA (US)	Mark Zdeblick, Andrew Thompson, Aleksandr Pikelny, Timothy L. Robertson, Belmont, CA (US)	[69]
US 7083837 B1	Registered microperforated films for modified/controlled atmosphere packaging	June 08, 2001	Marc Patterson	Elizabeth Varriano-Marston, 18 Wilson rd., Windham, NH (US)	[88]
US 7948381 B2	Reversibly de-activating a radio-frequency identification data tag	December 19, 2006	Binforma group limited liability company, Dover, DE (US)	Jeffrey Dean Lindsay, Herb Flores Velazquez, Neenah, Fung-Jou Unen, Eric Francis Wagner, Cary, John Christian Onderko, Appleton, WI (US)	[89]
EP 1477423 B1	Sealing wrapper for food products, corresponding process and installation	May 13, 2003	Bosotti, Luciano et al.	Mansuino, Sergio 12084 Mondovi (Cuneo) (IT)	[90]
US 2020/0051015 A1	Shipping package tracking or monitoring system and method	August 08, 2019	Bryan Jonathan Davis, Ronald Eugene Fisher, James Mark Oakley, David Fraser,	Bryan Jonathan Davis, San Francisco, Ronald Eugene Fisher, James Mark Oakley, David Fraser, Andrew David	[91]

Continued

Table 14.4 List of few patented products of smart package.—cont'd

Patent no.	Title	Filing date	Assignee	Inventor(s)	Reference(s)
US 11,317,480 B2	Smart packaging, systems and methods	March 14, 2019	Andrew David Cater, Mark Joseph Meyer, and Bradley Brian Bushard Inductive intelligence, LLC, Grand-Mother Rapids, MI (US)	Cater, Mark Joseph Meyer, Bradley Brian Bushard, Minneapolis (US) Gregory L. Clark, Ada, David W. Baarman, Fennville (US)	[92]
US 2015/0353236A1	Smart packaging and related technologies	February 06, 2015	TInk, Inc., NY (US)	John gentile, Terrance Z. Kaiserman, steven Martin, Anthony gentile, Tayler kaiserman, Adam joffee, NY (US)	[93]
US 9536449 B2	Smart watch and food utensil for monitoring food consumption	March 23, 2013	Medibotics LLC, st. Paul, MN (US)	Robert A. Connor, Forest Lake, MN (US)	[94]
US 2017/0178072A1	System, apparatus & method for transferring ownership of a smart delivery package	December 22, 2015	Intel corporation, santa clara, CA (US)	Rajesh Poornachandran, Portland, Ned M. Smith, Beaverton, (US)	[95]

think about. One of the most pressing issues is the safety of smart gadgets that come into contact with food, in case if they may migrate to food. As a result, the compounds used must pass all applicable food safety regulations. Some nations have food and medicine regulations, as well as rules governing food packaging and suppliers. As a result, getting regulatory permission for the use of innovative materials in labels is a time-consuming and difficult procedure. As a result, businesses may be cautious to engage in this sector. As a result, there will be a greater need for collaboration between governments and private firms.

Smart sensing systems must be exceedingly accurate and resilient in order to assure safety of public and welfare. The false positives should be denied at all costs, as they declare an item as safe instead of contamination can lead to serious health and legal consequences. Since a false negative is considered an economic loss, it should be avoided as much as possible. New validation procedures and methodologies must be developed to make testing of such systems easier. In order to reduce the number of errors and raise the bar for intelligent packaging systems, we need to investigate how existing systems can be improved.

Smart packaging system having high cost is a major barrier to their widespread implementation. If the unit price of packaged veggies is raised, most consumers will be dissatisfied. Due to the low return on investment, manufacturers may be hesitant to engage in smart packaging culture. Large quantities of product wastage in sales and distribution can be prevented with the use of these technology. After the outbreak of *E. coli*, most retailers, manufacturers, and distributors discarded the entire batch to avoid microbial contamination. This has proven to be a significant loss that could have been avoided by applying smart packaging technology. Information gathered and collected using real-time monitoring devices that give additional information on food habits may help risk manages better manage their risks. According to systems in development, any price increment should be within a few percent per bundle.

Data and cyber security are also major issues that must be resolved. The difficulty of dealing with massive data generated by real-time package monitoring has increased as traceability and tracking technologies have improved along the packaged vegetable supply chain. To do so, the network must be upgraded to accommodate the growing volume of data collected while also keeping the data private and secure from theft. This is because the collected data may contain user data or information such as location and preferences. In smart packaging systems, cryptography can help resolve this problem. With these precautions in place, collecting more precise data about each package step along the distribution network may give additional insight into customer preferences and current market trends while preventing hackers from misusing the data.

New production techniques that comply with existing packaging laws are necessary to spread and advertise the use of smart packaging systems. Printable electronics is a promising technology that has the capacity or power to overcome some of the industry's challenges. For instance, inkjet printing, is considered as a low-cost, straightforward way of creating films that may then be included into packaging. Nanotechnology's application in the packaging industry offers a lot of potential. This is due to the easy integration of the NP into the package, the rarity or uniqueness of the features it can provide, the extremely high accuracy, the communication capacity, and the inexpensive equipment [96].

8. Conclusion

Smart packaging technologies can alleviate issues, like food deterioration and microbial contamination that are common with traditional packaging. The most important factors to consider as part of a package item quality assessment are the carbon dioxide to oxygen ratio in the package, the ethylene reaction, the presence of pathogens, and the temperature that has a significant impact on the rate of pathogen development, there is humidity and all of these properties can be controlled using the active package. By capturing or releasing objects such as oxygen or carbon dioxide scavengers, dehumidifiers or release materials, antibacterial agents. Sensors like optical, bio, gas, and humidity sensors and various indicators are examples of intelligent systems, all used to track such properties. However, the difficulties in terms of regulation, cost, privacy issues and the long-term viability of the technologies utilized are preventing mass-production of smart devices. This can be resulted in the collaboration between researchers, industries and the regulatory agencies for the successful development of smart package system of required core medicinal food supplements as user's friendly, long life easy to handle stable system. Authorities and consumers will be able to further develop smart packaging technologies to deploy them in a cost-effective manner for the good storage of all types of medicinal food supplements in all conditions.

References

[1] S.K. Amit, M. Uddin, R. Rahman, S.M. Islam, M.S. Khan, A review on mechanisms and commercial aspects of food preservation and processing, Agric. Food Secur. (2017) 1–22.
[2] W.M. Cheung, J.T. Leong, P. Vichare, Incorporating lean thinking and life cycle assessment to reduce environmental impacts of plastic injection moulded products, J. Clean. Prod. 167 (2017) 759–775.
[3] K.L. Yam, P.T. Takhistov, J. Miltz, Intelligent packaging: concepts and applications, J. Food Sci. 77 (1) (2005) 1–10.
[4] J.P. Kerry, M.N. O'grady, S.A. Hogan, Past, current and potential utilisation of active and intelligent packaging systems for meat and muscle-based products: a review, Meat Sci (2006) 113–130.
[5] S. Otles, B. Yalcin, Intelligent food packaging, LogForum 4 (4) (2008) 3.
[6] Food Safety and Standards (Health Supplements, Nutraceuticals, Food for Special Dietary Use, Food for Special Medical Purpose, Functional Food and Novel Food) Regulations, 2016.
[7] E.B. Lehmann, B. Liebscher, C. Bendadani, Food Supplements: Definition and Classification, Drug Discovery and Evaluation: Methods in Clinical Pharmacology, Springer, Switzerland, 2011, pp. 625–636.
[8] E.A. Slawik, Food Supplements or Medicinal Products? Borderline Products in the EU, Wageningen, 2016.
[9] A Guide to what Is a Medicinal Product, MHRA, 2020, pp. 1–23.
[10] S. Thakkar, E. Anklam, A. Xu, F. Ulberth, J. Li, B. Li, M. Hugas, N. Sarma, S. Crerar, S. Swift, T. Hakamatsuka, V. Curtui, W. Yan, X. Geng, W. Slikker, W. Tong, Regulatory landscape of dietary supplements and herbal medicines from a global perspective, Regul. Toxicol. Pharmacol. (2020) 1–11.
[11] H. Cheng, H. Xu, D.J. McClements, L. Chen, A. Jiao, Y. Tian, M. Miao, Z. Jin, Recent advances in intelligent food packaging materials: principles, preparation and applications, Food Chem 375 (2022) 131738, 1–13.

[12] T. Bolumar, M.L. Andersen, V. Orlien, Antioxidant active packaging for chicken meat processed by high pressure treatment, Food Chem 129 (2011) 1406−1412.
[13] D. Schaefer, W.M. Cheung, Smart packaging opportunities and challenges, Procedia CIRP 72 (2018) 1022−1027.
[14] S. Pirsa, I.K. Sani, S.S. Mirtalebi, Nano-biocomposite based color sensors: investigation of structure, function, and applications in intelligent food packaging, Food Packag. Shelf Life 31 (2022) 100789, 1−17.
[15] P. Singh, A.A. Wani, S. Saengerlaub, Active packaging of food products: recent trends, Nutr. Food Sci. 41 (4) (2011) 249−260.
[16] S. Yildirim, B. Rocker, M.K. Pettersen, J.N. Nygaard, Z. Ayhan, R. Rutkaite, T. Radusin, P. Suminska, B. Marcos, V. Coma, Active packaging applications for food, Compr. Rev. Food Sci. Food Saf. 17 (1) (2018) 165−199.
[17] B.P. Day, Active Packaging, Food Packaging Technology, Blackwell Publishing Ltd., U.K., 2003, pp. 282−302.
[18] C. Vilela, M. Kurek, Z. Hayouka, B. Rocker, S. Yildirim, M.D.C. Antunes, J. Nilsen-Nygaard, M.K. Pettersen, C.S.R. Freire, A concise guide to active agents for active food packaging, Trends Food Sci. Technol. 80 (2018) 212−222.
[19] M. Soltani, R. Alimardani, H. Mobli, S.S. Mohtasebi, Modified atmosphere packaging: a progressive technology for shelf-life extension of fruits and vegetables, J. Appl. Packag. Res. (2015) 33−59.
[20] Anon, Up and Active—Pira's Latest Market Report Plots a Healthy Future for Active Packaging, Active & intelligent pack news, 2005, pp. 1−17.
[21] R. Coles, D. McDowell, M.J. Kirwan, Food Packaging Technology, CRC press, 2003, pp. 65−91.
[22] G.L. Robertson, Food Packaging − Principles and Practice, second ed., CRC Press, Boca Raton, FL, USA, 2006, pp. 1−8.
[23] M.L. Rooney, Active Food Packaging, first ed., Chapman & Hall, London, U. K, 1995.
[24] M.L. Rooney, Introduction to active food packaging technologies, in: J.H. Han (Ed.), Innovations in Food Packaging, Elsevier Ltd., London, UK, 2005, pp. 63−69.
[25] T.P. Labuza, W.M. Breene, Applications of "active packaging" for improvement of shelf-life and nutritional quality of fresh and extended shelf-life food, J. Food Process. Preserv. (1989) 1−69.
[26] A. Nura, Advances in food packaging technology-a review, Int. J. Postharvest Technol. Innov. (2018) 55−64.
[27] K.A. Mane, A review on active packaging: an innovation in food packaging, Int. J. Environ. Agric. Biotech. (2016) 544−549.
[28] S. Upasen, P. Wattanachai, Packaging to prolong shelf life of preservative-free white bread, Heliyon (2018), e00802.
[29] P. Suppakul, J. Miltz, K. Sonneveld, S.W. Bigger, Active packaging technologies with an emphasis on antimicrobial packaging and its applications, J. Food Sci. 68 (2) (2003) 408−420.
[30] N.D. Kruijf N, M.V. Beest, R. Rijk, T. Sipilainen-Malm, L.P. Paseiro, B. De Meulenaer, Active and intelligent packaging: applications and regulatory aspects, Food Addit. Contam. (2002) 144−162.
[31] J.H. Hotchkiss, Safety considerations in active packaging, Active Food Packaging (1995) 238−255.
[32] N. Church, Developments in modified-atmosphere packaging and related technologies, Trends Food Sci. Technol. (1994) 345−352.
[33] G.C. Otoni, J.P. Paula, R.J. Espitia, A. Bustillos, H.M. Tara, Trends in antimicrobial food packaging systems: emitting sachets and absorbent pads, Food Res. Int. (2016) 60−73.
[34] P. Appendini, J.H. Hotchkiss, Review of antimicrobial food packaging, Innov Food Sci Emerg Technol (2002) 113−126.
[35] H. Karleigh, Active and intelligent packaging: innovations for the future, in: Book: Food Processing and Technology: New Emerging Areas, 2014, pp. 10−21. Edition.
[36] M. Anderoost, P. Ragaert, F. Devlieghere, B. De Meulenaer, Intelligent food packaging: the next generation, Trend. Food Sci Technol (2014) 47−62.
[37] M. Ghaani, C.A. Cozzolino, G. Castelli, S. Farris, An overview of the intelligent packaging technologies in the food sector, Trends Food Sci. Technol. 51 (2016) 1−11.

[38] P. Prasad, A. Kochhar, Active packaging in food industry: a review, IOSR J. Environ. Sci. Toxicol. Dood Technol. (2014) 1–7.
[39] C.E. Realini, B. Marcos, Active and intelligent packaging systems for a modern society, Meat Sci (2014) 404–419.
[40] S.J. Lee, A.T. Rahman, Intelligent Packaging for food products, in: J.H. Han (Ed.), Innovation in Food Packaging, second ed., Academic Press UK, 2014, p. 171.
[41] A.A. Kader, S. Ben-Yehoshua, Effects of superatmospheric oxygen levels on postharvest physiology and quality of fresh fruits and vegetables, Postharvest Biol. Technol. (2000) 1–3.
[42] K.K. Kuorwel, M.J. Cran, J.D. Orbell, S. Buddhadasa, S.W. Bigger, Review of mechanical properties, migration, and potential applications in active food packaging systems containing nanoclays and nanosilver, Compr. Rev. Food Sci. Food Saf. (2015) 11–30.
[43] K.R. Sharrock, Advances in freshness and safety indicators in food and beverage packaging, in: Emerging Food Packaging Technologies, Woodhead Publishing, U.K., 2012, pp. 175–197.
[44] E. Mohebi, L. Marquez, Intelligent packaging in meat industry: an overview of existing solutions, J. Food Sci. Technol. (2015) 47–64.
[45] B.M. Coll, Manufacturers Services Ltd, Assignee. Solder Paste with a Time-Temperature Indicator. United States Patent US 6,331,076, December 18, 2001.
[46] A.A. Mirza, M. Masoomian, M. Shakooie, K.M. Zabihzadeh, M. Farhoodi, Trends and applications of intelligent packaging in dairy products: a review, Crit Rev. Food Sci. Nutr. (2020) 1–5.
[47] N.B. Borchert, J.P. Kerry, D.B. Papkovsky, A CO2 sensor based on Pt-porphyrin dye and FRET scheme for food packaging applications, Sens. Actuators B. Chem. (2013) 57–65.
[48] Anon, European Expansion for Moisture Absorber, Active & intelligent pack news, 2003.
[49] C. Chatterjee, A. Sen, Sensitive colorimetric sensors for visual detection of carbon dioxide and sulfur dioxide, J Mater. Chem. A (2015) 2–7.
[50] G.L. Robertson, Active and Intelligent Packaging, Food Packag, 2016, pp. 429–458.
[51] F. Formisano, E. Massera, D.S. Vito, A. Buonanno, G. Di Francia, P.D. Veneri, Auxiliary smart gas sensor prototype plugged in a RFID active tag for ripening evaluation, in: 2015 XVIII AISEM Annual Conference, 2015, pp. 1–4.
[52] G. Wang, H. Jinsong, Q. Chen, X. Wei, H. Ding, Z. Jiang, J. Zhao, Verifiable smart packaging with passive RFID, IEEE Trans Mob Comput (2019) 1217–1230.
[53] H. Beshai, et al., Freshness monitoring of packaged vegetables, App. Sci. (2020) 1–41.
[54] F. Formisano, E. Massera, S. De Vito, A. Buonanno, G. Di Francia, P.D. Veneri, RFID tag for vegetable ripening evaluation using an auxiliary smart gas sensor, Sens (2014) 2026–2029.
[55] A.V. Quintero, F. Molina-Lopez, E.C. Smits, F. Danesh, J. van den Brand, K. Persaud, A. Oprea, N. Barsan, U. Weimar, N.F. De Rooij, D. Briand, Smart RFID label with a printed multisensor platform for environmental monitoring, Flex. Print. Electron. (2016) 1–12.
[56] Z.M. Tahir, E.C. Alocilja, Fabrication of a disposable biosensor for *Escherichia coli* O157: H7 detection, IEEE Sens. J. (2003) 45–51.
[57] V. Velusamy, K. Arshak, O. Korostynska, K. Oliwa, C. Adley, An overview of foodborne pathogen detection: in the perspective of biosensors, Biotechnol. Adv. (2010) 32–54.
[58] Y. Yi, L. Sanchez, Y. Gao, Y. Yu, Janus particles for biological imaging and sensing, Analyst 141 (12) (2016) 3526–3539.
[59] M. Tahir, Z. Alocilja, A conductometric biosensor for biosecurity, Biosens. Bioelectron. (2003) 813–819.
[60] D.W. Unfricht, S.L. Colpitts, S.M. Fernandez, M.A. Lynes, Grating-coupled surface plasmon resonance: a cell and protein microarray platform, Proteomics (2005) 32–42.
[61] B.K. Oh, W. Lee, B.S. Chun, Y.M. Bae, L.H. Lee, J.W. Choi, The fabrication of protein chip based on surface plasmon resonance for detection of pathogens, Biosens. Bioelectron (2005) 47–50.
[62] C. Griesche, A.J. Baeumner, Biosensors to support sustainable agriculture and food safety, Trends Analyt Chem (2020) 1–57.
[63] Y. Liu, J. Ye, Y. Li, Rapid detection of *Escherichia coli* O157: H7 inoculated in ground beef, chicken carcass, and lettuce samples with an immunomagnetic chemiluminescence fiber-optic biosensor, J. Food Prot. (2003) 512–517.

[64] H. Wang, Y. Zhou, X. Jiang, B. Sun, Y. Zhu, H. Wang, Y. Su, Y. He, Simultaneous capture, detection, and inactivation of bacteria as enabled by a surface-enhanced Raman scattering multifunctional chip, Angew. Chem. Int. Ed. Engl. (2015) 1—5.

[65] J.L. McKILLIP, L.A. Jaykus, M. Drake, Influence of growth in a food medium on the detection of *Escherichia coli* O157: H7 by polymerase chain reaction, J. Food Prot. (2002) 5—9.

[66] S. Nakano, T. Kobayashi, K. Funabiki, A. Matsumura, Y. Nagao, T. Yamada, PCR detection of Bacillus and Staphylococcus in various foods, J. Food Prot. (2004) 1—7.

[67] M. Yoshida, K.H. Roh, S. Mandal, S. Bhaskar, D. Lim, D. Nandivada, X. Deng, J. Lahann, Structurally controlled bio-hybrid materials based on unidirectional association of anisotropic microparticles with human endothelial cells, Adv. Mater. (2009) 4920—4925.

[68] N.A. Luechinger, S. Loher, E.K. Athanassiou, R.N. Grass, W.J. Stark, Highly sensitive optical detection of humidity on polymer/metal nanoparticle hybrid films, Langmuir (2007) 3—7.

[69] M. Zdeblick, P. Valley, et al., Inventors; Proteus Digital Health, Inc., Redwood City, CA (US), Assignee; Pharma—Informatics System, US 10,610,128 B2, 2017.

[70] T.V. Duncan, Applications of nanotechnology in food packaging and food safety: barrier materials, antimicrobials and sensors, J. Colloid Interface Sci. (2011) 1—24.

[71] F. Mustafa, S. Andreescu, Chemical and biological sensors for food-quality monitoring and smart packaging, Foods 168 (2018) 1—20.

[72] T.C. Le, J. Zhai, W.H. Chiu, P.A. Tran, N. Tran, Janus particles: recent advances in the biomedical applications, Int. J. Nanomedicine (2019) 6749—6777.

[73] P.A. Suci, S. Kang, M. Young, T. Douglas, A streptavidin—protein cage janus particle for polarized targeting and modular functionalization, J. Am. Chem. Soc. 131 (2009) 4—5.

[74] L.F. Wang, J.W. Rhim, Preparation and application of agar/alginate/collagen ternary blend functional food packaging films, Int. J. Biol. Macromol. (2015) 460—468.

[75] M.R. De Moura, L.H. Mattoso, V. Zucolotto, Development of cellulose-based bactericidal nanocomposites containing silver nanoparticles and their use as active food packaging, J Food Eng 109 (2012) 520—524.

[76] A. Mills, Q. Chang, Fluorescence plastic thin-film sensor for carbon dioxide, Analyst. 118. 39—43.

[77] H. Sahoo, Forster resonance energy transfer—A spectroscopic nanoruler: principle and applications, J. Photochem. Photobiol. C. (2011) 20—30.

[78] F. Molina-Lopez, A.V. Quintero, G. Mattana, D. Briand, N.F. De Rooij, Large-area compatible fabrication and encapsulation of inkjet-printed humidity sensors on flexible foils with integrated thermal compensation, J Micromech Microeng (2013) 1—11.

[79] C. Boniello, T. Mayr, J.M. Bolivar, B. Nidetzky, Dual-lifetime referencing (DLR): a powerful method for on-line measurement of internal pH in carrier-bound immobilized biocatalysts, BMC Biotechnol (2012) 2—10.

[80] W. Gopel, J. Hesse, J.N. Zemel, Sensors: A Comprehensive Survey, 1989, pp. 79—106.

[81] Lindsay, Jeffrey, Heriberto, Eric Inventors; Callahan Cellular, Assignee; Activating a Data Tag by Load or Orientation or User Control, EP 1 743 282 B1, 2005.

[82] Withers, Inventor; International Consolidated Business Group, Assignee; Antioxidant Active Food Packaging, AU 2016325872 B2, 2016.

[83] Bell, Dippo, Inventors; Tda Research inc., Self-Heating Flexible Packages, WO 01/04548 A1, 2001.

[84] J. Foster, C. Albert, Hyung, Inventors; AGILEQR, INC. US, Assignee; System and Method for Product Track and Trace Utilizing Unique Item-specific Urls and Quick-Response (Qr) Codes, WO 2014/126654 A1, 2014.

[85] C. Lenexa, J. Faragher, H. Richfield, inventors Knebel, DuPont Nutrition Biosciences APS, Copenhagen K (DK), Assignee; Encapsulated Antimicrobial Material, US 8,337.923 B2, 2004.

[86] A.J.M. Garwood, Mercer Island, WA, SafeFresh Technologies, LLC, Mercer Island, WA (US), Packages and Methods for Processing Food Products, US 7,205,016 B2, 2003.

[87] Mingerink, B. Corey, et al., Inventors; Binforma Group Limited Liability Company Dover, Assignee; Packaging Systems Integrated with Disposable Rfid Devices, EP 2 047 413 B1, 2007.

[88] E. Varriano-Marston, inventors, Marc Patterson, Assignee; Registered Microperforated Films for Modified/controlled Atmosphere Packaging, US 7,083,837 B1, 2001.

[89] J.D. Lindsay, et al., Inventors; Binforma Group Limited Liability Company, Dover, DE (US), Assignee; Reversibly Deactivating a Radio Frequency Identification Data Tag, US 7,948,381 B2, 2003.
[90] S. Mansuino, inventor, L. Bosotti, et al., c/o Buzzi, Notaro & Antonielli d'Oulx Srl, via Maria Vittoria, Sealing Wrapper for Food Products, Corresponding Process and Installation, EP 1 477 423 B1, 2003.
[91] B. Jonathan Davis, S. Francisco, et al., Inventors, Assignee; Shipping Package Tracking or Monitoring System and Method, US 2020/0051015 A1, 2019.
[92] G.L. Clark, M.I. Ada, et al., Inventors; Inductive Intelligence, LLC, Grandmother Rapids, MI, Assignee; Smart Packaging, Systems and Methods, US 11,317,480 B2, 2019.
[93] J. Gentile, Montclair, et al., Inventors; TInk, Inc., New York, NY (US), Assignee; Smart Packaging and Related Technologies, US 2015/0353236A1, 2015.
[94] R.A. Connor, Forest Lake, MN, Inventors; Medibotics LLC, St. Paul, MN (US), Assignee, Smart Watch and Food Utensil for Monitoring Food Consumption, US 9,536.449 B2, 2013.
[95] R. Poornachandran, et al., Inventors; Intel Corporation, Santa Clara, CA, Assignee; System, Apparatus and Method for Transferring Ownership of a Smart Delivery Package, US 2017/0178072 A1, 2015.
[96] Y. Wyser, M. Adams, M. Avella, D. Carlander, L. Garcia, G. Pieper, M. Rennen, J. Schuermans, J. Weiss, Outlook and challenges of nanotechnologies for food packaging, Packag. Technol. Sci. (2016) 15–48.

CHAPTER 15

Smart packaging to preserve fruit quality

Pinku Chandra Nath[a], Biswanath Bhunia[a], Tarun Kanti Bandyopadhyay[b] and Biplab Roy[b]

[a]Department of Bio Engineering, National Institute of Technology Agartala, Jirania, Tripura, India; [b]Department of Chemical Engineering, National Institute of Technology Agartala, Jirania, Tripura, India

Abbreviations

H_2S Hydrogen sulfide
$KMnO_4$ Potassium permanganate
MAP Modified atmosphere packaging
NDRI Non dispersive infrared
RFID Radio frequency identification device
TTIs Time temperature indicators

1. Introduction

Fruits have always been an important part of a healthy diet because they contain a wide range of vitamins, minerals, fibers, phytochemicals, and flavonoids, all of which have been shown to be good for you (Table 15.1). Fruits are believed to reduce the incidence of some cancers, cardiovascular disorders, and strokes through increasing antioxidant capability [9]. In most cases, fruit contains water-emulsifiable vitamins including B and C vitamins, along with fat-dispersible vitamins like A, E, and K. Apart from specialized minerals such as Cu, Se, and iodine, the majority of fruits are rich in Ca, K, Na, K, and Fe, among other nutrients.

Fruit fibers are classified as soluble or insoluble, with the former referring to primarily indigestible components of fruit contained within the fruit surface that behave as 'roughage' and add bulk. Cellulose, hemicellulose, lignin, mucilage and pectin are the constituents of fruit fibers [10]. This type of fiber acts as sponges for its considerable water ability, and it has been shown to help regulate bowel motions and also decrease cholesterol concentrations in blood. Middle-aged persons have a significant benefit from fruit fibers and antioxidants, including the presence of monounsaturated fatty acids, when it comes to lowering inflammatory response and associated coronary heart disease events [11].

Phytochemicals called fruit flavonoids have a wide range of medical and pharmacological applications [12]. Citrus fruits in particular are high in flavonoids such as tangeretin, rutin, and nobiletin, which aid in the treatment of numerous forms of lungs, mouth,

Table 15.1 Nutritional features and dietary implications of fruits.

Fruit name	Chemical name	Active chemical name	Mode of action	Applications	Reference
Acai	Flavonoids, omega-6 and omega 9 fatty acids	Oleic and linolenic acids	Antioxidants	Cholesterol level low in blood, skin regenerate, anti aging properties	[1,2]
Apples	Polyphenols, vitamin C, flavonoids, malic acid, tartaric acid	Catechin, quercetin, chlorogenic acid, pectin	Antioxidants	Heart function improves, promotes prevention against cancer	[3,4]
Grapes	Flavonoids and polyphenol	Resveratrol, pterostilbene	Anti-initiating, anti-promoting	Reduction in prostrate and colon cancer	[5,6]
Citrus fruits	Flavonoids, flavonol	Tangeretin, rutin, nobiletin, naringenin	Antioxidants and antibiotic	Protection against Rheumatoid Arthritis	[7,8]

cancers and stomach [13]. Increasing the intake of beta-carotene-rich fruits and vegetables, as well as C and E vitamins, has been shown to minimize the chances of pharyngeal, oral, and breast cancer [14]. The flavonoids and polyphenols found in fruits behave as signal substances, affecting the gut micro ecology in two ways: straightforwardly by efficiently hindering pathogenic microbial species binding to gut exterior and obliquely by expanding the quantity of advantageous gut microbes through augmented probiotic microbial species adhesion to the gut surface. While fruits are emphasized in today's good dietary plans, the molecular (genomic, proteomic, and metabolomic) processes behind their beneficial benefits must be explored. Smart packaging technologies will be discussed in this chapter, as well as their interoperability with freshness detectors and the issues connected with integrating them with freshness detectors. We will also address the most up-to-date applications of freshness detectors as smart packaging systems for fruit protection and quality control, as well as their limitations.

2. Fruit freshness is related to classification, stages, and harvesting

Fruits' freshness is determined by a number of factors, including their classification, stage of development, and harvesting process. Fruits are natural plant structures that offer nourishment for a wide variety of animals, including humans. The taxonomy and basic growth of fruit are described in this part to serve as a foundation for the remainder of this review. Fruits are classified as reproductive components of plants when they are classified anatomically. Plants create seeds for their future generation in their fruits, which are made from the tissues of their flower structures. A plant's ovulation undergoes pollination, which involves the transfer of pollen to fertilize it. When the pericarp, or the tissue around the ovary, grows into a thick coating, the seeds are protected. In majority of cases, pericarp is the portion of a fruit which is intended for consumption. As illustrated in Fig. 15.1, the fruit is categorized into three classifications: simple fruits, aggregate fruits, and numerous fruits.

3. Smart packaging technologies for fruit preservation and freshness

A fruit's freshness indicator can identify and review on the fruit's juiciness, ripeness, and constancy. On-package signals that detect the surroundings both within and without the package and provide consumers with information on the state of the fruits are known as fruit freshness detectors. Smart packaging from another end refers to an integrated technology which is made up of electrochemical sensing devices which are placed adjacent to packaged fruits in order to improve their quality. As demonstrated in Fig. 15.2, this system enables continual monitoring of packaged fruits quality from the time they left the plant till they reach customers. The usage of freshness detectors in fruit packing is highly variable depending on the fruit variety and physiology.

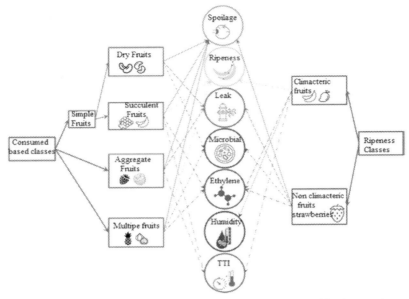

Fig. 15.1 Fruit categorization framework and its relationship to several freshness characteristics.

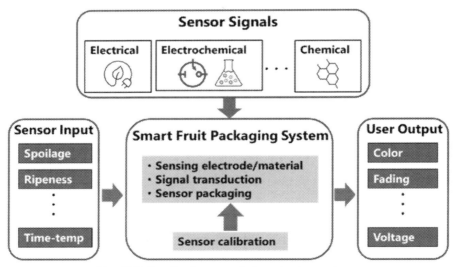

Fig. 15.2 Details of a smart fruit packaging systems.

Therefore, as explained in the next subsections, it is critical to have a comprehensive taxonomy of fruits as well as a basic concept of their physiology for developing and deploying freshness indicators and smart packaging technologies.

3.1 The significance of smart packaging and its connection to freshness sensors

Consumer and transportation packaging are the two primary categories of fruit packaging used in the shipment and selling of produce [15]. The term "Consumer packaging" refers to a group of fruits that are individually packaged and marketed as a single item. In this packaging, plastic films are frequently used as the packing material. Aside from ensuring the effective and secure transportation of large quantities of goods, transport packaging has been designed to make it easier to carry goods in bulk. Shipments are usually made in bulk in crates of various sizes made of plastic or cardboard; although, plastic containers too are favored because of their excellent longevity and simplicity of cleaning [16]. It is critical for both kinds of packing to ensure that the goods included within them are shielded from damages and to convey authentic and complete details on the source, grade, nutritional value, and other characteristics of the goods contained within them [15]. The properties of plastic packing containers, such as their opacity, mechanical durability, and chemical stability, are crucial for the integration of smart packaging indicators. This will have a negligible effect on the fruit packing industry's standards.

Fruits are packaged in a variety of ways and sold to consumers. For instance, frozen fruits that are fat-free and do not decompose often have a far longer shelf life than their fresh counterparts. However, other studies indicate that when frozen fruit is thawed after prolonged storage, nutrients are lost and the texture of fruit is distorted [17]. Freezer burn can occur if frozen food gets exposed to air, which causes the food to lose moisture. This changes the taste and texture [17]. The nutritional content of canned fruit is diminished when contrasted to the nutritive benefits of genuine organic fruit, despite the fact that canned fruit has a prolonged shelf life. In spite of its limited shelf life and high price, fresh food is still regarded to become the greatest flavorful and nutritionally rich form of produce available [18]. As a result, it continues to be one of the most common methods of produce delivery, accounting for a sizable percentage of the fruit and vegetable industry [18]. The quality of fresh fruits, nutritional content, and safety of fresh fruits must be monitored with smart packaging in order to meet the needs of customers [19]. As a result, the sector is increasingly driven by the demand for better quality control and maintenance systems. After the product has been examined, processed, and packaged, it receives almost no additional scrutiny during transportation or by re-sellers until it reaches its final destination. Fresh produce across most marketplaces is kept at a temperature of roughly 2–4°C in order to slow the ripening process. While keeping an eye on the storeroom's temperature and humidity can help keep things from going bad, certain fruits, like bananas, which turn black when cooled, can't be preserved in this way. Additional environmental controls are lacking in grocery shop display cases, where fruit is frequently exposed to air [20]. Because of a deficiency of monitoring infrastructures at this stage of supply chain, it is challenging task to authenticate the freshness of fruits are preserved before they are delivered to customers. To solve these issues, a number of companies [19] have started implementing smart packaging technologies which may be utilized to assure quality control and maintenance from the very beginning of the manufacturing process.

3.2 Intelligent packaging technologies for fruit

Fruit packaging problems can be detected and remedied using a number of sensor techniques and smart packaging approaches. Depending on the sort of fruit being observed, different sensors may be utilized.

3.2.1 Direct freshness

Freshness sensors are classified into two types: direct sensors and indirect sensors. A direct freshness sensor, as the name implies, detects a specific analytes straight from fruit itself in order to determine its level of freshness. In order to determine whether or not fruits are still fresh, special markers or compounds are used in direct fruit freshness detectors. Among the direct fruit freshness detectors that are often employed are rotten, ripeness, leakage, microbiological infection indicators, as well as ethylene and senescence detectors. The sensors include a color indicator that allows the freshness level to be easily recognized with the naked eye.

3.2.1.1 Spoilage

Fruits decay when their quality deteriorates to the point where they are no longer edible. While fishes and meats are much more commonly linked with spoiling indicators, fruits can also show certain signs of decomposition due to a variety of outside stimuli. External elements such as temperature, humidity, gas, and the surrounding atmosphere all contribute to the rotting of fruits. Food rotting can be detected utilizing chemical indicators, such as pH indicator which monitors the freshness of fruit noninvasively and in real time. A pH sensor can be used to determine how the pH of an encapsulated fruit package varies inside the headspace. Typically, pH indicator dyes are employed, which change color when exposed to acids or bases. When a fruit degrades, it emits a variety of volatile organic chemicals that are detectable with pH indicators. pH sensors are a low-cost, extremely precise, and simple-to-use method of determining the freshness of food on-package. Colorimetric pH sensors for the detection of aldehydes have already been described, for example. A pH detector was designed to detect glutaraldehyde in liquid state using chemical reaction principle of the H_2 bond generated among glutaraldehyde oligomer generated by aldol process and the detector [21]. Another research implemented the chemical interaction among formaldehyde and primary amines to identify the presence of formaldehyde in environment [22]. Researchers found that both experiments used fruit with different levels of basicity (Fig. 15.3).

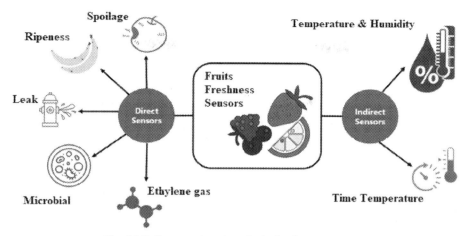

Fig. 15.3 Represent various fruits freshness sensors.

3.2.1.2 Ripeness

Ripeness is a critical sign of fruit freshness, yet it is frequently difficult for consumers to evaluate. Typically, these indications are geared toward detecting the initial type of alteration (pH, gas content, etc.). The indicators detect these alterations and convert them to a response, typically a color reaction that could be conveniently quantified and connected to food freshness. This class of sensors detects metabolites by color signals (e.g., pH) or biosensors. Fresh-Tag®, a colorimetric sensor that notifies customers to volatile amines production in fisheries commodities, was launched by COX Technologies but abandoned in 2004 [17]. Methyl red has been employed to construct the label detector; the cellulose (metyl/red) membranes rely on an elevation in pH brought about by decomposition of volatile amines in order to function. Real-time surveillance of raw broiler chicken meat was made possible with this sensor label [23]. Ripe-Sense™ is the nation's foremost smart sensor label that alters color in response to the fruit's ripeness [24]. It functions by responding to fragrances generated by fruit as it reaches maturity; initially red, it progresses to orange followed by yellow color, depending on the amount of ripeness desired for eating the fruit (Fig. 15.4).

3.2.1.3 Leak

Fruit ripeness is a measure of the freshness of uncooked fruits. However, another critical aspect is the freshness of processed/packaged fruits. The package integrity can jeopardize the freshness of packed fruits. As a result, leaks can be a substantial impediment to preserving the freshness of packaged food. Because of this, a leak detector could be employed to evaluate the durability of packaging fruits which affects fruit freshness (Fig. 15.5). The hazardous content of gases released during the breakdown of packed fruit provides a concern to customer health and wellbeing when leak indications, which are often gas sensors, are activated [25].

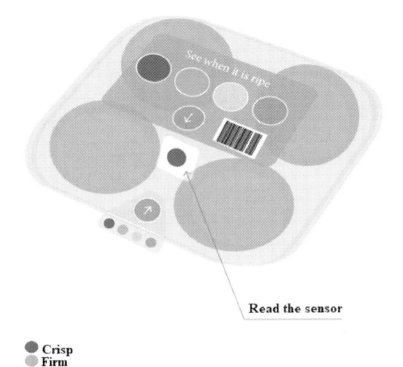

Fig. 15.4 Graphic representation of Ripe-Sense detector.

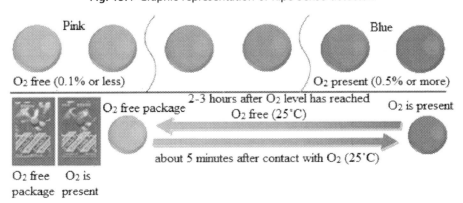

Fig. 15.5 Schematic representation of the leak indicators.

MAP and equilibrium MAP have been commonly used in the packing of processed fruits, which necessitates the employment of gas detectors to monitor fruit freshness [26]. In MAP, the atmosphere often has a lower O_2 content (2%), but a greater CO_2 content (20%–80%). MAPs can be divided into two categories: passive and active

MAPs. In order to ensure a stable environment inside the MAP, the product's respiration and the opacity of the packaging substance must be taken into consideration. For instance, when pure cut apples were maintained at 2 °C, O_2 concentration declined from 21% (first hour) to 6.89% (190 h), while the CO_2 content grew from 0.02% to 7.98%; after that, the concentrations reached nearly constant or equilibrium [27]. Product purity is maintained or extended by using active MAP, which includes ingredients that aid in the packaging process. As a result, sensors that detect leaks may keep tabs on the condition of packaged food and supply valuable feedback on fruit's freshness. Fruits benefit greatly from reduced oxygen levels in packing. In apples, 0.5–5 kPa O_2 causes diminished ethylene sensitivity as well as slowed respiratory and metabolic activities [28]. A leakage in MAP packaging considerably enhances the quantity of O_2 obtained from the environment and reduces the level of CO_2, permitting aerobic microbial development. Since MAP leak detectors are able to identify O_2 instead of CO_2, they are more accurate [29].

3.2.1.4 Microbial pathogens

Fruit freshness requires the identification of microbiological pathogens. In many circumstances, a sample must be extracted in order for the sensors to detect it. Because of this, incorporating sensors into packaging may provide an easily visible variation in color while maintaining a low manufacturing cost is difficult. Following on from the preceding point, the most straightforward approach of identifying microbiological contamination seems to be to observe variations in gas content within packaging that are induced by bacterial activity. However, diseases are uncommon in fruit packaging, and there are rare specific examples of packaging sensors applied to identify pathogens [30]. Detecting microbiological degradation in a fruit packaging with elevated CO_2 concentrations is a challenging task. Microbial pathogen detectors that rely upon CO_2 gas identification may not work in packaging that include any CO_2 gas [31]. Conductive polymers are sometimes used to make biosensors that detect gases. Microbial metabolism produces gases, which can be detected with biosensors [32,33]. A matrix of insulating nanomaterials and conducting polymer nanomaterials was employed to create the biosensors. The matrix's resistance varies in proportion to the amount of gas emitted [34].

3.2.2 Indirect freshness

Indirect freshness sensors as it should be reproduce the trade in precise first-class signs of food exposed to the identical indirect freshness markers. The sensors price of trade need to be connected with the price of decay of the packed food due to temperature/humidity versions through the years at some stage in transit, distribution, and garage. When subjected to aberrant garage temperatures, the oblique detectors must suggest the results of freshness monitoring by alteration of color and digital sign output. Indirect freshness sensors fall into different basic classes which are described below:

3.2.2.1 TTIs

Indirect freshness sensors, such as TTIs, detect changes in freshness by a combination of physi0-chemical, and biological means [35]. Chemical and physiological techniques are entirely based on chemical reactions induced by temperature and timing adjustments, respectively. Acid—base interactions, polycondensation, and melting are all examples of such mechanisms. Microorganisms, spores, and enzymes all showcase organic hobby in response to variations in temperature and timing. TTI detectors alternate color while subjected to better than acceptable garage temperatures for a sustained duration of time and in addition they indicate while a product have reached the quit of its shelf lifestyles. Thus, TTIs can be used to screen the physical, chemical, and organic interest of end result, imparting a clear, accurate indication in their freshness in phrases in their first-rate, protection, and shelf existence. Numerous industrial TTIs were developed and are generally used for perishable commodities tracking. For instance, as proven in Fig. 15.6, a few examples of TTIs consist of Fresh-Check®, Monitor-Mark®, OnVu®, eO®, Time-strip®, Check-point®, and Tempix®. Fruit programs can regularly gain from these business indicators, which are primarily based on chemical or enzymatic processes.

3.3 Container-based smart packaging technology

3.3.1 RFID sensors

The quality of the fruits or fruit products has been the primary focus of earlier sensing systems, but the fruit container itself can provide a wealth of information as well. RFID tags can also be used to track and identify the product quality while it is being stored and transported in containers. An RFID tag is an electronic barcode that may be read by a computer network to identify a single packed product [36]. A RFID tag including an antenna and a RFID scanner are necessary for these technologies, which transmit data via radio waves. Protective material covers an embedded circuit and antenna that are encased in the tag. Information regarding the package could be obtained by scanning RFID tag including an antenna and receiver which transform radio waves emitted by the labels toward proper data structure [37]. These scanners could operate alone or in conjunction with a central network capable of storing, processing, and further propagating the received data.

3.3.2 CO_2 non-dispersive infrared (NDIR) sensor

The concentration of specific gases can be detected using an NDIR sensor, which measures the absorption of infrared light. Light having wavelengths varying between 700 nm and 1 m is commonly used to illuminate an optical detector while passing through

Smart packaging to preserve fruit quality   277

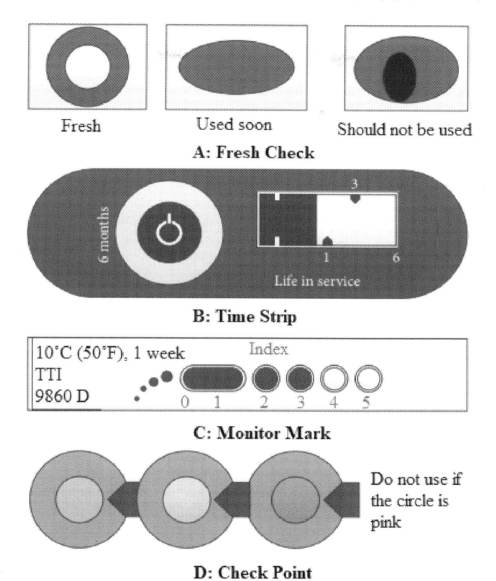

Fig. 15.6 Schematic diagram of the TTIs.

atmosphere containing target gas [38]. Certain wavelengths of light are absorbed by atmospheric gases, while others are allowed to flow through. Consequently, the amount of light absorbed by the target gas particles is recorded by optical sensor device, which turns it into a detectable electrical output based on the amount of light absorbed [39]. For measuring atmospheric gas concentrations, NDIR sensors make use of infrared sensors to analyze light absorption properties (Fig. 15.7). Light is absorbed by certain gases at varying wavelengths, whereas others are allowed to flow through. Light absorption or light transmission via the optical sensor can be used to determine how many target gas molecules are present. The optical detector detects the amount of light and converts it to a quantifiable electrical output [40–42].

3.4 Fruit packaging with active technologies

Active packaging is a step forward in the development of the notion of smart packaging. The goal is to implement a continual quality assurance strategy for each product that is separately wrapped. Through the use of regulatory elements that are integrated into the packaging substances, the package themselves has the ability to extract substances as well as adjust interior air conditions to satisfy the demands of the packing material. Sensors will identify undesirable components in the package environment and will take appropriate action to mitigate them. In situations when perishable goods cannot be maintained in optimal environments, this is ideal for prolonging their shelf life [43].

3.4.1 Ethylene scavengers

Ethylene levels should be kept as low as possible when the fruit is being transported or stored. Ethylene scavengers could sometimes be delivered to counteract the effects of ethylene if the sensors detect higher than permissible levels. An effective scavenger is one that has the ability to react with the target compound and transform it into something else that is safe for the environment [44]. In the context of ethylene, $KMnO_4$ is the best choice because it breaks down ethylene into CO_2, H_2O, and MnO_2 in stages [45].

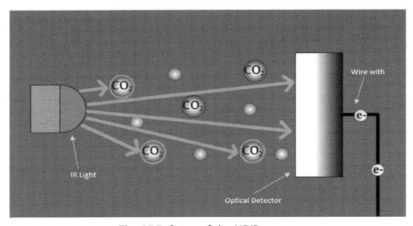

Fig. 15.7 Setup of the NDIR sensor.

Table 15.2 Commercially available intelligent and active packaging systems.

Smart packaging	Sensors	Trade name
Intelligent packaging	Freshness	Freshtag®, Sensorq®
	Temperature	Timestrip®
	TTIs	Thermax, Onvu™
		Checkpoint®, Monitormark™
		Fresh-Check®
	Leak	Novas®, Timestrip®
		Novas®, O₂ Sense™
	RFID	CS8304, Temptrip
		Easy2log®
Active packaging	Ethylene scavenger	Neupalon, Peakfresh
		Evert-fresh
	Antimicrobial packaging	Biomaster®, Agion®
	Moisture absorber	Dri-Loc®, Tenderpac®

3.4.2 Carbon dioxide emitter

By increasing carbon dioxide in the package rather than decreasing it like the active packaging technology does, carbon dioxide emitters can help protect fruits from the harmful effects of ethylene gas exposure [46]. To illustrate how this technology is implemented, it is used to slow down fruit respiration by exploiting the interaction among $NaHCO_3$ and H_2O, which leads to the creation of CO_2.

3.4.3 Moisture absorbers for humidity control

In order to maintain humidity levels low, moisture absorbers contain substances that are normally included in the packaging of the product. These compounds remove the extra moisture from the air in the packaging, making it easier to control. Silica gels, that can hold approximately 35% of its own water weight, and CaO_2, that could be utilized to pack dry foods including raw fruit, seem to be two such examples among these substances [47]. Table 15.2 provides a full overview of widely viable active and smart packaging solutions.

4. Conclusion

Freshness assessment and smart packaging technologies for fruit protection are discussed. Traditional packaging problems such as damaged products and wastes can be addressed by these new technologies. Various aspects of fruit biology and classifications, growth, processing, and harvesting were covered. This background information were analyzed in order to build smart packaging which potentially assist in reducing fruit wastes through harvesting, post-harvesting, and packaging segments of the fruit production cycle. For implementation of intelligent packaging technologies for fruit freshness monitoring, cross-disciplinary collaboration between companies, academics, and consumers may produce more sustainable solutions.

References

[1] E.M. Collins, An A–Z Guide to Healing Foods: A Shopper's Reference, Mango Media, 2010.

[2] L.A. Pacheco-Palencia, S.T. Talcott, S. Safe, S. Mertens-Talcott, Absorption and biological activity of phytochemical-rich extracts from acai (*Euterpe oleracea* Mart.) pulp and oil in vitro, J. Agricult. Food Chem. 56 (10) (2008) 3593–3600.

[3] S. Ahmad, T. Mahmood, R. Kumar, P. Bagga, F. Ahsan, A. Shamim, et al., A contrastive phytopharmacological analysis of gala and fuji apple, Res. J. Pharm. Technol. 13 (3) (2020) 1527–1537.

[4] R.P. Guiné, M.J. Barroca, T.E. Coldea, E. Bartkiene, O. Anjos, Apple fermented products: an overview of technology, properties and health effects, Processes 9 (2) (2021) 223.

[5] V. Georgiev, A. Ananga, V. Tsolova, Recent advances and uses of grape flavonoids as nutraceuticals, Nutrients 6 (1) (2014) 391–415.

[6] D. Tagliazucchi, E. Verzelloni, D. Bertolini, A. Conte, In vitro bio-accessibility and antioxidant activity of grape polyphenols, Food Chem. 120 (2) (2010) 599–606.

[7] F. Alam, K. Mohammadin, Z. Shafique, S.T. Amjad, M.H.H. Asad, Citrus flavonoids as potential therapeutic agents: a review, Phytother. Res. (2021).

[8] I. Alexander, Exploitative Beneficial Effects of Citrus Fruits. Citrus—Health Benefits and Production Technology, IntechOpen, London, UK, 2019, p. 31.

[9] N.M.A. Hassimotto, M.I. Genovese, F.M. Lajolo, Antioxidant capacity of Brazilian fruit, vegetables and commercially-frozen fruit pulps, J. Food Compos. Anal. 22 (5) (2009) 394–396.

[10] P. Ramulu, P.U. Rao, Total, insoluble and soluble dietary fiber contents of Indian fruits, J. Food Compos. Anal. 16 (6) (2003) 677–685.

[11] A. Basu, S. Devaraj, I. Jialal, Dietary factors that promote or retard inflammation, Arterioscler. Thromb. Vasc. Biol. 26 (5) (2006) 995–1001.

[12] D. Tungmunnithum, A. Thongboonyou, A. Pholboon, A. Yangsabai, Flavonoids and other phenolic compounds from medicinal plants for pharmaceutical and medical aspects: an overview, Medicines 5 (3) (2018) 93.

[13] M. Zhang, S. Zhu, W. Yang, Q. Huang, C.-T. Ho, The biological fate and bioefficacy of citrus flavonoids: bioavailability, biotransformation, and delivery systems, Food Funct. 12 (8) (2021) 3307–3323.

[14] J. Lin, N.R. Cook, C. Albert, E. Zaharris, J.M. Gaziano, M. Van Denburgh, et al., Vitamins C and E and beta carotene supplementation and cancer risk: a randomized controlled trial, J. Natl. Cancer Instit. 101 (1) (2009) 14–23.

[15] H.R. El-Ramady, É. Domokos-Szabolcsy, N.A. Abdalla, H.S. Taha, M. Fári, Postharvest management of fruits and vegetables storage, Sustain. Agricult. Rev. (2015) 65–152.

[16] J.A. Watson, D. Treadwell, S.A. Sargent, J.K. Brecht, W. Pelletier, Postharvest Storage, Packaging and Handling of Specialty Crops: A Guide for Florida Small Farm Producers, University of Florida, Florida, 2015.

[17] A.U. Alam, P. Rathi, H. Beshai, G.K. Sarabha, M.J. Deen, Fruit quality monitoring with smart packaging, Sensors 21 (4) (2021) 1509.

[18] J.C. Rickman, D.M. Barrett, C.M. Bruhn, Nutritional comparison of fresh, frozen and canned fruits and vegetables. Part 1. Vitamins C and B and phenolic compounds, J. Sci. Food Agricult. 87 (6) (2007) 930–944.

[19] G. Fuertes, I. Soto, R. Carrasco, M. Vargas, J. Sabattin, C. Lagos, Intelligent packaging systems: sensors and nanosensors to monitor food quality and safety, J. Sens. 2016 (2016).

[20] N. Donthu, A. Gustafsson, Effects of COVID-19 on Business and Research, Elsevier, 2020, pp. 284–289.

[21] E. Vo, D.K. Murray, T.L. Scott, A. Attar, Development of a novel colorimetric indicator pad for detecting aldehydes, Talanta 73 (1) (2007) 87–94.

[22] L. Feng, C.J. Musto, K.S. Suslick, A simple and highly sensitive colorimetric detection method for gaseous formaldehyde, J. Am. Chem. Soc. 132 (12) (2010) 4046–4047.

[23] P.V. Mahajan, O.J. Caleb, Z. Singh, C.B. Watkins, M. Geyer, Postharvest treatments of fresh produce, Philos. Trans. Royal Soc. A 372 (2017) (2014) 20130309.

[24] B. Kuswandi, Y. Wicaksono, A. Abdullah, L.Y. Heng, M. Ahmad, Smart packaging: sensors for monitoring of food quality and safety, Sens. Instrument. Food Quality Safety 5 (3) (2011) 137–146.
[25] S. Matindoust, M. Baghaei-Nejad, M.H.S. Abadi, Z. Zou, L.-R. Zheng, Food quality and safety monitoring using gas sensor array in intelligent packaging, Sens. Rev. (2016).
[26] Y. Fang, M. Wakisaka, A review on the modified atmosphere preservation of fruits and vegetables with cutting-edge technologies, Agriculture 11 (10) (2021) 992.
[27] C. Fagundes, B.A.M. Carciofi, A.R. Monteiro, Estimate of respiration rate and physicochemical changes of fresh-cut apples stored under different temperatures, Food Sci. Technol. 33 (1) (2013) 60–67.
[28] R. Beaudry, MAP as a Basis for Active Packaging. Intelligent and Active Packaging for Fruits and Vegetables, 2007, pp. 31–55.
[29] A. Dodero, A. Escher, S. Bertucci, M. Castellano, P. Lova, Intelligent packaging for real-time monitoring of food-quality: current and future developments, Appl. Sci. 11 (8) (2021) 3532.
[30] F. Mustafa, S. Andreescu, Nanotechnology-based approaches for food sensing and packaging applications, RSC Adv. 10 (33) (2020) 19309–19336.
[31] X. Meng, S. Kim, P. Puligundla, S. Ko, Carbon dioxide and oxygen gas sensors-possible application for monitoring quality, freshness, and safety of agricultural and food products with emphasis on importance of analytical signals and their transformation, J. Kor. Soc. Appl. Biol. Chem. 57 (6) (2014) 723–733.
[32] J.R. Retama, D. Mecerreyes, B. Lopez-Ruiz, E. Lopez-Cabarcos, Synthesis and characterization of semiconducting polypyrrole/polyacrylamide microparticles with GOx for biosensor applications, Coll. Surf. A 270 (2005) 239–244.
[33] T. Ahuja, I.A. Mir, D. Kumar, Biomolecular immobilization on conducting polymers for biosensing applications, Biomaterials 28 (5) (2007) 791–805.
[34] K. Arshak, C. Adley, E. Moore, C. Cunniffe, M. Campion, J. Harris, Characterisation of polymer nanocomposite sensors for quantification of bacterial cultures, Sens. Actuat. B 126 (1) (2007) 226–231.
[35] G.L. Robertson, Food packaging, in: Textbook of Food Science and Technology, 2009, pp. 279–298.
[36] L. Ruiz-Garcia, L. Lunadei, P. Barreiro, I. Robla, A review of wireless sensor technologies and applications in agriculture and food industry: state of the art and current trends, Sensors 9 (6) (2009) 4728–4750.
[37] P. Guenther, A Case Study of Perceptions of Asset Tracking and Inventory Management Technology in a Small Construction Company, Wilmington University, Delaware, 2020.
[38] X. Yin, H. Wu, L. Dong, B. Li, W. Ma, L. Zhang, et al., Ppb-level SO_2 photoacoustic sensors with a suppressed absorption–desorption effect by using a 7.41 μm external-cavity quantum cascade laser, ACS Sensors 5 (2) (2020) 549–556.
[39] J. Hodgkinson, R.P. Tatam, Optical gas sensing: a review, Measur. Sci. Technol. 24 (1) (2012) 012004.
[40] M. Vanderroost, P. Ragaert, F. Devlieghere, B. De Meulenaer, Intelligent food packaging: the next generation, Trends Food Sci. Technol. 39 (1) (2014) 47–62.
[41] X. Tang, C. Tan, A. Chen, Z. Li, R. Shuai, Design and implementation of temperature and humidity monitoring system for small cold storage of fruit and vegetable based on Arduino, J. Physics (2020).
[42] B. Kuswandi, Freshness Sensors for Food Packaging, 2017.
[43] D. Schaefer, W.M. Cheung, Smart packaging: opportunities and challenges, Proc. CIRP 72 (2018) 1022–1027.
[44] L.M. Magalhães, M.A. Segundo, S. Reis, J.L. Lima, Methodological aspects about in vitro evaluation of antioxidant properties, Analyt. Chim. Acta 613 (1) (2008) 1–19.
[45] J. Boerman, M. Bauersfeld, K. Schmitt, J. Wöllenstein, Detection of gaseous ethanol by the use of ambient temperature platinum catalyst, Proc. Eng. 168 (2016) 201–205.
[46] A.L. Brody, E. Strupinsky, L.R. Kline, Active Packaging for Food Applications, CRC press, 2001.
[47] S. Yildirim, B. Röcker, M.K. Pettersen, J. Nilsen-Nygaard, Z. Ayhan, R. Rutkaite, et al., Active packaging applications for food, Comprehen. Rev. Food Sci. Food Saf. 17 (1) (2018) 165–199.

CHAPTER 16

Evaluating the sustainable metal packaging for cooked foods among food packaging materials

Figen Balo[a] and Lutfu S. Sua[b]

[a]Department of METE, Firat University, Elazig, Turkey; [b]Department of Management and Marketing, Southern University and A&M College, Baton Rouge, LA, United States

1. Introduction

Packaging of food is a significant aspect of contemporary civilization. Without packaging, the food produced commercially could not be treated and transported effectively and securely. Food packaging is described as a coordinated mechanism to prepare food for transit, storage, distribution, end-use, and retailing to accomplish the potential consumer's requirements at the lowest possible cost [1]. It is impossible to design high-attribute packaging of food if the packaging does not execute its tasks. Poor packaging is estimated to squander more than 25% of food, according to the World Packaging Organization [2]. As a result, it is apparent that proper packing can help decreasing food waste. Furthermore, the present users' desire for quick, high-attribute food has enhanced the food packaging influence.

Packaging systems are divided into four categories. The "primary package" is the first-level package that has direct contact with the product. Primary packaging must be compatible with the food, it should be non-toxic, and should not create any modifications in food characteristics like flavor, unwanted chemical processes, color, and so forth. The secondary package preserves the first-level packages from harm in the course of storage and transportation by containing two or more primary packages. Secondary type packages are utilized to keep grime and pollutants out of primary shipments and to group primary packages together. The shipping container is the tertiary package, which usually comprises several primary or secondary shipments. The "distribution package" is another name for it. Its primary purpose is to safeguard the goods throughout distribution while also allowing for efficient handling. A unit load is made up of corrugated boxes that have been put on a stretch wrapped and pallet for mechanical usage, storage, and shipment. The goal is to help with the automated handling of huge quantities of merchandise [3].

There are two types of packaging in general. A package that will be delivered to the end customer at a retail store is referred to as consumer packaging. This category usually includes major and secondary packages. A package for storage and delivery to a retail outlet is referred to as industrial packaging. This category includes tertiary packages and unit loads.

Food packaging serves four primary purposes. Convenience is a concept used to describe the utility function of packing. Consumers want products to match their lives, and the packaging business has to adapt. As a result, the utility function comprises all packaging qualities that bring value and convenience to product and/or package. All packaging operations rely heavily on confinement. The confinement function makes a substantial contribution to product protection and preservation throughout distribution. For distribution from the manufacturing site to the last destination, all items must be confined. All the same, it must be cost-effective [3,4].

Fresh and processed foods are subjected to two forms of harm during storage and transit. Physical damage, like vibration, shock, compressive pressures, and so on. The other is environmental damage caused by water, light, gases, scents, bacteria, and other factors. The contents of the package will be protected or reduced by a proper packing strategy. Nonetheless, in the event of fresh-food goods, packaging alone rarely provides enough protection. Because temperature has such a large impact on food deterioration, it is more cost-effective to modify the supply chain to regulate temperature (refrigeration, freezing, etc.). Packaging, on the other hand, can provide some protection by slowing down temperature swings [4].

Therefore, the selection of the appropriate packaging material is important. Many scientific methods have been used to select materials suitable for their use in research. Among these methods, MCDM methods were frequently used in the selection of the right and suitable material. Some studies on this subject are as follows.

For the best phase change material for household air conditioning, ventilation, and heating, Rastogi et al. employed a combined TOPSIS and Ashby method [5]. Chatterjee et al. investigated nine potential materials and five selection criteria for choosing the best material for a gear: core hardness, surface hardness, tensile strength, limit of surface fatigue, and limit of bending fatigue. The authors used the ARAS and COPRAS approaches to address the decision problem. They evaluated COPRAS and ARAS methodologies for material analysis to close the gap [6]. Senyigit et al. utilized a combined SAW and AHP method for the choice of material for dental implants in health-related study [7]. Kiong et al. utilized the AHP approach to define the best material for screw manufacturing from a variety of options, including cast iron and low-C steel. The researcher used the AHP method since it is possible to analyze both qualitative and quantitative information [8]. Chakraborty et al. used the PROMETHEE, VIKOR, and TOPSIS methodologies to choose the best material for less-load wagon external walls from a variety of materials, including aluminum, aluminum alloys, low-C steel, nickel alloys, copper alloys, titanium alloys, and zinc alloys. Material expense, density, wear resistance, corrosion resistance, and specific stiffness were used as the selection criteria. According to the authors, the methodologies were chosen because of their ease of implementation, contrary to techniques like the ELECTRE methodology, which are computationally complicated. They also investigated a challenge involving the choice of the proper material for a high-velocity nautical ship. The authors used the PROMETHEE, VIKOR, and TOPSIS techniques to choose the best material from six alternatives based on nine selection criteria [9]. For a cryogenic storage warehouse, Chatterjee et al. utilized an combined EVAMIX and COPRAS techniques to resolve the material selection challenge. Since the methods were novel and had restricted use in the engineering field, they used the EVAMIX and COPRAS methodologies to resolve the materials choice problem [10].

Chothani et al. demonstrated the application of PROMETHEE and AHP methodologies to evaluate five alternative hacksaw blade materials based on 6 decision criteria including young's modulus elasticity, yield strength, and hardness. The PROMETHEE approach was utilized to rank the various materials, while the AHP methodology was employed to analyze choice criteria weigh [11]. Dweiri et al. utilized the AHP model to define the best material for the keys fabrication. The authors investigated high-C steel, cast iron, low-C steel, copper, composite material, and aluminum as alternative materials for key production, with the following decision criteria: appearance, environment, strength, hardness, expense, and availability. One of the grounds for the researchers' adoption of the method is the pairwise comparison's consistency measurement of the options, which helps to reduce decision maker inconsistency [12]. Ilangkumaran et al. evaluated the use of the PROMETHEE and Fuzzy-AHP methods in combination to analyze material selection for vehicle bumpers [13]. Cavallini et al. employed the HOQ and VIKOR methods to determine the optimal coating for protecting an aluminum alloy substrate (Al-7075), taking into account numerous factors like fill density and fill as the foundation for assessing the coating performances [14]. Soni et al. implemented COPRAS, MOORA, TOPSIS, SAW, and ELECTRE methodologies for the magnesium alloy choice to comply the car implementation while considering elements like thermal conductivity, density, and stiffness to demonstrate the MCDM methodologies' effectiveness in material choice in the car sector [15]. For the manufacturing cylindrical caps without heat treatment, Rao et al. presented an integration of TOPSIS and AHP approaches for material choice. For a tank of cryogenic storage, the researchers also used the technique to choose the best material, utilizing factors including thermal expansion, toughness index, and essence heat as the foundation for determining the best choice. Since TOPSIS methodology is insufficient of calculating decision criteria weights, they combined AHP with TOPSIS method in order to analyze the aforementioned challenges [16].

This study is divided into two primary parts. The first part of the study discusses the qualities and kinds of food packaging systems and materials in order to aid appreciation and knowledge of the primary packaging materials that might impact food quality and shelf life, such as paper, plastic, glass, and metal. The second part of the chapter includes a multi-criteria decision-making analysis to assess all types of packing criteria by explaining the advantages and disadvantages amongst other packaging technologies of metal packaging.

2. Major food packaging materials and metal packing

On a global scale, the most widely used packaging materials are paperboard and paper. Over half of all packaging is paper-sourced, including packing of food. Plant fibers are used to make paper. Agricultural by-products like straw (of rye, wheat, rice, and barley), cotton, bagasse of sugar cane, flax, maize husks, bamboo, and other agricultural by-products make up the remaining sources of paper. Creating papier-mache is the first step in making paperboard or paper, and the papier-mache quality influences the quality of the finished product. Mechanical, chemical, or a mix of processes can be used to pulp wood. Mechanical pulp generates papers with a low strength and large bulk, in addition to a less expense. Their application is quite restricted in packaging. Chemical pulp provides a higher-attribute, stronger paper, but it is also expensive. In terms of cost and properties, combination processes are in the middle. The pulp can be bleached or unbleached to different stages, and diverse agents of sizing and other added substances are employed to adjust appearance and functions. Fig. 16.1 illustrates the process of paper manufacturing.

Plastics are a subclass of polymers, however the names are often used interchangeably in the packaging industry. At very low temperatures, plastics are a kind of polymer that, unlike metals and glass, can be molded into a broad variety of forms utilizing controlled pressure and heat. There are hundreds of discovered synthetic polymers species, but solely several are commonly utilized for packing of food. Plastics have grown in popularity faster than any other product, and it is presently the second most commonly utilized packaging material. Depending on its chemical makeup, each plastic has its own set of characteristics. Each material's performance and interaction with a range of meals is unique. As a result, the plastic material for a given food's packaging is chosen to perform effectively within the application's constraints.

Glass is an amorphous inorganic product that has been chilled to a hard form. No-crystallizing jars or bottles are the most common forms of glass packaging utilized for food packing [17]. Bottles make up 75% of overall glass food-containers. The glass is mostly composed of silica, which comes from sandstone or sand. Silica is blended with other basic materials in varied amounts to make most glass. On the other hand, glass has drawbacks owing to its great fragility and weight. The food packaging's fragility has raised certain security issues, like the danger of glass chipped in items of food. Recently, the utilization of glass for packaging of food has decreased, with metal cans and, progressively, plastics gaining market dominance. It does, however, continue to play a significant part in packaging.

Fig. 16.1 Paper production process.

Active packaging is a significant and quickly expanding industry. For active packaging, there are various alternative definitions in the literature. This packaging form frequently includes interplay between the food product and the packaging components that goes plus the packaging material's inactive obstacle operate [18,19]. Moisture absorbers, oxygen scavengers, releasers of antimicrobial agent, packaging of temperature control, odor/flavor absorbers, and ethylene scavengers are among the most common active packaging technologies. The key deteriorative elements for food goods must be understood in order to utilize this technique. For example, water activity, acidity, oxidation, respiration rate, temperature, microbial spoilage, and other variables all impact the shelf life of packed foods. Active packaging may be devised and used to preserve product quality and/or increase shelf life by carefully examining all of these criteria. Sachets or direct incorporation of active agents into packing containers are used.

To sustain product quality and expand shelf-life, modified environment packaging involves adjusting or changing the air inside the box. Food degradation responses like oxidation, enzyme activity, post-harvest metabolic actions, and moisture loss, in addition to microbe development, are all postponed with the optimal gaseous environment. O_2, CO_2, and N_2 are the three major gases utilized in modified environment packaging. They can be used alone or in combination in most circumstances. Commercially, Ar and CO mixtures are also used.

Filling a sterilized commercial product into a container sterilized under aseptic circumstances then stamping it hermetically to avoid contagion throughout the handling and distribution process is known as aseptic packaging. Aseptic packaging allows for less thermal influence for the nutrient than traditional heat disinfection procedures like retorting and canning, resulting in typically better food quality than traditional thermal disinfection. The container design and material choice can be more versatile than with characteristically heat disinfection since the packaging and food product are sterilized separately. The utilized aseptic packaging material's type is influenced by the product's nature, the expense of both packaging and product, and the consumer choice. The paperboard laminated carton is the most common aseptic packaging. This material's usual structure includes bleached or unbleached paperboard, aluminum foil, and polyethylene. Gas, liquid, and light cannot pass through the layered framework. Cans, bottles, pouches, trays, and cups are all examples of aseptic packaging. Environmentally friendly packaging material's comparison is given in Table 16.1 [20,21].

2.1 Metal packaging materials

Life has altered dramatically since the beginning of the 21st century as a result of its widespread consumer acceptance and adaptability. Despite changes in market trends, metal packaging material has been present in market statuses. The sector has been continually on the drive toward research and innovation in order to test out sizes and new designs in

Table 16.1 Environmentally friendly packaging material's comparison.

	Metal (stainless)	**Paper**	**Glass**	**Plastic**
Availability	Easy to get, abroad	Easy to get, domestic	Easy to get, domestic and aboard	Difficult to get, abroad
Safety (rating scale 1–4)	4	3	4	3
Compatibility material and product	Cold, hot, solid, liquid, dry, wet product	Cold, hot, solid, dry product	Hot, cold, liquid, wet, dry product	Cold, hot, solid, liquid, dry, wet product
Price rate	18	0.8	1.3	1
Portability (rating scale 1–4)	3	2	2	2
Durability	>1 year, reuse	Disposable	For multiple uses	Disposable, reuse >1 year
Recycling alternatives	None	Recycled paper, crafts	Recycled	Recycled plastic, plant's receptacle

order to suit the demands of clientele. Metal packaging is the nutrient preservation process's complementary section. In the last ten years, the metal packaging use in the beverage and food industries has more than doubled.

Food packaging made of metal comprises a wide variety of goods from closures to cans. From tuna cans to rack systems, metal is employed in packaging in a range of implementations. Aluminum, steel, chromium, and tin are the most frequent metals utilized in packaging of food. Steel and aluminum are the major materials for metal packaging and are often utilized in the fabrication of food cans. For cans of food, steel is the most popular material, whereas aluminum is used for beverage cans. When steel is exposed to moisture and oxygen, it oxidizes, resulting in rust. Therefore, chromium and tin are utilized as preservative coatings for steel.

The majority of metal packaging contains an interior nutrient contact covering, which is critical to the package's efficiency. Failure of the coating might lead to packing failure and, as a result, spoiling of the food. As a result, a large variety of coatings based on a small number of authorized compounds are employed. Properties and types of used resins in interior coatings for cans are given in Table 16.2 [18–20].

To prevent corrosion, tin-plate is a tin and steel combination created by covering sheet steel with a thin tin sheet. Tin free steel or electrolytic-chromium coated steel is the outcome of using chromium instead of tin to give corrosion protection. Electrolytic chromium coated steel has a lower corrosion resistance than tinplate, but it is less costly and more heat resistant.

Table 16.2 Features and types of used resins in interior coatings for cans.

	Flexibility	Main end-uses	Nature	Pack resistance
Oleo-resinous	Variable	Too limited uses	Oils that exist naturally but have been synthetically modified	Dependent on the pack
Organosol	Too good	Cans that have been drawn ends that are simple to open Frequently used over an epoxy-phenolic basecoat.	PVC dissolved in a varnish and traditionally stabilized using a low molecular weight epoxy resin or novolac epoxy resin. Oils that have been epoxidized can also be utilized.	Too good
Epoxy-phenolic	Good	3-Piece cans universal gold lacquer cans with a shallow draw	Epoxy resins with a high molecular weight phenolic resole resins are used to crosslink the material.	Too good
Epoxy-amino	Good	Beer and beverage can universal lacquer (water reducible) Striped side seams some food production methods	Cross-linking of high molecular weight epoxy resins with amino resins. Internal epoxy acrylic water-based sprays for B&B DWI are also available.	Limited

Continued

Table 16.2 Features and types of used resins in interior coatings for cans.—cont'd

	Flexibility	Main end-uses	Nature	Pack resistance
Epoxy-anhydride	Good	For 3-piece cans, internal white is used.	Cross-linked high molecular weight epoxy resins with anhydride hardeners.	Too good
Phenolic	Too poor, however, film quality is weight-dependent.	Pails and drums where flexibility isn't a need	Self-crosslinking phenolic resin(s) (cure)	Excellent, especially with abrasive foodstuffs.
Polyester	Too good	May not be proper for aggressive foods and too acidic	Polyester resins that have been cross linked with amino or phenolic resins. Epoxy resin with a lower molecular weight may be present.	Dependent on the pack

The word "canning" is mostly utilized to define a technique. It is especially useful in global places where there is limited or no cooling for nutrient storage. It is a method of protecting goods without causing microbial degradation. During the last century, canned nutrient has become a major element of the dietary products in industrialized nations. The metal packaging material is commonly utilized in the canned goods' packaging. Metal packaging has a dual purpose: it protects the food from external influences throughout thermal storage and treatment storage, in addition to serving as data pack and sales. The container's hermetic tightness is the most fundamental criterion for such a package.

Metal was formerly solely a packaging's technical component; nevertheless, metal has now become a whole novel packaging material in the cosmetic and fragrance industries. The cause for this change is the recent advancements in metal decoration. Marketing strategists investigate the benefits of metal packaging in expediting the marketing of new items. Metal packaging is also employed in cap manufacturing since it is suited for a wide variety of materials.

The nutrient that has been sterilized through the thermal treatment should be kept facing reinfection with microbes or another external effect. Container integrity is a term used to represent this somewhat difficult requirement. Metals are utilized in a variety of food-contact applications, including saucepans, coffee receptacles, and packaging. This monograph focuses solely on metal food packaging. Coatings have been given special attention since most of them comprise a natural sheet (called as a covering) on the metal superficies from the metal to the foodstuff. Food is packaged in a variety of receptacles, some of which are entirely made of metal and others which incorporate metal elements. Food cans, lids (for butter and yoghurt containers, open trays, aerosol containers, soft drink and beer cans), closures (for bottle tops and glass jars) and caps, pails and drums, and tubes are metal packaging's all examples.

Numerous metal packaging (usually containers, cans, lids, and closings) have one or both sides coated. Internal coatings, lacquer, or enamel are used on the inside (food contact) while exterior coatings, enamel, ink, or varnish are used on the outside (non-food contact). Can coatings, different numerous other implementations, are usually thermally treated (baked or dried). Coatings are used on metal containers for a variety of reasons. Carbonated beverage and beer cans are packaged under compression, whereas nutrient cans are packed at ambient pressure or vacuum. Another distinction is the processing that takes place after the can has been filled and the lid has been sealed. Soft drinks are usually not treated any further. Beer in cans is commonly pasteurized. Many meals are heated when they are cooked and packed in the can under a variety of circumstances. This allows nutrient to be kept for lengthy durations of time without the use of preservatives. The sterilizing operations are carefully monitored to ensure microbiological safety during the expected shelf life.

2.1.1 Metal packaging alternatives

There has been a cost-driven and ecologically friendly development of metal packaging toward thinner metal gauges, which is currently supported by package waste laws that emphasize material minimization. This must be accomplished without jeopardizing the package's security features. Engineering or design improvements, such as beading on nutrient cans and the switch to less diameter inserts for various beverage cans, are made to achieve this. As a result, the coatings, inks, and lacquers utilized must perform to higher standards. Lids, closures, trays, tubes, aerosol containers, drums, pails, and cans are all examples of metal packaging used for food. Commonly used metal packaging materials are listed below [21–28].

Pails and drums are 3-section steel cans that are big in size. They are delivered unfilled, with plugs inside the lid to be filled by the purchaser. When they are filled, they are not processed in any way. Drums often refer to containers with a capacity of 100–220 L, whereas pails typically refer to containers with a volume of 5–25 L. Drums come in a variety of grades, depending on the intended contents and mode of delivery.

There are several different types of cans and lids. After filling the can with food, the lids are always attached, thus fillers and packers buy unfilled lids and cans and sew the lids on boxes. Cans are made up of 2 or 3 independent elements; 3-section cans have a top, a cylinder, and a bottom final, whereas 2-piece boxes have one piece for the bottom and wall and a separate top. They come in a variety of sizes, from extremely tiny to purveyance package extents (usually for 2–10 kg of contents).

Extruded metal tubes are made from a slug of metal (generally Al). Not all of them, especially toothpaste tubes, are polished on the inside. Metal is no longer used in many tubes. Metal-based tubes are only those with contents that require minimum oxygen contact.

Cosmetics, personal care products, pesticides, and lubricants are among the most common uses for aerosol cans. Only a few goods are packaged in aerosol cans, such as canned whipped cream. Three-piece or two-piece aerosol cans are available. In essence, 2- or 3-piece can manufacturing principles imply, while the ends different and are suited as one unit. Steel is used to construct all three-piece aerosols. The 2-piece aerosol, called as one mono-bloc, is the major distinction between aerosol and can manufacture. Impact extrusion is used to make the aerosol container out of aluminum. The open end's diameter is decreased to accommodate the spray head, and an interior varnish is sprayed on.

Sealing containers is done with closures. They can be re-sealable/reusable, like a pickle jar lid, ketchup bottle lid, or whiskey bottle lid, or a way, like one top bottle peak for beer. Shutdowns don't have to be composed of mine, and polymers are rapidly gaining traction in this industry. Various people, especially those in the organizing field, consult to some shutdowns as caps, especially those for jars.

Semi-rigid and rigid Al trays for nutrient implementations are made from Al rolled with a thickness of 70–300 m. To offer retort resistance, containers with polyurethane adhesives and polyolefin laminate constructions are sometimes utilized for the nutrient touch the tray's.

Whilst caps and closures refer to the lids on glass jars and bottle tops the types of caps utilized on butter and yoghurt receptacles. Sealed caps hermetically shut a tray and keep outside influences out of the food. In most situations, these lids make it simple to open the food pack. After the rolling procedures, 20–70 m aluminum foil is utilized for this purpose. The metal may or may not be subjected to a pre-treatment procedure.

The classic bottle caps that are removed with a bottle opener are known as crowns. As a substrate, either ETP or TFS can be employed. Old or modern technology may be met depending on the geographical area. The sealing insert, which caps the crown's beverage touch area, is the main distinction. It's either a compound derived from PVC that's or a plastic liner (vinyl acetate copolymer, an ethylene or polyethylene) inflated in a foam rubber to signet the bottle.

2.1.2 Metal packaging disadvantages and advantages

Metal packing materials are among the most suggested techniques for packaging foods like soups, spices, and soft beverages. They are also widely used in the packaging of medications, cosmetics, and other consumer goods. Metal materials may be molded by current manufacturing processes and advanced machinery, resulting in different superficies plans, easy to open caps, tightness, and high solidity.

The metal packing creates a barricade that is impervious to light, water, and air. It is pest and rodent resistant. Its use is widespread since it can be chilled and heated for disinfection, and if correctly treated through employing appropriate lacs, it doesn't react negatively with the food within. This is due to the significant advantages that metal packaging materials for meals have over alternative approaches [21–28].

Metal packaging is constructed of 2 different materials: bin and Al steel layers. Steel layers are coated with organic and tin lac, which prevents direct contiguity with nutrient. As a result, rust-resistant metal packaging are created. It is manufactured as rust-resistant metal packing.

Metal cans are classified into two types: those with a double-stitching that are utilized to create tinned goods and those with push-on screw-caps or lids that are utilized to store cooking oils or dried foods (e.g., dried yeast, coffee or milk powder). Double-stitched boxes are constructed of aluminum or can coated steel and tinned with varnishes that are appropriate to the type of nutrient. Boxes have several advantages over other kinds of containers: they give entire protection of the scopes with a double-stitch; they are tamper-proof; and they may be constructed in a range of sizes and forms.

Cans, on the other hand, are expensive in comparison to other receptacles because of the considerable expense of metal and the considerable production expenses. They are heavier than plastic containers and hence incur greater transportation expenses. Because of these difficulties and/or a lack of availability, there are several box-making companies in undeveloped nations. Small scale nutrient process normally do not employ metal boxes. Bigger metal cans are commonly utilized as transportation receptacles for juices, oils, and other fluid goods, while less expensive plastic barrels are gradually replacing them.

Aluminum foil trays and cups, foil pouches laminated as options to jars or cans, foldable Aluminum tanks for pastes, and Al casks are examples of metal containers. Aluminum has the benefit of being impermeable to humidity, microorganisms, light, and odors, in addition to being a great gas barrier. It features an excellent weight to strength rate and a good-quality surface for decorating or printing.

2.1.2.1 Metal packaging advantages

Metal cans composed generally of tin, steel, or aluminum can be a terrific assets and investment when it is concerned with keeping nutrient and storing it over time. On the other hand, metal receptacles are created using a can-making machine. When it comes to metal packaging, it has numerous benefits over other packaging techniques, like resistance and durability to elevate temperatures, which give steadiness in the course of the operation.

Metal packaging is becoming increasingly popular among businesses since it is a strong yet elastic material that lots of manufacturers and designers like working with. In terms of packaging, tin cans are incredibly versatile. Because of the numerous features of tin material, such as strong rigidity and strength, these cans provide a great level of elasticity in packaging. It may organize these boxes anyway it see fit, and they can be carried through any methods of transportation while retaining their original state. It is an important feature that would not be seen in other kinds of boxes since they can simply bend or harm the contents interior while in transit [26]. Furthermore, even in typical shapes, metal packaging frequently includes characteristics like pull tabs, volume variance, or protrusions, which provide greater elasticity than other material kinds [25]. Furthermore, even in characteristic types, metal packaging often contains properties like pull tabs, volume variance, or protrusions, which supply larger elasticity than other material kinds [23].

Because of their rigidity and robustness, tin cans provide remarkable packing versatility. These are the type of boxes that can retain their original form even after being subjected to a strong effect. As a result, it will not have a trouble with distorted boxes, considerably reducing losses correlate with return stock because of formlessness. These qualities also allow employing different modes of transportation without worrying about the boxes' condition. This is another attribute that both vendors and buyers search for in tin boxes [23].

Some items must be stored in away from other light resources or sunlight and dark-packaging. The package, whether made of steel or aluminum, is opaque and prevents solar light from harming the contents within. Moreover, metal as material is durable that may protect the contents from damage. For example, metal stores CO_2, resists O_2 entry, and keeps out light, emerging in a long shelf life for the beverage.

Tin has a longer shelf life than most other metals, including aluminum and steel due to the fact that the tin substance does not react chemically with the nutrient. On the other hand, steel and aluminum can readily oxidize relying on the acidity of the nutrient wrapped and ruining it. As a result, these containers cannot be used to store soups, beverages, or medications. Tin cans, on the other hand, are not corrosive, which is why they can preserve food for lengthy periods of time without spoiling [23].

Despite the limitations of metal packaging, there are certain positives, such as the fact that metal packaging is more robust than other materials used for food packing. Some packing materials deteriorate over time. Paper, for example, will be deconstructed over time and is easily affected through humidity. The plastics become sticky and degrade over time. Steel and Aluminum are much more durable than plastic and paper. The metal is planned to be reusable and long-lasting. Metal may be scuffed and repositioned repeatedly without sustaining major exterior damage. It will keep such objects secure and safe as much as possible.

Green projects are spreading which is good for both the environment and people. Because metal receptacles are re-useable, they fit in well with waste-reduction strategies. It can be argued that no metal receptacle is completely worthless. In excellent condition, utilized metal receptacles are just as useful as novel ones. Used containers are a terrific money-saving solution for businesses who cannot afford to buy brand new containers. While secondhand receptacles are just as functional as novel ones, there is nothing incorrect with renting or purchasing trademark novel receptacles as a longtime investing if you have the finances.

If a metal receptacle is no longer functional, it can be reused and recycled to make another metal receptacle. On one hand, having metal receptacles is an investment, and it can be sold quickly. It will still generate money from the container regardless of its state. The majority of metals are recycled simply. Steel and Aluminum are two of the world's most recyclable commodities. Instead of extracting fresh metal from the ground, most industries that utilize metal packaging employ materials that are at minimum partly recovered; actually, it is believed that 80% of all metal ever created in the globe is still in use today.

Metal packaging materials offer a unique service that allows for the creation of any size or shape tradition tin, depending on trademark objectives and planned around the produce. When it comes to beverages, though, a metal can have its own set of characteristics. Metal has a high ranking in terms of security. One of its weaknesses is the inability to retain unique shapes. Cans can now be shaped thanks to new processing processes, although they are pricey and confined to smaller market uses. Metal packaging, like glass packaging, is wonderfully attractive.

Most of us associate metal containers with canned vegetables or soda. Popular objects like these have a beneficial appearance; nevertheless, as production industry improves, it will see more variation in metal packaging. Metal may be formed into any shape, even entirely ornamental ones. In order to provide a cool temperature for customers, a metal container is frequently the preferred option.

One of the primary advantages of the metals is its long-term worth. Metal mining is frequently less harmful to the ecology than oil puncture. Once metal obtained from the soil. It has different composites that may be transformed into any form to fulfill the demands of all types of providers. Furthermore, metal is quite significant on the basis of continuity.

Metal lasts a long time before it needs to be recycled because it is naturally durable to rust, especially in rust-proof composite types. Metal's high melting-point and resilience allow firms to bend and reshuffle it without losing its key properties. In general, recycling metal generates minimal more waste and allows it to take on each new shape without deterioration [21−28].

2.1.2.2 Metal packaging disadvantages

The metal packaging material includes everything from Al soft drink boxes to can biscuit receptacles. The metal is important since it is long-lasting, inexpensive, and non-toxic, making it ideal for food storage. Metal packing issues differ depending on the type of metal utilized. However, there are a few drawbacks to this type of packing.

Metal manufacturing involves a number of complex operations in the production line, as well as a number of input materials. Different machinery, including as the splitting machine, seaming machine, and others, are also used to produce the can producing machine.

The metal packaging material may keep a receptacle's ingredients fresh and secure, but it has the drawback of being not opaque, so buyers cannot see inside the package to examine the ingredients or further scrutinize a possible purchase. This restricts the metal packaging's usage in the retail trade industry, as alternative packaging materials such as plastic-based ones are preferable sometimes. Canned meals, for instance, are usually sealed, making it difficult for anybody to see what is within. That makes it the appropriate option though since one might simply select a tin with undesirable nutrient composition, posing a significant obstacle to end customers. This is why transparent packaging is chosen.

Metallic cans cannot be used to package excessively acidic goods. Since metal is corrodent, the acidic items' packaging would result in a chemical reaction that would devastate both the container and the product. As a result, acidic items cannot be packaged in these cans. This is one of the primary reasons behind the demise of tin cans. Metallic metals can potentially create serious health problems in food. Aluminum, for example, is another popular metal packing material. While aluminum is not corrosive when utilized to keep nutrient, it does react with acidic nutrients like tomatoes and rhubarb. These nutrients are particularly acidic due to chemical structure and can be harmed through Al if it is utilized to keep them. For these reasons, the flavor and quality of the dish are ruined. As a result of employing metal packaging for certain goods, the food will taste like aluminum.

When it comes to packing and shipping items across great distances, weight is crucial. Metal containers, particularly bigger ones for shipping products, can be significantly more costly than plastic-based boxes of the identical size. When budgeting for storage, it is important to consider the costs and advantages of each [25]. Most service providers charge a transportation cost depending on the load's weight being sent. That is, if it is searching for methods to lower manufacturing costs, decreasing transportation costs is a great place to start. Tin cans are lightweight, which helps to cut production costs.

Because of this trait, they are also easier to handle than their predecessors [24]. One frequent insight that businesses have reached is that they may avoid the increased cost of metal by acquiring products in bulk, such as wholesale tin containers.

Steel, unlike other packing materials such as plastic, is relatively hefty. Because of the building process, this is what makes it a costly packing option. It raises the price of cans based on the grade of the material used. Steel's biggest drawback is that it is prone to rust, which is what causes the metal to decay. Rust consists when a metal begins to come back to its initial state; for instance, steel comes back to the iron mine from which it welded. Rust is produced through oxidization, which consists when metals are subjected to water and air. Rust is one type of rust that happens on steel packing and creates it to flake off. Metal containers must be treated to prevent rusting; otherwise, moisture will cause them to corrode. To avoid corrosion, metal packaging material is frequently covered in other materials like chromium.

Rust degrades product quality and has a negative influence on nutrient ingredient. This is, what creates nutrient to spoil [25]. Furthermore, a significant difficulty for the soft drinks business is rust of beverage in aluminum boxes, which concludes in financial losses [24].

Can is frequently utilized for specific sorts of receptacles, notably cookie containers. The receptacles are hard to warehouse adequately, both after and before utilize, because the metal packing is not readily bent or squished by hand. A paper or plastic container, on the other hand, may be easier to squash or fold up and hide in a cabinet or another warehouse plant.

3. Results and discussion

Multi-criteria decision-making is a methodological strategy that uses decision-criteria, as well as advantage information and decision makers' opinions, to select the best option from a list of options [29]. For analysis of material, AHP, MAUT, VIKOR, TOPSIS, and ELECTRE are some of the widespread multi-criteria decision support systems that have been used in the scientific researches. The framework ranks alternatives based on selection criteria that are typically measured in different units. Scientific research revealed many MCDM methodologies that were utilized to select the best materials depending on various criteria. These methodologies differ in their properties and the extent to which data can be included in the estimations.

The most widely used AHP method, aids in the solution of multi-criteria tasks by employing a pairwise comparison scale [30]. This method's calculation technique is quite efficient, and computational results are found comparatively rapidly when compared to alternative methodologies. The methodologies are simply implemented in diverse areas [31], and are sensible and depend on a hierarchical structure, focusing on overall criteria selected. Nonetheless, it should be considered that decision makers' experimentation information plays a critical role in determining the criteria weights. If there is more than one decision-maker, this can complex the assessment operation [32–34].

In this study, the selected packaging materials and their performance scores based on the list of factors are presented in Table 16.3 [35].

Depending on the pair-wise comparison, the decision-matrix given in Table 16.4 is constructed. The table displays the used characteristics' relative priorities.

Table 16.3 The selected packaging materials and their performance scores based on the list of factors.

	Paper packaging material	Plastic packaging material	Glass container	Metal packaging material
Barrier performance	2	5	4	4
Mechanical strength	2	5	2	4
Chemical stability	3	5	4	3
Processing adaptability	4	4	5	4
Convenience	4	4	3	4
Decoration	4	5	5	4
Economical	4	5	4	3
Sanitary	5	5	4	5
Disposability	4	1	4	4
Shielding	0	0	0	1

Table 16.4 Decision matrix.

Matrix	Barrier performance 1	Mechanical strength 2	Chemical stability 3	Processing adaptability 4	Convenience 5	Decoration 6	Economical 7	Sanitary 8	Disposability 9	Shielding 10	Relative priorities
Barrier performance	1	3	1	2	1/5	9	7	1/6	3	1/4	9.07
Mechanical strength	1/3	1	1/2	1/3	1/7	2	3	1/7	1/3	1/4	3.24
Chemical stability	1	2	1	3	1/5	9	2	1/3	2	1/2	8.22
Processing adaptability	1/2	3	1/3	1	1/6	3	2	1/5	1/3	1/7	3.97
Convenience	5	7	5	6	1	9	3	1/2	2	1/2	19.07
Decoration	1/9	1/2	1/9	1/3	1/9	1	1/2	1/9	1/7	1/9	1.43
Economical	1/7	1/3	1/2	1/2	1/3	2	1	1/9	1/5	1/9	2.42
Sanitary	6	7	3	5	2	9	1	1	5	2	25.21
Disposability	1/3	3	1/2	3	1/2	7	5	1/5	1	1/3	7.24
Shielding	4	4	2	7	2	9	9	1/2	3	1	9.07

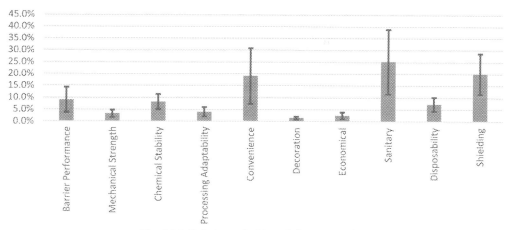

Fig. 16.2 Relative priorities of the criteria (%).

According to the resulting relative priorities given in Fig. 16.2; Sanitary, Shielding, and Convenience are obtained to be the criteria with the maximum effect on the packing materials.

For the purpose of this study, four packing material alternatives are determined. The table below displays the normalized values of the alternatives, their priority values for each characteristic, and the overall scores, which are indicated in the table's last row. These normalized values are shown in Table 16.5.

These normalized values are then multiplied by the relative weights of each of factors to produce the weighted-priorities of the alternatives shown in Table 16.6 below.

The conclusions in Table 16.6 and Fig. 16.3 display that Metal packaging material has the maximum overall score (0.408) amongst other options.

Table 16.5 Normalized priorities.

	Normalized			
	Paper packaging material	Plastic packaging material	Glass container	Metal packaging material
Barrier performance	0.1333	0.3333	0.2667	0.2667
Mechanical strength	0.1538	0.3846	0.1538	0.3077
Chemical stability	0.2000	0.3333	0.2667	0.2000
Processing adaptability	0.2353	0.2353	0.2941	0.2353
Convenience	0.2667	0.2667	0.2000	0.2667
Decoration	0.2222	0.2778	0.2778	0.2222
Economical	0.2500	0.3125	0.2500	0.1875
Sanitary	0.2632	0.2632	0.2105	0.2632
Disposability	0.3077	0.0769	0.3077	0.3077
Shielding	0.0000	0.0000	0.0000	1.0000

Table 16.6 Resulting scores of packing materials.

	Priorities			
	Paper packaging material	Plastic packaging material	Glass container	Metal packaging material
Barrier performance	0.012	0.030	0.024	0.024
Mechanical strength	0.005	0.012	0.005	0.010
Chemical stability	0.016	0.027	0.022	0.016
Processing adaptability	0.009	0.009	0.012	0.009
Convenience	0.051	0.051	0.038	0.051
Decoration	0.003	0.004	0.004	0.003
Economical	0.006	0.008	0.006	0.005
Sanitary	0.066	0.066	0.053	0.066
Disposability	0.022	0.006	0.022	0.022
Shielding	0.000	0.000	0.000	0.201
Scores	0.192	0.214	0.186	0.408

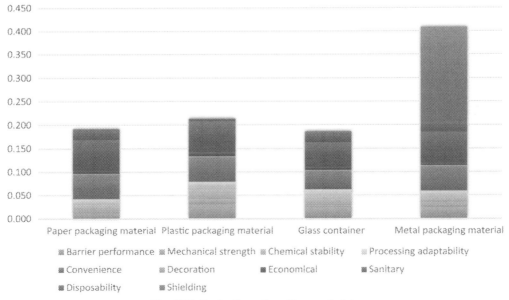

Fig. 16.3 Evaluation of packing materials.

4. Conclusions

Recent innovations in metal packaging for nutrient products include tools to supply a foam head to coffee chilled and beer, big opening stay on ends of tab for drink boxes, whole aperture nutrient box ends that are easier to open, self-chilling and self-heating drink boxes, foursquare part treated nutrient boxes for more performance shelf warehouse, 2-part wall and draw iron, and 2-part draw redraw boxes, ends of peelable membrane for treated nutrient boxes. The packaging's primary goal in metal boxes is to protect the product from chemical and physical damage.

In all of these ways, a perfectly lacquered can is an ideal food container. This will supply that metals proceed to play a vital role in the nutrients' expense-effective packaging for long or short-term atmospheric safekeeping area circumstances. Metal containers' innate resistance and the fact that they are light resistant contribute to preserving a high standard for the product contained over long shelf-life durations.

The aim of this research is to help to identify the most ideal option packaging material for foods. Within the scope of this study, Analytical Hierarchy Process is used for analysis. The result shows metal packaging material as the best packaging material for particularly cooked nutrients. The primary aim of this research is to make use of qualitative evaluations concurrently in evaluating complicated material selection factors. As a result of determining the list of evaluation criteria and their relative priority values based on expert opinions, packaging materials and their performance scores based on this list of factors are calculated. Computational results reveal that metal packaging material has the maximum overall score among other options considered in this research.

References

[1] R. Coles, D. McDowell, M.J. Kirwan, Introduction, in: R. Coles, D. McDowell, M. Kirwan (Eds.), Food Packaging Technology, CRC Press, Boca Raton, FL, 2003, pp. 1–31.
[2] World Packaging Organization (WPO), Packaging Is the Answer to World Hunger, 2009.
[3] S. Joongmin, S. Selke, Food packaging (chapter 11), in: S. Clark, S. Jung (Eds.), Food Processing: Principles and Applications, second ed.Buddhi Lamsal, John Wiley & Sons, Ltd, 2014.
[4] T.S. Lopez, D.C. Ranasighe, B. Patkai, D.M. Farlane, Taxonomy, technology and applications of smart objects, Information of Systems Frontiers 13 (2011) 281–300.
[5] M. Rastogi, A. Chauhan, R. Vaish, A. Kishan, Selection and performance assessment of phase change materials for heating, ventilation and air-conditioning applications, Energy Convers. Manag. 89 (2015) 260–269.
[6] P. Chatterjee, S. Chakraborty, Gear material selection using complex proportional assessment and additive ratio assessment-based approaches: a comparative study, Int. J. Mater. Sci. Eng. 1 (2013) 104–111.
[7] E. Senyigit, B. Demirel, The selection of material in dental implant with entropy based simple additive weighting and analytic hierarchy process methods, Sigma J. Eng. Nat. Sci. 36 (3) (2018) 731–740.
[8] S.C. Kiong, L.Y. Lee, S.H. Chong, M.A. Azlan, M. Nor, N. Hisyamudin, Decision making with the analytical hierarchy process (AHP) for material selection in screw manufacturing for minimizing environmental impacts, Appl. Mech. Mater. 315 (2013) 57–62.
[9] S. Chakraborty, P. Chatterjee, Selection of materials using multi-criteria decision making methods with minimum data, Decis. Sci. Lett. 2 (3) (2013) 135–148.

[10] P. Chatterjee, V. Athawale, S. Chakraborty, Materials selection using complex proportional assessment and evaluation of mixed data methods, Mater. Des. 32 (2) (2011) 851–860.
[11] H. Chothani, B. Kuchhadiya, J. Solanki, Selection of material for hacksaw blade using AHP-PROMETHEE approach, Int. J. Innov. Res. Adv. Eng. 1 (15) (2014) 26–30.
[12] F. Dweiri, F.M. Al-Oqla, Material selection using analytical hierarchy process, Int. J. Comput. Appl. Technol. 26 (4) (2006) 182–189.
[13] M. Ilangkumaran, A. Avenash, V. Balakrishnan, S.B. Kumar, M. Raja, Material selection using hybrid MCDM approach for automobile bumper, Int. J. Ind. Syst. Eng. 14 (1) (2013) 20–39.
[14] C. Cavallini, A. Giorgetti, P. Citti, F. Nicolaie, Integral aided method for material selection based on quality function deployment and comprehensive VIKOR algorithm, Mater. Des. 47 (2013) 27–34.
[15] A. Soni, D. Gautam, A. Dwivedi, Implementation of multi-criteria decision-making method for the selection of magnesium alloy to suit automotive application, Int. J. Adv. Res. Dev. 3 (6) (2018) 4–12.
[16] R.V. Rao, J.P. Davim, A decision-making framework model for material selection using a combined multiple attribute decision-making method, Int. J. Adv. Manuf. Technol. 35 (7–8) (2008) 751–761.
[17] ASTM, Standard Terminology of Glass and Glass Products, ASTM International, West Conshohocken, PA, 2003.
[18] W. Soroka, Packaging terms, in: W. Soroka (Ed.), Illustrated Glossary of Packaging Terms, Institute of Packaging Professionals, Naperville, IL, 2008.
[19] T.P. Labuza, W.M. Breene, Applications of "active packaging" for improvement of shelf life and nutritional quality of fresh and extended shelf life of foods, Journal of Food Processing and Preservation 13 (1989) 1–69.
[20] N.R. Mukarromah, A.P. Hendradewa, D.T. Utari, Expert system for selecting product packaging materials as a means to achieve sustainable development goals, in: Advances in Social Science, Education and Humanities Research, Volume 474, Proceedings of the 2nd International Seminar on Science and Technology (ISSTEC), 2019.
[21] https://donusumdernegi.org/yesile-donuyoruz-projesi/?yesile-donuyoruz-projesi/. (Accessed 18 March 2022).
[22] (Directive 94/62/EC, 1994).
[23] https://bizfluent.com/list-7601569-disadvantages-metal-packaging.html.
[24] https://environmental-conscience.com/cans-pros-cons/.
[25] https://www.haulaway.com/blog/2012/08/advantages-and-disadvantages-of-metal-storage-containers/.
[26] http://www.canmakingmachine.net/news/291.html.
[27] https://packagingoptionsdirect.com/understanding-the-advantages-of-metal-packaging.
[28] https://tinwaredirect.com/blogs/tips-advice/five-advantages-of-metal-packaging.
[29] G. Kabir, R. Sadiq, S. Tesfamariam, A review of multi-criteria decision-making methods for infrastructure management, Struct. Infrastruct. Eng. 10 (9) (2014) 1176–1210.
[30] T.L. Saaty, The Analytic Hierarchy Process, McGraw-Hill, New York, NY, USA, 1980, pp. 11–29.
[31] I. Kaya, M. Çolak, F. Terzi, Use of MCDM techniques for energy policy and decision-making problems: a review, Int. J. Energy Res. 42 (2018) 2344–2372.
[32] T.L. Saaty, Decision making-the analytic hierarchy and network processes (AHP/ANP), J. Syst. Sci. Syst. Eng. 13 (2004) 1–35.
[33] A. Ishizaka, A. Labib, Analytic hierarchy process and expert choice: benefits and limitations, Or Insight 22 (2009) 201–220.
[34] K. Shahroodi, A. Keramatpanah, S. Amini, K.S. Haghighi, Application of analytical hierarchy process (AHP) technique to evaluate and selecting suppliers in an effective supply chain, Kuwait Chapter Arab. J. Bus. Manag. Rev. 1 (2012) 119–132.
[35] https://www.jxblet.com/advantages-of-metal-cans-what-is-the-difference-between-metal-cans-glass-bottles-and-plastic-bottles/.

CHAPTER 17

Smart packaging products and smart showcase design

Mustafa Kucuktuvek[a] and Caglar Altay[b]

[a]Department of Interior Architecture, Faculty of Architecture, İskenderun Technical University, İskenderun, Hatay, Turkey;
[b]Department of Interior Design, Aydın Vocational School, Aydın Adnan Menderes University, Aydın, Turkey

1. Introduction

Smart packaging is used in a wide range of areas in our daily lives, from fragrance bottles to the box we empty our cereal. Initially, the packaging had to be fully functional, and its task was simply to preserve and protect the products contained within, primarily during shipping. After the industrial revolution, mass production increased rapidly, and new packaging began to be designed in the first half of the 20th century. Cardboard and plastic manufacturers understood the importance of using packaging in these products to store the materials and reflect the value of the businesses. Later, these companies added more innovations to the packaging they used to sell these products. As a result, a concept called smart packaging, which has more functionality beyond the original use of packaging, has emerged.

The concept of smart packaging is further divided into two parts. These are respectively; 1. Active packaging is divided into 2. Intelligent packaging. The first of these is active packaging; it is defined as packaging that only improves the product itself. The second, smart packaging, is a system that has the capability to detect that the manufactured goods or packaging has changed in some way. If we give an example of this; They are packages that can monitor and show the freshness of the food. When we switch to a subset of smart packaging during transportation and storage, so-called linked packaging is used, packaging that integrates technologies such as NFC tags or QR codes to perform a specific function. In addition, many brands can leverage these technologies in their packaging to deliver branded digital content to their consumers. Consumers who can scan an NFC or QR code with their phone can greatly benefit from enabling digital experiences that allow them to receive exclusive content or even verify a product, creating a strong and direct sense of connection between many more brands. As we mentioned before, if we talk about how smart packaging is used today to better interact with consumers; For example, thanks to the NFC application, users can communicate with each other. In this way, they can share their experiences with the products and make suggestions.

Smart packaging also enables consumers to buy something online and even unlock exclusive content. For example, in a sneaker brand, high-speed sneakers appear to have an NFC chip embedded in the shoe when we touch the phone in our hand. These systems can also be placed on a website, products, and packaging that includes custom content training. Another example is learning how to use this product properly by touching a phone directly to the makeup product. In addition, you can also get makeup tips. The smart packaging system can also include product validation feedback and storytelling. With this innovative and dynamic system, important values are added to the products and brands can connect with their consumers on a level like never before. Packaging has been used to package things that people need to survive, such as food, health, and personal care. However, since none of these products have much intelligence, there is a concern that they will spoil.

There are two methods by which packaging can communicate with different types of intelligence and detect these packages. First, it is known that the newly purchased yogurt container, which should last for two weeks, actually reaches a temperature that causes it to spoil and does not spoil. This system tells you how to make a cocktail while making a cocktail and notifies you of the ingredients needed for it while making your favorite cocktail. So actually this system is a personalized helper for us. For example, when a jar of jam or something is purchased, it notifies you of these steps. This type of personalization can be important for driving the sale of these products and the packaging will change to you. Another example is vegetable-based packaging. For example, when we buy a burger, you can heat its box and put it in your soup, or if it is protein-based, it can be reused later, but you can use it for a different purpose, not as a package.

In the future, this smart package can be eaten or a dog can eat it. In other words, it is thought that these will happen in the next 10, or 20 years. Many packaging companies are waiting for their customers to ask for a different innovation. The challenge, however, is that the demand for a new converter often increases when these customers want something different. Going forward, it needs to be much closer to a partnership with brand owners, retailers with upstream recyclers, and even new types of customers downstream, which will make it easier for retailers to think about how this packaging will work and improve the value chain.

IBCSGate solution allows you to verify people or incoming/outgoing products without the need for physical control or counting. The system, which takes less time than passing through the door, can simultaneously identify your sources by using Radio Frequency Identification (RFID) devices and can be transferred to the system where the data is recorded. At the same time, the control of logistics processes in enterprises becomes possible. In this system, everything can be done remotely and automatically and can work without human intervention. This system incomparably increases work efficiency without making mistakes. It also provides precise information about the products in stock. For example, RFID door installation is now very simple and employees can install it in less than 2 h. At the same time, RFID technology will work well in industrial buildings and warehouses and the esthetic version will fit in any office space, allowing to eliminate white spots in logistics processes. It optimizes and automates the recording of constantly moving resources. Thus, it increases the efficiency of the organization. This means saving time and money.

2. Smart packaging

Traditional food packaging is used for protection, communication, convenience, and preservation. Packaging refers to a product in the existence or nonexistence of heat, light, humidity, weight, bacteria, gas releases, etc. It is used to protect it starting the detrimental possessions of outdoor ecological conditions. At the same time, it provides the user with better comfort of usage and reduces the amount of time, and contains manufactured products in several dimensions and forms [1,2]. The primary safety objective of conventional packaging materials for food contact is to be as passive as possible. The smart packaging structures, such as the concepts of dynamic and smart packing, depend upon the beneficial interface between the packing space and the nutrition to offer dynamic conservation to the nutrition [3].

Smart packaging technology, also known as smart labels, are indicators used inside or outside the packaging to monitor the quality and freshness of the packaged foods, such as temperature changes, microbial deterioration, and packaging integrity during the distribution and storage processes from production to consumption [1]. Theoretically, the expiry date of a food is determined by considering that the product is kept under optimal conditions. However, there is a long process between the production and consumption of food. In this process, manufacturers can only control the optimal conditions in the production and shipping stages. However, other stages develop beyond their control, and the shelf life of the product may vary depending on the retailer's way of keeping and selling the product on the shelf, and the way consumers preserve the product [1,4].

Smart tags provide many benefits for users; for example, reducing food-related safety risks for manufacturers, while for retailers; It reduces temperature-induced perishable food waste by improving sales of fresh and perishable food. For consumers, it provides the opportunity to choose the freshest product and eliminates hesitations about food safety by allowing it to control its cold chain [5]. In addition, the smart tags to be used; must be accurate, reliable and reproducible, cost-effective, and easy to implement or activate. Again, these systems should be stored at room temperature, safe in indirect contact with food, and versatile, that is, they should be suitable for printing and can be included in the existing label or package [6,7].

3. Smart showcase design

Especially in businesses established for sales purposes, the design processes of showcases are very important. In general, each business provides services in different sectors. Such that, in these businesses that sell different products according to the sector, the showcase design should be made in the most appropriate way for the products. For example, the window work done in the markets directly affects the sales. A design project to be prepared should be a project that will provide solutions to many important points. In other words, businesses must be designed and organized with a professional study. Designs are generally grouped into two. One of them is knowledge-based design and the other is random design. Scientific data and behavioral characteristics of people must be considered in the design. Therefore, knowledge-based design always comes to the fore. Concrete results of esthetic and technical studies in knowledge-based design will be seen as soon as possible. This design should be in harmony with the general appearance, theme, and area of interest of the business. Accordingly, people's shopping behaviors play an important role [8]. A smart showcase design is shown in Fig. 17.1.

3.1 Smart nanosensors for intelligent packaging

Nanotechnology is expressed as the design, manufacture, and implementation of configurations and schemes at the nanoscale through the controller of their form and dimension. This know-how is practical in many diverse arenas, counting natural, electrical, materials, and farming production [9–11].

3.2 Smart nutrition packet methods and graphene-based nanosensors for nutrition protection and quality checking

Food consumption is one of the elementary requirements of people, and its feature and quality are fundamental to living a vigorous lifetime. Due to the growing glob of shops,

Fig. 17.1 Smart showcase design example (designed by Mustafa Kucuktuvek).

the processing, distribution, and consumption of food where it is created are being less preferred. These developments are known to cause significant changes in the food supply chain. When designing new systems and approaches to ensure food safety and quality, factors in food quality, transport, harvest, processing, and storage should be considered. Food quality and safety need to be controlled in different parts of the supply chain. Biosensors have been developed and it has become possible to detect various harmful analytes in foods in the last decade. The sensitivity of biosensors has increased in parallel with the developments in nanotechnology. This technology and nanomaterials have many special properties. These properties let to ideal conditions for usage in other devices. Mechanical flexibility, versatile surface functionalization, high conductivity, biocompatibility, and high conductivity can be counted among the pre-important properties [12].

3.3 Improvement of the consumer/packaging interface

Although consumer packing has remained unchanged for years, efforts are being made to develop more of it today. There is a growing demand from consumers to support their busy lifestyles by improving the convenience and functionality aspects of packaging while also demonstrating positive environmental credentials. Brand owners are requested to make packaging, to make a brand that makes a difference, to differentiate it from similar systems, and to make unique systems [13].

3.4 Temperature-sensitive smart packaging for meat

In meat imports, the cold air chain in transportation significantly affects the product quality. In order not to affect the freshness and quality of the meat, temperature fluctuations should not occur [14]. Deterioration of the shelf life of meat affects the growth and reproduction of bacteria. Food companies have a responsibility to ensure the integrity and safety of refrigerated products from meat production to serving on the plate [15,16]. Therefore, to develop the safety and quality of high-value food products, it is important to have good cold chain and storage standards [17—23]. In order to prevent temperature fluctuations in the cold chain, cold stores, refrigerated trucks, and small freezers are used [11]. Smart showcase can help the product reach the customer at the required temperatures, thanks to advanced air-conditioning features and high technology. It can provide detailed information about the cold chain and the stages followed by the products from the manufacturer to the customer, with the help of the NFC system.

3.5 Smarter packaging for consumer food waste reduction

Presently a main nutrition safety issue worldwide is food waste. Worldwide nutrition leftover has become an economic burden, exceeding 1.3 billion tons in 2011 and above $990 billion in 2019 [24]. The main causes of food waste throughout the supply chain can be listed as follows. Microbial spoilage is a result of packaging damage, errors in agricultural production, problems in the food processing process, and problems in the storage, transportation, and sales process [25—27]. Besides toxins, pathogenic organisms and contamination can create the most important food safety problems. It has been determined by the World Health Organization that food contamination causes more than 200 diseases [28]. For this reason, it is important to take some measures, monitor and improve the safety of food, and develop important methods to improve its quality and sustainability. Traditional food packaging has some basic features. Considering these parameters, smart packaging systems have been developed [29]. By using sensors for longer shelf life, foods can be tracked in smart packages throughout the supply chain [30]. In this way, the quality, safety, and freshness of the food can be assured. Smart showcase systems can be used as a complement to the smart packaging system.

3.6 Smart control systems and showcase design (smart control design for complex flexible aerospace vehicles)

A low-cost ventilation recording and control system is made possible by recent advances in wireless sensor networks (WSN). These developments eliminate the inefficiency of the heating, ventilation, and air conditioning (HVAC) system. In the HVAC system, wide temperature distributions occur due to thermal loads in different areas, and this not only creates additional costs for the user but also has a negative effect on global warming. WSN can be a good solution to reduce the total amount of energy used. Data from wireless sensors are controlled by a master controller. These sensors transmit information such as occupancy, temperature, and humidity to the main control system. The main control system provides to balance the cold/hot airflow and humidity according to the incoming data. Smart thermostats are also used in the WSN system [31].

Thanks to the WSN used in the smart showcase design, the shelf life of the products can be extended and healthy and fresh products can be delivered to the customers. This system not only provides energy efficiency but also prevents waste by preventing the products from shrinking to a large extent.

4. Showcase design principles

Because showcases reflect the image and atmosphere of the store, they are an important communication tool. The two key properties of the showcases are; the product identity of the stock, promoting its image, and being effective, inviting people to enter the store [32]. Chain businesses, small and medium-sized businesses use showcases as an effective sales tool to introduce their brand identities to the target audience, attract the attention of the product, and attract the customer to the store, in the globalizing competitive environment. Showcases; It is designed on special occasions such as New Year's Day and Mother's Day, can give messages with social content, draw attention to seasonal changes, give hints about seasonal trends, and increase consumption. Expanding with large glass surfaces, storefronts and showcases are designed with very different materials to impress and attract the customers with their displays. In this sense, fast consumption, the change that renewed seasonal showcases go through simultaneously with style, and showcase design basis and methods draw regard. Showcases, which are the creative scenes of today's design, are shaped as artistic creations and graphic arrangements, which are creative installations with design values. Although shop windows are considered leisure entertainment, they have important communication values. With this feature, it can fascinate customers and draw them toward itself while describing the brand image [33].

4.1 Shelf life and showcase design

Thanks to smart labels, product quality can be monitored effectively in terms of physicochemical, chemical, and biological properties of cold chain beverages and foods. Freshness indicators in products directly affect product quality. The temperature history of the product can be measured over time through integrators. Smart packaging provides info about the excellence and security of the nutrition and the food, lend to extending the shelf life of the product, better control of the cold chain, reducing nutrition leftover, protecting the consumer, and sustainability [34].

4.2 Color in showcase

When choosing the color in the showcase design, the store image, brand identity, marketing principles of the brand, consumer class, the category of the products displayed, and the material used are an element that is decided together. Colors have psychological effects on consumers. Colors are one of the design principles that contribute to the communication between the consumers and the store [35]. Colors; It is possible to divide them into three different classes vivid, inanimate (natural), and pastel colors. Among these color classes, those that fall into the vivid color class and the pastel color class have two different tones, which are separated as warm and cold. Warm colors are more exciting, attractive, and stimulating, cold colors create a relaxing, calming, and distant effect. Vivid colors; red, orange, and yellow (warm colors); blue, green, and purple (cold colors), muted (natural) colors; white, off-white, beige, cream, light brown, gray, black, pastel colors; pink, peach (warm); They are classified as lavender, light blue, light green (cold). Colors are known as elements that can communicate with consumers. It is desired to cause consumption as a result of the showcase design with the right color use, which attracts the attention of the consumer, together with the reason for the consumption desire of the consumer. Colors, which have the effect of increasing the consumption desire of consumers, should contact the consumer in the right places and in the right way. The use of colors in the wrong places and in the wrong way reduces the consumer's desire to consume and causes the consumer to feel uncomfortable [36]. For example, using the natural color of wood material in a place can transform that space into a warmer and more interesting one.

4.3 Use of light in the showcase

With the light used in the design of the showcase, consumers can perceive the products more accurately and comfortably [37]. For example, the light in the showcase designs of the ready-to-wear merchandising, which includes a fashion element; the product must be at the forefront, to reflect the theme in the best way and to reflect the colors in the showcase correctly. Light helps to explain what the class of products is [38].

5. Showcase accessories

Accessories; It should not be in a way that precludes the products displayed in the showcase. They are classified as objects that are used in the showcase but cannot be owned by the consumer to meet their consumption desire. The background used when setting up the showcase; are objects used on the necessity of the theme, such as walls, floor coverings, shelves, and steps [32].

6. Showcase materials

The material options and application methods that will be determined according to the periodical showcase designs of the shop windows vary according to the area that the shop window designers want to use in the showcase [32]. When determining the material type; According to the determined showcase concept, the price is determined as being easy to apply in the showcase and ensuring that the message the brand wants to give to the consumer reaches the consumer [35].

For example, the use of wooden materials in a showcase gives that area a warm image. Especially the fact that wooden (solid) furniture is natural, harmless to human health, and long-lasting when viewed carefully, being resistant and easily repairable is one of the most preferred reasons for this material. It can be shown that one of the negative features of solid furniture is the humidity that is, taking and giving water to its body. In order to eliminate this negative feature, various surface treatments can be applied to the wood material. These are the impregnation method, which is generally referred to as the penetration of chemicals into the wood, the heat treatment method, and varnishing methods that prevent the wood from deforming (working), that is, taking in water and swelling over time. Among these methods, it is recommended to prefer the varnishing method to ensure that the wood shows its natural texture, especially in the design of the showcase. It is recommended to use water-based varnishing systems, which reflect the natural texture of wood more, while varnishing is performed in the showcase design.

6.1 Wood-based showcase materials

The wood material is a natural material obtained from the trunk of the tree. Due to its fibrous structure, it retains a high amount of water in its body. If the water content is more than 30%, it is called wet wood, if it is between 30% and 20%, semi-dry wood, and less than 20%, it is called dry wood. Wood is made suitable for processing by undergoing certain drying processes. Hardwood; is obtained from broad-leaved trees such as walnut, oak, and beech, while softwood is obtained from coniferous trees such as pine, fir, and juniper. Wood types are examined under two headings massive and composite;

6.2 Solid wood

This type of wood material is wood that exists with the pure properties of the material without undergoing any processing. Since it is completely natural, it is a material that is harmless to the environment and human health. Due to its natural properties, expansion, water intake, color change, infestation, and so on.

6.3 Composite wood

They are products whose raw material is natural but their qualities have been changed by supporting them with adhesives and filling materials. Wood waste and non-functional small parts can be recycled into production with composite products. In other words, it can also be called recycling material.

6.3.1 Particleboard (particleboard, chipboard, and OSB)

Hardboard; sawdust formed by chipping and processing wood raw material with a synthetic resin; It is obtained as a result of bonding to each other using heat treatment and pressing methods. It is a hollow structure, light, low cost, but not high strength sheet product. It is generally preferred in furniture manufacturing and decoration products. It can be produced in various thicknesses and densities, and accordingly, the impact resistance varies. The chipboard is also not resistant to water and fire. However, nowadays it is possible to obtain chipboard with increased water resistance. In other words, it can also be called raw chipboard.

My chipboard; It is obtained by laminating the particle board with a surface resistant to water and heat. Thanks to this coating, it is possible to choose the desired color and model, and an affordable and higher-strength material can be obtained compared to chipboard.

OSB (Oriented Strand Board); that is, oriented particleboard is produced from small-diameter logs that cannot be used in plywood production. The production technique is similar to plywood. Screw resistance and flexibility increase thanks to the staggered layers. Ease of application, variety of usage areas, and reasonable cost are among the reasons for preference. It is a material resistant to external weather conditions. It is generally preferred in exterior furniture such as camellia roofs.

6.3.2 Fiberboard (MDF, MDF lam, and HDF)

The production process of MDF (Medium Density Fiberboard), which means medium density fiberboard, and chipboard is almost the same. The difference between them is that the raw material of MDF is fiber instead of sawdust. The fact that the fiber has a smaller structure compared to the chips and sawdust pieces ensures that the MDF is homogeneous, void-free, heavy, and durable. It is preferred because it is resistant to heat and humidity, suitable for coating and painting, and easy to shape. It is priced higher than MDF chipboard and lowers priced than plywood sheet. Suitable for moldings, furniture, and cabinets. It also finds a place in production as a raw material for laminate flooring.

MDF Lam is formed by combining raw MDF with decor paper of various colors and patterns. Since these coatings increase the resistance of the material to water, MDF Lam is more robust than MDF and is more preferred. The fact that the density of MDF is the same at every point increases the smoothness of its surface and its screw holding ability. Thanks to its smooth surface, clean results are obtained in lacquer paint application. Fiberboards with a density of more than 800 kg/m^3 are called HDF. Due to its high density, HDF is thinner and therefore its usage area is narrower than MDF. MDF is suitable for furniture and decoration, while HDF; It is preferred for laminate flooring, wall coverings, and door wings. The high moisture resistance and durability of HDF cause it to be costly.

6.3.3 Plywood

Plywood is obtained by gluing together 3 or more thinned wood layers crosswise. Thanks to the cross-linking of the layers, both sides of the product are of equal strength, and expansion and contraction are minimized. It is more stable than other processed woods and has a high screw retention rate. 3-Layer plywood is generally preferred indoors, 5-layers are preferred for exteriors of temporary structures and interior wall coverings, and multi-layers are preferred for roof coverings and exteriors. In plywoods with the same thickness but a different number of layers, the strength increases as the number of layers increases [39].

6.3.4 Metal-based showcase materials

The use of metal in joinery is reaching an increasing rate. The most important reason for this event is the concept change in building production. The industrial age has brought to the fore "Metal joinery", which shows the characteristics of the product of this age. When comparing metal with wood as a material, its advantages in various aspects are also valid when comparing joinery made with these materials. Positive results have been obtained in terms of the cost of raw material and the high investment requirement for the establishment of the workshop, which are the most restrictive factors for metal joinery. Instead of construction methods that require a large investment, new techniques were used (bending or drawn profile instead of casting), type standardization and mass production were made and new marketing opportunities were obtained [40].

6.3.5 Aluminum (al) and light alloy-based showcase materials

Aluminum is a precious metal material with superior qualities. Ideal solutions are obtained with light alloys obtained by the combination of other metals, especially Magnesium and Silicium. The most important features of this material are; lightness, resistance, corrosion under the influence of the atmosphere, long life, easy processing and bonding ability, and ease of maintenance are shown [41].

6.3.6 PVC and plastic based showcase materials

Between 1950 and 1957, France, USA England, Italy, Russia, etc. The successive exhibition of houses built with plastic materials with very different techniques in foreign countries has heralded the opening of domestic roads in the field of construction. Plastics; light, long-lasting, insulating power, easy processing, and propensity for mass production are the prominent features [41].

6.3.7 Glass showcase materials

It is a material obtained as a result of melting an oxide mixture containing silicium in glass at high temperature and cooling it with a subsequent process. If the mixture is completely homogeneous (equal), transparent glass is obtained. The presence of substances in the main shaper and the density of these substances form the series of "translucent" (light-transparent - non-sight) or "non-transparent" (light and vision-impermeable) glasses. Glass is a hard material that is highly resistant to external influences, does not get dirty, does not rot, does not deteriorate, and is easy to clean. The most important problem with glass is its fragility. Normal glass cannot resist mechanical impact. This important issue led to a special production series and the construction of high-resistance glasses was realized. The thermal conductivity value determines that the glass poses a significant problem. The resolution of heat and sound issues such as fragility has also led to the acquisition of new glass types.

7. Smart showcase design proposal

The smart packaging system has revealed the need for a smart showcase design. The smart showcase design proposal is shown in Fig. 17.2. This showcase adjusts climatic factors such as humidity, temperature, and light according to the products in it, and ensures that the products in it can be stored for a healthier and longer time. In addition, information exchange and interaction are provided with the smart showcase by using NFC and RFID technologies.

Besides, the smart showcase can more fully display the display features of the products, promotions, advertisements, and other information content to the customer. At the same time, it not only reduces labor costs but also improves the store's service image. The advantages of the smart showcase are displayed in Table 17.1.

Fig. 17.2 Smart showcase design proposal.

Table 17.1 Advantages of smart showcase.

Advantages for customers	Advantages for the store
Advice to customers	Stock information on the showcase
Access to training	Climatization conditions of the showcase
Product verification	Follow-up of stock supply processes
The story of the product	Automatic product counting in showcase
Correct use of the product	Active stock control against theft
	Promotions, advertisements, and other information content to the customer

Smart showcases can be controlled remotely with applications installed on mobile phones. These showcases are a design proposal that appropriately interacts with people, machines, systems, and the environment. In this way, business and energy efficiency can be increased, time can be saved, and sustainability and contribution to the country's economy can be made.

References

[1] K.L. Yam, P.T. Takhistov, J. Miltz, Intelligent packaging: concepts and applications, J. Food Sci. 70 (2005) 1–10.
[2] K. Marsh, B. Bugusu, Food packaging: roles, materials, and environmental issues, J. Food Sci. 72 (2007) 39–55.
[3] K.B. Biji, C.N. Ravishankar, C.O. Mohan, T.S. Gopal, Smart packaging systems for food applications: a review, J. Food Sci. Tech. 52 (2015) 6125–6135.

[4] J.H. Han, New Technologies in Food Packaging: Overview, in: J.H. Han (Ed.), Innovations in Food Packaging, Elsevier Academic Press, London, 2005, pp. 3—10.
[5] S.F. Schilthuizen, Communication with your Packaging: Possibilities for intelligent functions and identification methods in packaging, Pack. Technol. Sci. 12 (1999) 225—228.
[6] J.H. Han, in: J.H. Han (Ed.), Innovations in Food Packaging, Elsevier Academic Press, London, 2005, pp. 138—153.
[7] S. Otles, B. Yalçın, Intelligent Food Packaging, LogForum 4 (3) (2008). İzmir.
[8] S. Otles, B. Yalçın, intelligent food packaging, LogForum 4 (3) (2008). İzmir, https://www.dekomika.com/market-dizayn/. Accessed date: 25.01.2022.
[9] M.T. Bohr, Nanotechnology goals and challenges for electronic applications, IEEE Trans. Nanotechnol. 1 (2002) 56—62.
[10] S. Mobasser, A.A. Firoozi, Review of nanotechnology applications in science and engineering, J. Civ. Eng. Urban. 6 (2016) 84—93.
[11] T. Singh, S. Shukla, P. Kumar, V. Wahla, V.K. Bajpai, I.A. Rather, Application of nanotechnology in food science: perception and overview, Front. Microbiol. 8 (2017) 1501.
[12] A.K. Sundramoorthy, T.H.V. Kumar, S. Gunasekaran, Graphene-based nanosensors and smart food packaging systems for food safety and quality monitoring, in: Graphene bioelectronics, Elsevier, 2018, pp. 267—306.
[13] B.P.F. Day, Active packaging of food, in: J. Kerry, P.J. Butler (Eds.), Smart packaging technologies for fast moving consumer goods, Wiley and sons Ltd, West Sussex, England, 2008.
[14] K.T. Lee, C.S. Yoon, Quality changes and shelf life of imported vacuumpackaged beef chuck during storage at 0_C, Meat Sci 59 (2001) 71e77, https://doi.org/10.1016/S0309-1740(01)00054-7.
[15] K.K. Gaikwad, J.Y. Lee, Y.S. Lee, Development of polyvinyl alcohol and apple pomace bio-composite film with antioxidant properties for active food packaging application, J. Food Sci. Technol. 53 (2016) 1608—1619, https://doi.org/10.1007/s13197-015-2104-9.
[16] C. Wallace, W. Sperber, S.E. Mortimore, Food Safety for the 21st Century: Managing HACCP and Food Safety throughout the Global Supply Chain, John Wiley & Sons, 2018.
[17] S. Singh, I. Park, Y.J. Shin, Y.S. Lee, Antimicrobial properties of polypropylene films containing AgSiO2, AgZn and AgZ for returnable packaging in seafood distribution, Journal of Food Measurement and Characterization 10 (2016a) 781—793.
[18] S. Singh, Y. Shin, Y.S. Lee, Antimicrobial seafood packaging: a review, Journal of food science and technology 53 (2016b) 2505—2518, https://doi.org/10.1007/s13197-016-2216-x.
[19] S. Singh, K.K. Gaikwad, M. Lee, Y.S. Lee, Thermally buffered corrugated packaging for preserving the postharvest freshness of mushrooms (Agaricus bispours), J. Food Eng. 216 (2018a) 11—19, https://doi.org/10.1016/j.jfoodeng.2017.07.013.
[20] S. Singh, M. Lee, K.K. Gaikwad, Y.S. Lee, Antibacterial and amine scavenging properties of silver-esilica composite for post-harvest storage of fresh fish, Food Bioprod. Process. 107 (2018b) 61—69, https://doi.org/10.1016/j.fbp.2017.10.009.
[21] K.K. Gaikwad, S. Singh, Y.S. Lee, A pyrogallol-coated modified LDPE film as an oxygen scavenging film for active packaging materials, Prog. Org. Coating 111 (2017) 186—195.
[22] W.S. Choi, S. Singh, Y.S. Lee, Characterization of edible film containing essential oils in hydroxypropyl methylcellulose and its effect on quality attributes of "Formosa"plum (Prunus salicina L.), LWT - Food Sci. Technol. (Lebensmittel- Wissenschaft -Technol.) 70 (2016) 213—222.
[23] B.J. Ahn, K.K. Gaikwad, Y.S. Lee, Characterization and properties of LDPE film with gallic-acid-based oxygen scavenging system useful as a functional packaging material, J. Appl. Polym. Sci. 133 (2016).
[24] P. Morone, A. Koutinas, N. Gathergood, M. Arshadi, A. Matharu, Food waste: Challenges and opportunities for enhancing the emerging bio-economy, J. Cleaner Prod. 221 (2019) 10—16, https://doi.org/10.1016/j. jclepro.2019.02.258.
[25] C. Chauhan, A. Dhir, M.U. Akram, J. Salo, Food loss and waste in food supply chains: A systematic literature review and framework development approach, J. Clean. Prod. 295 (2021) 126438, https://doi.org/10.1016/j. jclepro.2021.126438.

[26] Y. Omolayo, B.J. Feingold, R.A. Neff, X.X. Romeiko, Life cycle assessment of food loss and waste in the food supply chain, Resour. Conserv. Recycl. 164 (2021) 105119, https://doi.org/10.1016/j.resconrec.2020.105119.
[27] X. Zhang, Y.A. Zhao, Q. Shi, Y. Zhang, J. Liu, X. Wu, Z. Fang, Development and characterization of active and pH-sensitive films based on psyllium seed gum incorporated with free and microencapsulated mulberry pomace extracts, Food Chem 352 (2021) 129333, https://doi.org/10.1016/j.foodchem.2021.129333.
[28] X. Zhai, Z. Li, J. Shi, X. Huang, Z. Sun, D. Zhang, X. Zou, Y. Sun, J. Zhang, M. Holmes, Y. Gong, M. Povey, S. Wang, A colorimetric hydrogen sulfide sensor based on gellan gum-silver nanoparticles bionanocomposite for monitoring of meat spoilage in intelligent packaging, Food Chem 290 (2019) 135–143. https://doi.org/10.1016/j.foodchem.2019.03.138.
[29] E. Drago, R. Campardelli, M. Pettinato, P. Perego, Innovations in smart packaging concepts for food: an extensive review, Foods 9 (11) (2020) 1628. https://doi.org/10.3390/foods9111628.
[30] T.N. Tran, B.T. Mai, C. Setti, A. Athanassiou, Transparent bioplastic derived from CO2-based polymer functionalized with oregano waste extract toward active food packaging, ACS Appl. Mater. Interfaces 12 (41) (2020) 46667–46677. https://doi.org/10.1021/acsami.0c12789.
[31] A. Redfern, M. Koplow, P. Wright, Design architecture for multi-zone HVAC control systems from existing single-zone systems using wireless sensor networks 6414, SPIE, 2007, pp. 239–246. Smart Structures, Devices, and Systems III.
[32] A. Bayraktar, Görsel Mağazacılıkta Vitrinlerin Önemi. İstanbul: Beta Yayınları, 2011.
[33] N.C. Ataoğlu, Vitrin Tasarımında Güncel Yaklasımlar. 2. Ulusal İç Mimari Tasarım Sempozyumu, Karadeniz Teknik Üniversitesi İç Mimarlık Bölümü, Bildiri Özetleri Kitabı, Trabzon, 2018.
[34] P. Taoukis, T. Tsironi, 5 - Smart Packaging for Monitoring and Managing Food and Beverage Shelf Life, The Stability and Shelf Life of Food, Second Edition, Woodhead Publishing Series in Food Science, Technology and Nutrition, 2016, pp. 141–168.
[35] S. Melikoğlu, Marka Mekanı Olarak Vitrin Tasarımının Önemi: Tüketici Davranısları Üzerinden Deneysel Bir Arastırma, Karadeniz Teknik Üniversitesi, Yüksek Lisans Tezi, Trabzon, 2008.
[36] E.C. Korkmaz, Hazır giyim sektöründe üretici firmaların gözünden vitrin tasarımı (Master's thesis, Fen Bilimleri Enstitüsü), 2014.
[37] L. Pırılkan, Tüketici Satınalma Davranısları Bağlamında Giyim Mağazalarının İç Mekan Tasarımı, Hacettepe Üniversitesi, Ankara, 2005.
[38] Y. Sabuncuoğlu, Mimari Mekanda Yapay Aydınlatma Ve Mağazalar, Dokuz Eylül Üniversitesi, İzmir (2002).
[39] Y. Sabuncuoğlu, Mimari Mekanda Yapay Aydınlatma Ve Mağazalar, Dokuz Eylül Üniversitesi, İzmir, 2002. https://cesmimarlik.com/ahsap-esasli-malzemeler-nelerdir/. Accessed date: 26.01.2022.
[40] N. Sanıvar, İ. Zorlu, Z. Isık, İç mimari ve dekorasyonda konstrüksiyon, Milli Eğitim Bakanlığı, 1982.
[41] G. Çelikbas, Mağazalarda Kurumsal Kimlik ve Vitrin Tasarımı İliskisi: İstanbul Louis Vuitton Mağazalarının Vitrin Tasarım Analizi, Maltepe Üniversitesi, Yüksek Lisans Tezi, İstanbul, 2013.

CHAPTER 18

Biodegradable polymers- a greener approach for food packaging

Bably Khatun[a], Jonali Das[a], Shagufta Rizwana[b] and T.K. Maji[a]
[a]Department of Chemical Sciences, Tezpur University, Tezpur, Assam, India; [b]Department of Food Engineering and Technology, Tezpur University, Tezpur, Assam, India

1. Introduction

In layman's words, packaging may be described as something that encloses a product. But the term packaging has many layers to it. As defined by G. L. Robertson, a material to be employed for 'packaging', should satisfy the four functions-containment (contains product), protection (protects from damage), convenience (facilitates storing and transport), and communication (carries information) [1]. Food is shielded from the environment, its nutrients are preserved, and its shelf-life is preserved via food packaging [2]. Also, a package is the only product part that is initially presented to the customer communicating about the details of the product inside. Packaging serves as a convenience for handling, disposal, product visibility, reseal ability, and microwavability. Packaging offers a defence against physical, biological, and chemical risks (Fig. 18.1) [3]. There are different levels in a packaging system. The different levels of packaging are:

Primary: The material is directly in contact with the product, usually serving as a barrier, e.g., plastic bags, paperboard boxes, metal cans, and glass bottles.

Secondary: The material holding the primary packaging, working as a carrier, e.g., corrugated boxes.

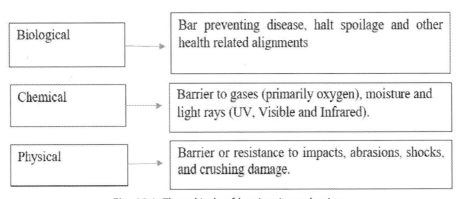

Fig. 18.1 Three kinds of barriers in packaging.

Tertiary: The material holding several secondary packages, e.g., stretch-wrapping over corrugated cases.

Quaternary: In a few cases, a big metal container holding many tertiary packages, usually used in ships and trains.

To keep the food supply chain safe, food packaging plays a tremendous role. Food packaging contributes to 50 % of the packaging sales, due to its high consumption [4]. Food packaging is mostly based on non-renewable and non-degradable derivatives of plastic, thus contributing to a major amount of pollution. Therefore, this chapter focuses on how biodegradable packaging can help to minimize waste generation and sustainable sources for the environment.

2. History of food packaging

Packaging materials had been used for ages. Earlier, naturally available products like leaves, skin, and gourds, were used as food packaging materials. Baskets were weaved from grasses and fibers to store dry harvested products. Clay amphorae and wooden barrels were used for storing wines and oils. It was until the 18th century; packaging applications were limited. Afterward, packaging industries started, and the idea of consumer packaging came into existence [5]. The progress of packaging industries can be explained through different time groups (Table 18.1) [5–9].

Table 18.1 Description of the evolution of packaging industry through ages.

Timelines (Year)	Packaging evolution age	Manufacturing type
1800–1890	Emerging consumer packaging technologies	Uses of glass, paper, and metal cans in smaller quantities [5].
1890–1920	Packaged and branded	Mechanization of packaging industries for mass production, development of corrugated boxes, and printing of brand names [6].
1920–1940	Package and sell	Packages itself serves as an advertisement. Manufacture of packaging materials in compliance with food safety regulations [7].
1950–1970	Growth of packaging management	Consideration of packaging as a discipline and profession, introduction of plastics as a cheaper alternative, and widening the range of

Table 18.1 Description of the evolution of packaging industry through ages.—cont'd

Timelines (Year)	Packaging evolution age	Manufacturing type
1980–2020	Increasing functionality and technological packaging	packaging types like flexible packaging [8]. Introduction of tailor-made packaging materials according to consumer requirements. Development of tetra pack, modified atmosphere packaging, vacuum packaging, etc. [9]

3. Characteristics and criteria of packaging materials

The most used packaging materials with their consumption percentage is displayed in the Fig. 18.2 [10]. The aspects of packaging material include labelling and designing. A food packaging engineer's job is to deliver optimized packaging systems in terms of price, ease, preservation, trade, and sales.

The packaging materials and their characteristics are mentioned below [11]:

Paper. Papers are made up of wood pulp consisting of hemicellulose and lignin. Paper and paperboard are made of natural fibers of bleached and unbleached cellulose and can be recycled. For years, bamboo and mulberry barks have been used for paper making. For food packaging, papers are combined with oil, grease, and waxes for achieving better

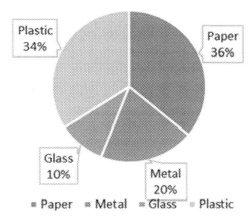

Fig. 18.2 Packaging material consumption rate.

barrier properties. It is lightweight, easily shape-able, and rigid but its barrier property is less. The types of different paper boards used for food packaging are:
1. White-lined chipboard (WLCB): It is made up of 60–100% of recycled fiber and is arranged in a multilayer structure. The top layer is high-grade quality bleached paper having better printability. The mid-layer and end layers are formed from wastepaper. These are generally used as outer cover which doesn't remain in direct contact with food.
2. Uncoated recycled board (URB): It contains nearly 20% virgin fibers without any coating. Mostly, it is combined with different paper or paperboards and utilized for general packaging applications and set-up boxes. These types of boxes are generally used as secondary or tertiary packaging and not in direct contact with food.
3. Solid bleached sulphate (SBS): It is often used for aroma and flavor-sensitive products. It is made of entirely premium-grade bleached pulps.
4. Coated unbleached kraft (CUK) board: It is molded from sulphate consisting of the unbleached pulp of higher strength. The upper layer consists of bleached pulp. It is used as a beverage carrier for being inexpensive.

The different types of papers utilized are kraft paper (usually coarse with much strength), bleached paper (soft and bright), glassine paper (high density and transparency), and grease-proof paper (translucent and heavily hydrated).

Plastic: Plastics are generally derived from petroleum resources. They are classified as thermosetting or thermoplastic. They are non-biodegradable in nature because it takes millions of years to decompose. Some common characteristics of plastics used in food packaging are good resistance to chemical substances and temperature; permeability of gas; low water vapor transmission rate, accurate shearing resistance, and mild thermal and mechanical behavior, and so on. Ethylene, the simplest monomer, can be considered as the building block of most of the food packaging plastic materials. Processing ethylene involves mixing it with additional monomers to create the copolymers ethylene vinyl acetate (EVA), ethylene methacrylic acid (EMAA) & ethylene acrylic acid (EAA). The various other types of polymers used for packaging are discussed as follows:
- Low density polyethylene (LDPE)
- High density polyethylene (HDPE)
- Polyethylene terephthalate (PET).

Due to its flexibility and other properties, LDPE is used to manufacture films, garment bags, bubble wraps, and foams. HDPE is used for freezer bags, bottles to carry milk, juice bottles, etc. PET is the most frequently employed polyester-class thermoplastic polymer resin.

Metals: The most used metals in food packaging are steel, tin, aluminium, and chromium. Tinplate and steel that have been electrolytically coated with chromium are used in composite materials for construction (ECCS). Aluminium is utilized in purified alloy forms of magnesium and manganese in a very small amount. Metal containers are preferred for their superior barrier properties to gases, moisture & light; mechanical strength, low toxicity, and heat resistance, and hence are ideal for decorating and lacquering purposes.

Metals are mostly used for the purpose of canning. The canning process is adopted for the preservation of many food materials for a longer duration. It involves hermetically sealing the already sterilized containers and creating a vacuum inside. For canning, two types of lacquering are considered: acid-resistant and sulphur-resistant. Acid-resistant can use R-type enamel. Basically, R enamel is a phenolic resin varnish. Sulphur resistant (C-type enamels) are used for non-acidic fruits like most vegetables. The C-type enamels are pigmented with zinc oxide or zinc carbonate.

Glass: It is made up of sand, flint, and quartz. Glass has very little tensile strength and so it is fragile in nature. Glass containers are generally preferred for their thermal resistant property. But thermal shock resistance must be taken into consideration while working with glass bottles. Glasses tend to darken and lose much of their ability to transmit light when bombarded by high-energy radiations such as those used in food irradiation.

Glass is employed because it is inert, does not react with or migrate into food items, and is resistant to moisture, gases, odors, and germs. The filling speeds of glass are comparable to those of cans.

In essence, testing packaging materials entails examining their physical characteristics [12].

a) Paper and paperboard: "Technical Association of the Pulp and Paper Industry" reports that the essential properties of paper to test are water vapor transmission rate, air resistance, basis weight, binding swiftness, brightness, opacity, roughness, tensile strength, burst strength, tear resistance, coefficient of friction, and thickness. Such attributes or properties primarily rely on the non-impurity, variety, & dimension of the materials such as pulps and fibers, and secondly on the chemical composition and moisture content of the pulp.

b) Plastic: While designing plastic for packaging, the properties that are taken into consideration are density, degree of polymerization, permeability, chemical resistance, etc. The coefficient of friction, elongation, modulus, elasticity, tensile strength, elongation, viscosity, plasticity, elasticity, etc. are also very important physical properties. Thermal properties that are very critical are heat expansion, enthalpy, and melting, crystallization, glass transition & heat deformation temperatures (HDT). Morphological properties of plastic packaging like crystallinity affect the physical and thermal properties. Since plastics are molded from monomers of different atoms, therefore, configuration, arrangement, conformation, the number of molecules and atoms are the most important factors characterizing individual plastics.

c) Metal: The primary and most crucial testing property for metals are corrosion and denting because they have a high chance of electrochemical reaction with food materials. Thermal resistance properties and pressure resistance properties are also to be checked.

d) Glass: Here, physical properties like impact resistance, resistance to vertical load, internal pressure resistance, resistance to impact, and shearing resistance are to be checked. Thermal shock resistance is also an important criterion. Since glass is optically isotropic, the light transmission rate needs to be measured.

4. Types of food packaging

Based on technology and materials, food packaging is mainly grouped as aseptic, active, modified atmosphere, and biodegradable packaging.

4.1 Aseptic packaging

Aseptic means non-presence or omission of non-desirable organisms from food substances as well as packaging materials. It is also referred to as hermetically sealed because it restricts the entrance of air or water or any such environment that can promote organisms to grow. It is used to package pre-sterilized products (e.g., milk products, fruit juices, and sauces) or non-sterile products (e.g., fresh-cut fruits) by creating an aseptic environment in the package itself. High-temperature-short-time (HTST) and ultraheat or ultrahigh temperature (UHT) are the sterilization processes used in aseptic processing.

4.2 Active packaging

The idea of smart packaging comes from concepts of active packaging that allow indicators for packed food quality control. It is essential for evaluating the shelf-life of packaged items because it functions as temperature-time indicators for frozen foods, oxygen indicators to check the gaseous concentration, carbon dioxide indicators, which are primarily used in perishable foods such as fish, meat, & poultry, and microbial indicators to find pathogens. Table 18.2 provides some illustrations of active packaging systems [13–16].

4.3 Modified atmosphere packaging (MAP)

To slow down fruit respiration, minimize microbiological amplification, and prevent enzymatic deterioration for self-life extension, the internal gaseous ambiance must be adjusted or modified before sealing. MAP packing can be done in either an active or passive manner. By creating a small vacuum, the active MAP packaging technology fundamentally substitutes the interior ambient conditions in the container with the desired atmosphere. In the case of passive MAP, depletion of oxygen and release of carbon dioxide occurs by the commodity which matches with film permeability properties, because of which proper airspace develops (Table 18.3).

Table 18.2 Active packaging system and their application.

Packaging type	Absorbing or releasing compounds	Purpose of applications	Food material
Oxygen absorbers [13]	Ferro-compounds, ascorbic acid, metal salts, glucose oxidase, alcohol oxidase	Prevention of mold, yeast and aerobic bacteria, oxidation of fats, oils, vitamins, colors, etc.	Cheese, meat products, ready-to-eat products, bakery products, coffee, tea, nuts, milk powder, etc.
Ethylene absorbers [14]	Aluminum oxide and potassium permanganate (sachets), activated carbon + metal catalyst (sachet), zeolite (films), clay (films), Japanese soya	Prevention of too fast ripening and softening, inhibition of mold growth.	Fruits like apples, apricots, banana, mango, cucumber, tomatoes, avocados, and vegetables like carrots, potatoes, etc.
Carbon dioxide emitters [15]	Ascorbic acid, sodium hydrogen carbonate, and ascorbate	Growth inhibition of gram-negative bacteria and molds.	Vegetables and fruits, fish, meat, poultry item, etc.
Antimicrobial preservative releasers [16]	Sorbic acid or other organic acids, silver zeolite, spice, and herb extracts, allylisothiocyanate enzymes, e.g., lysozyme	Growth inhibition of spoilage and pathogenic bacteria.	Meat, poultry, fish, bread, cheese, fruit, and vegetables.

4.4 Biodegradable packaging

It refers to the production of packaging materials by using natural polymers. These are environment friendly because they disintegrate into carbon dioxide, biomasses, water, and inorganic compounds, and as a result, discard no toxic degraded product in the surroundings [17]. Food has traditionally been packaged using non-renewable resources. Paper-based materials are excluded from biopolymers as trees cut for papermaking, by and large, have a renewal time of 25–65 years. Films and coatings that are edible are included as biodegradable polymers as they are developed from food-based materials. They may be eaten collectively with the food. Edible packing can be used in the form of pouches, films, coatings and sheets and are safe-to-eat [1].

Table 18.3 Conventional plastic materials, their bio-based alternatives, and uses.

Conventional plastic materials	Alternate biobased materials	Manufactured items
PET (polyethylene terephthalate)	Bio-PET	Container for carbonated beverages
PP (polypropylene), PS (polystyrene)	PLA or its blends or Bio-PET	Trays, caps, etc.
PE (polyethylene)	Bio-PE	Transparent films for packaging of vegetables and fruits
PE, PP	Blends of PHA/PLA/PBAT	Carry bags, shopping bags

5. Why is biodegradable packaging important?

The management of packaging waste is a big environmental concern these days. Regulations are amended by countries to control pollution caused by packaging waste. According to recent reports, plastic consumption around the globe is over and above 200 million tonnes, with a yearly increase of roughly 5%, which consists of the biggest area for crude oil application. It is quite perturbing how oil and natural gas dependency on the plastic industry can have an economical influence on countries [18]. Biodegradable packaging promotes the utilization of renewable and potentially more sustainable sources of organic material and facilitates integrated waste management [19]. Waste management has several approaches, the first is the reduction and prevention of waste proposal by minimal utilization of resources and the second is the reusing, recycling, and composting of the material. The composition of matter depends on various factors like the environment and the physicochemical properties of the matter. Another way is the combustion of biomass to generate energy, but it is less feasible. The last option is the dumping of waste which is the least reliable option as it concentrates our landfills [20]. Packaging waste consists of a substantial portion of municipal garbage. Recently, biodegradable, and compostable polymers have evolved as a potential alternative for conventional plastics. Thus, we need to emphasize more on recyclable and sustainable sources of packaging. Biodegradable packaging is one such need of the hour.

6. Biopolymers used for food packaging

6.1 Polysaccharide based polymers

6.1.1 Cellulose

Cellulose is a bounteous biopolymer available in plants, sea animals, some ameba, bacteria, and fungi [21]. It is utilized as a good substitute for synthetic petroleum-based polymers and is commercially made on a big scale in the packaging industry all over the world [22]. But directly extracted native cellulose is unsuitable for packaging due to poor water solubility and highly crystalline structure [23,24]. Therefore, cellulose properties are modified by combining them with other biopolymers, plasticizers, and nanomaterials. Improved cellulose composite films' tensile strength and moisture vapor barrier qualities can be applied to packaging [21].

Each cellulose molecule contains long anhydro-D-glucopyranose units (AGU) with three hydroxyl groups/AGU except at the terminals. Naturally, cellulose exists in two different crystal forms, cellulose I and cellulose II. The cellulose generally found in marine algae (cellulose II) remains in crystalline form and it forms under the treatment of cellulose I with aqueous sodium hydroxide [25]. This biopolymer consists of several thousands of D-glucose residues linked linearly in a β-1, 4 forms (Fig. 18.3) [26].

Four different modified forms of cellulose (ether and esters) have practical importance. They are bacterial cellulose, cellulose acetate, ethyl cellulose, and hydroxypropyl cellulose. Bacterial cellulose is devoid of hemicellulose, lignin, and pectin [27,28]. A few drawbacks of bacterial cellulose consist of excessive production cost, low yield, use of steeply-priced culture media, working costs, and downstream processing. Another important ester of cellulose is a cellulose ester. Based on the way of processing, cellulose esters can be applied to membranes or fibers.

Ethyl cellulose is made from natural cellulose that has part of its hydroxyl groups reformed into ethyl ether groups [27,29]. Ethyl cellulose is a non-ionic material, non-reactive in nature [30]. A modified kind of cellulose called hydroxypropyl cellulose is dissolvable in water as well as organic solvents and water. The OH groups of repeating glucose units of cellulose are interplaced by hydroxypropyl groups via an etherification reaction in synthetic approaches to its manufacture.

The intramolecular & intermolecular hydrogen bonding among the chains make cellulose an odorless, tasteless, fibrous substance that cannot be dissolved in polar solvents like water or other common solvents [25]. It exhibits a range of significant structural properties like chirality, biodegradability, water-solubility, & variable reactivity due to the donor-OH group at the terminal end of the molecular structure. It has a unique ability to retain its semi-crystalline state of aggregation [27,29]. Although humans cannot

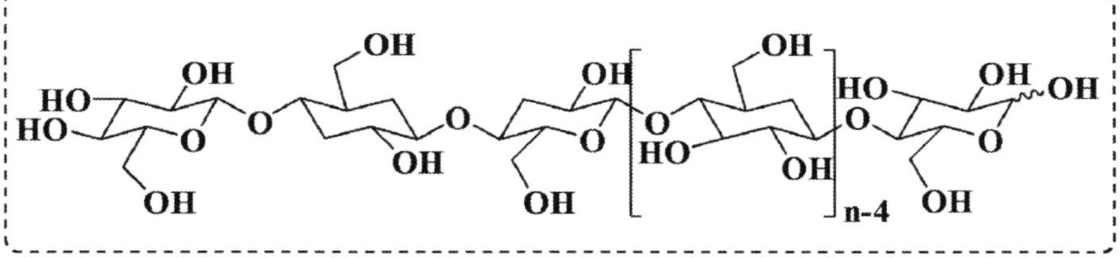

Fig. 18.3 Chemical structure of cellulose.

digest cellulose, it covers varied ways of applications in human life including wood for building, cotton, paper products, rayon for clothes, linen, cellulose acetate for films, nitrocellulose for explosives, etc. [25]. Cellulose's utilization in the formulation of food packaging materials is divided into three categories: coating materials, composites, and edible & non-edible films. Composites are derived directly by extraction from plants i.e., it is solely dependent on renewable raw materials. Cellulose derivatives are widely employed as a barrier coating in food items, edible, and biodegradable films. Edible coatings are protective coatings made of edible polymers applied to food surfaces. Fruits are submerged into active methylcellulose (MC) and MC-palmitic coatings. MC-based coatings are also used in the quality and storage of tomatoes. Cellulose ester derivatives including both organic and inorganic esters are highly used as a film former, as an additive to different film former, as a reactive curing coating, etc. [31]. Ether form of hydroxypropyl cellulose can be employed as a lubricant. Further, it can be utilized for treating corneal erosions, neuro paralytic keratitis, keratoconjunctivitis sicca, etc. [27,29]. Cellulose ethers have significant use in food, cosmetics, paints, and pharmaceutical industries. Commercially, carboxymethylcellulose (CMC) is the most usable significant cellulose ether derivative. It is widely utilized as an additive. It possesses several advantageous properties like high water solubility and organic solvency, responsible for its behavior as a multifunctional agent. Natural cellulose does degrade upon melting and can be regenerated to manufacture fibers known as rayon or viscose and the film is called by brand name 'cellophane'. Cellophane is widely used in the market as breathable packaging for baked goods, cheese products, and live yeasts as well as oven able & microwavable packaging [31,32]. Nanocellulose can be represented as biodegradable cellulose nanomaterial having significant properties like renewability, sustainability, nontoxicity, & recyclability, and it produces a low carbon footprint with approximately 35—50% natural fibers. Nanocellulose has been used as an important key constituent for the application of cellulose-derived food packaging. Different types of nanocellulose include nanofibrillated cellulose (NFC), bacterial nanocellulose (BNC), and cellulose nanocrystals (CNCs). Nanocellulose is available from agricultural waste such as sugarcane, wheat, corn, etc. [33].

6.1.2 Starch

Starch is a highly demanding environment-friendly renewable natural polymer. It is cheap, abundant, and has high production capability from renewable resources like water, sunlight, and carbon dioxide (CO_2) by photosynthesis of plants [34]. It is available in various plants such as wheat, potatoes, corn, beans, and rice as well as certain strains of algae and fungi. Starch granules exist in semi-crystalline form, 2—100 μm in size having wide-ranging morphologies and distributions. Chemically, Starch consists of polysaccharides, viz., amylose (β-amylose) and amylopectin (α-amylose) along with trace amounts of lipids and proteins. Both polymers are composed of the same monomeric units, i.e., α-D-glucose, linked by glycosidic bonds at two different positions. Starch contains amylopectin in a major amount (more than 80%) which is present in the outer part of the starch grains. Whereas only 20% of amylose present in amylose starch is hydrophilic which lies in the inner part of starch granules. Amylose has a linear structure containing α-1, 4 anhydroglucose units with a size range of 200–800 kg/mol. The film-forming properties of starch arise because of the presence of amylose. As opposed to that, amylopectin may be a strongly branched chemical molecule with a large molecular weight (5000–30,000 kg/mol) and short α-1, 4 chains connected by α-1, 6 glycosidic branching sites every 25–30 glucose units. Depending upon the diffraction pattern of starch grains, starch can be classified into two crystal types, viz. '**A-type**' and '**B-type**'. Starch derived from cereal grains is mostly of A-kind while that from root tubers is more of B-kind. C-kind of diffraction patterns are characterized by the presence of both A- and B-kind of diffraction patterns. Another type of crystal structure, known as the V-type, may arise upon plasticization which is due to the interaction of amylose with the plasticizer [35,36].

As a result of the intramolecular hydrogen bonding present in cold water, starch granules are not soluble; but, in hot water, some of the starch grains become soluble due to the small disruption of the hydrogen bonding. Starch fulfills all the chief criteria such as biodegradability, biocompatibility, high yield, chemical inertness, resistance to degradation, & easy availability thus making it edible and protective. However, there are a lot of limitations of starch to maintain safety and food quality. To circumvent these constraints, starch is combined with other plasticizers and polysaccharides to improve its film-forming ability. Starch-based films are semi-permeable to oxygen, carbon dioxide, moisture, liquids & flavoring agents. They are tasteless, odorless, biodegradable, non-toxic, and physiologically absorbable [37]. Strong intermolecular forces and hydrogen bonding resist it to act as a thermoplastic material. Therefore, despite the presence of water in starch, food-safe plasticizers like glycerol, urea, glycerine, sorbitol, etc. are added to form a deformable thermoplastic material commonly known as thermoplastic starch (TPS). Plasticizers increase flexibility and enhance processability by lowering the glass transition temperature (T_g). Because of the growing cost of petroleum products and environmental problems, the importance of plasticizers has increased [38]. Starch films become more rigid due to a high percentage of amylose indicating high elastic modulus (EM) & tensile strength (TS) with a reduction in reasonable elongation at break (EAB). Low amylose content prohibits the formation of films. Furthermore, various other characteristics such as protein content, granule size, granular morphology, granule size distribution, amylose/amylopectin molecular weight, and phosphate monoester content control the mechanical properties of starch films. However, those traits range from species to species relying upon the assets of extraction. Additionally, time, humidity levels, and film synthesis techniques all significantly influence the characteristics of films. Cassava (26.7% of amylose), corn (28.5 % of amylose), and wheat (27.4 % of amylose starch) form very good films. On the opposite hand, films made with potato starch (10% of amylose) crack upon drying in any respect concentrations. The physical alteration of starch also affects the characteristics of starch films. The starch "Babassu mesocarp" is extracted under acidic, alkaline, or neutral conditions. The crystal structures produced under acidic conditions are of B-type, while those under neutral or alkaline conditions are A-type structures. Starch modification strategies like acetylation, oxidation and crosslinking enhance the characteristics of starch films. Acetylation improves starch film degradability; crosslinking improves the strength of almost any polymeric materials and oxidation increases the aqueous solubility and moisture content [35].

A thin layer of packaging material known as "edible film" can be consumed in whole or in little amounts. Commercially available Bioplast GS2189, Bioplast 105, Bioplast GF106/02, Bioplast 200, Bioplast 900, Bioplast wrap 100, and Bioplast TPS are made of potato starch, most of which are free from any plasticizer. These films are composed of approximately 23—100% bio-based portions and can be applied as thermoformed food trays, flexographic, and offset printings. Both Pentafood Biofilm and Pentaform Biofilm TPS are composed of excessively hydroxy-propylated maize starch. They are utilized to provide a glossy surface, outstanding flavor, odor barrier, and grease resistance and comprise 85—90 wt percent renewable materials. Fresh fruit and vegetable packaging options include Mater-Bi and MagicNet [39].

6.1.3 Polyhydroxyalkanoates

Polyhydroxyalkanoates (PHA) are biodegradable polymers derived from renewable resources. It's a group of 3, 4, 5, and 6-hydroxyacid linear biopolyesters made from lipids, sugars, alkane, alkene, and alkanoic acids fermented by bacteria. Its easy biodegradability and excellent biocompatibility make it a promising bioplastic material and in recent times extensive research has been going on to search for an alternative to petroleum resources [40].

PHAs are spherical-shaped granules having diameters in the range of 0.1—0.2 μm stored in the cytoplasm of bacteria. PHAs are nature-based polyesters with different side-chain lengths and fatty acid-containing hydroxyl groups at the fourth and fifth positions. Based on the number of carbon atoms attached to their monomeric units, PHAs are divided into three distinct classes, viz. short-chain length PHAs (PHA_{SCL}), medium-chain length PHAs (PHA_{MCL}), and long-chain-length PHAs (PHA_{LCL}). PHA_{SCL} contains a monomeric unit with 3—5 carbon atoms and PHA_{MCL} contains 6—14 carbon atoms. Long-chain fatty acids with more than 14 carbon atoms are used to make PHA_{LCL} [41—43]. The structure of PHA is displayed in Fig. 18.4 [44].

In contrast to PHA_{MCL}, which is elastic and flexible and has a glass transition temperature below room temperature, PHA_{SCL} is brittle, stiff, and extremely crystalline having a high elongation at break. PHAs are employed as a food packaging material due to their favorable mechanical and physical characteristics, low permeability for O_2, H_2O, and CO_2 without leaving catalyst residue. The barrier quality of a film is crucial since the food shelf life is mostly affected by permeability. PHAs like PHBV, PHB, and PHBHHX

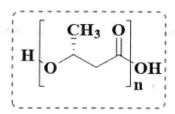

Fig. 18.4 Chemical structure of polyhydroxyalkanoates.

are favorable materials for food packaging. PHAs have brilliant film-forming and coating capabilities due to their hydrophobicity. But PHAs production and industrial application are still limited because of their high cost [45–47].

Some of the promising characteristics of PHAs are hydrophobicity, biodegradability, thermoplasticity, nonlinear optical activity, nontoxicity, gases or water resistivity & piezoelectricity [42]. The type of polymer generated, and the composition of the monomeric units affect the rate of thermodegradation, young's modulus, tensile strength, and water vapor & oxygen transport. PHAs have different melting and glass transition temperatures that range from 40 to 180 °C and -50 to 4 °C, respectively. One of the most significant thermoplastic polymers, polyhydroxy butyrate (PHB), has a melting point of 170–180 °C. It is an attractive candidate for food packaging because of its biodegradability in various environmental conditions [46]. Milk packaging uses PHA-coated film and paper to increase the gas barrier properties of conventional packages. In addition, PHAs are used as paper or paperboard coating material to make biodegradable water-resistant surfaces which are used as an alternative to polyethylene-based paperboard. PHB-coated and PHAs copolymer PHBV-coated paperboards have been utilized in the packaging of ready-made meals and for dairy products, dry products, and beverages, respectively. Besides their barrier properties, PHAs are also beneficial in producing modified paper and board with grease-resistant properties and sealability [45].

6.1.4 Pectin

Pectin is the fundamental structural element of a plant's cell wall. It provides flexibility & strength, structural integrity, and protection from the external environment. It is a polysaccharide-based branched polymer made up of neutral carbohydrates such as rhamnose, galactose, arabinose & xylose as well as extended α-chain galacturonan components. Pectin is not only found in the cell wall of plants (35–40%) but is also predominantly present in vegetables and fruits [47,48]. Pectin is mostly made up of α-(1, 4)-linked polygalacturonic acid structure. Based on their process of extraction, the carboxylic group present in the backbone is esterified with methanol [49,50]. The general molecular structure of pectin is displayed in Fig. 18.5 [51].

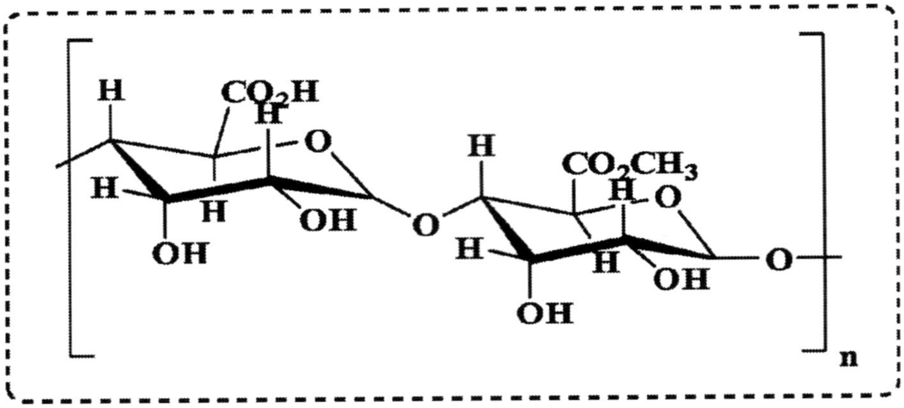

Fig. 18.5 Chemical structure of pectin.

The free carboxylic group provides acidic properties to the pectin solution. The percentage of the esterified carboxylic group, present in the pectin structure, is termed the degree of esterification (DE). Many properties of pectin including solubility, gelation, viscosity, and acidity are highly dependent on the degree of polymerization [47]. Esters are hydrophobic, which reduces the solubility of pectin. However, DE significantly speeds up pectin gelation. High methoxy pectin (HMP) gels can be formed by entanglement, self-aggregation, and different mechanisms that de-esterify. Electrostatic attraction causes monovalent cations (Na^+, K^+) to create salts with free carboxylic groups, which leads to the creation of 3D crystalline networks, also known as gel networks. In the presence of water or ethanol, these networks get collapsed and aggregated. The addition of a water-ethanol mixture to HMP shows emulsifying properties that are highly applicable in food processing. Smooth portions of pectin, an anionic biopolymer, are those without side chains, while hairy regions are those with non-ionic side chains. HMP is chiefly employed in the food industry as it forms gels at low pH because of hydrogen bonding as well as lipophilic interaction among the pectin chains. Neutral sugar such as sucrose actively participates in gelation. Both HMP and LMP (low methoxy pectin) have film-forming abilities under specific conditions by incorporating antioxidants, crosslinking agents, and active additives. In pectin-based films, calcium chloride can be used as a crosslinking agent. Furthermore, clove essential oil additionally improves the antioxidant properties, antimicrobial, and mechanical barrier. Pectin-based films are highly acceptable as they are yellowish, semi-transparent, visually homogeneous, water-resistant, and have high tensile strength. Moreover, pectin has good compatibility with protein, lipid, and other natural or synthetic biopolymers. Pectin is in high demand because it may be used to create an edible coating for food containers. Red cabbage extract, tea extract, and other plant extracts can be used to alter the structural and functional characteristics of pectin-based films [51]. Extract of pomegranate peel can be incorporated as an antimicrobial agent into the edible coating for emergency food products (EFP) [52]. Nanoparticles are added to pectin-based films to enhance their mechanical and barrier qualities, antibacterial activity, and antioxidant capabilities, making them acceptable for use in food packaging [51]. Pectin is a colloidal polymeric material widely used as hydrogels because of its soft and flexible nature, intrinsic biocompatibility, high water content, unique structure, and similarity to natural materials. Moreover, pectin gel-based packaging materials show high oxygen barrier properties, reduce oxidation and respiration rate of food, and thus are an appropriate choice for food packaging application [52,53]. Pectin has been used for a very long time as a coating material on fresh and cut fruits and vegetables as well as a gelling, thickening, and emulsifying agent, colloidal stabilizer, and texturizer in food products including jams, jellies, and marmalades. LMP is highly used to prepare low sugar content jams in the food industry. Pectin hydrogels have antibacterial & antioxidant properties that prolong the shelf-life of foods [51,53]. Pectin is a composite material having several qualities that make it a viable candidate for its future application in food packaging [54].

6.1.5 Alginate

Alginates are naturally available water-soluble polysaccharides derived from brown algae species. Alginates are biodegradable polymers that are commonly used in the creation of edible films because of their ability to form gel and films. Alginic acid which is isolated from *Phaeophyceae* plants can be converted to alginates in presence of alkali (i.e., the salt form of alginic acid called alginate). Alginate is composed of linear binary copolymers of D-mannuronic and L-guluronic acids joined by 1, 4 glycosidic links. The structure of alginate is portrayed in Fig. 18.6 [55].

Alginate-based edible films are used for active packaging. They actively serve as a vehicle for antioxidants, antimicrobials, and other ingredients that can increase the

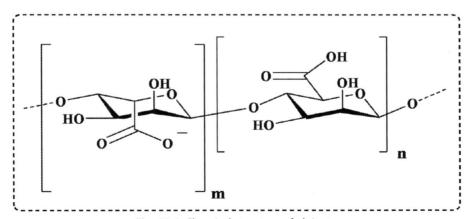

Fig. 18.6 Chemical structure of alginate.

shelf-life of food goods. Alginates and their calcium salts, calcium alginates, are hydrophobic, but monovalent alginate salts, such as sodium, ammonium potassium alginates, are hydrophilic, albeit only to a limited extent at lower pH levels. It is used as an emulsifier, gelling & thickening agent. As per the European Commission, alginic acid and its salts can be used as food additives. Charged states of alginate (Ca^{2+}, Sr^{2+}, Ba^{2+}, Ni^{2+}, Pb^{2+}, Zn^{2+}, Mn^{2+}, etc.) can form gel via ion exchange. Polyuronide is a well-known natural ion exchanger. Monovalent cations and Mg^{2+} are incapable of forming gels. The uncharged state of alginate is responsible for increasing the viscosity of the gel. The mechanical, organoleptic, functional & nutritional characteristics of alginate-based films can be reformed by incorporating several chemical and natural additives like plasticizers, surfactants, antimicrobials, antioxidants, anti-browning agents, pigments, and flavors. The plasticizers such as water, sorbitol, glycerine, sodium lactate, etc. are generally added for the formulation of alginate-based edible coating and films to modify the mechanical properties.

Surface active agents are important elements that improve the wettability and adherence of a hydrophobic surface, resulting in a smooth, uniform edible coating. Antimicrobial agents are applied to edible coatings and films to lessen the microbiological burden of alginate-based coated food items. The addition of antioxidant agents like ascorbic acid, citric acid, etc. to an edible coating decreases oxidation due to its gas barrier property. Anti-browning compounds like polyphenol oxidase, N-acetylcysteine, and glutathione are frequently used in crosslinking solutions & employed on the surface of fresh or cut fruits & vegetables to prevent enzymatic browning, oxidative rancidity & degradation, flavorings, sweeteners, spices, coloring agents, and seasonings are incorporated into the alginate-based coating matrix to develop the organoleptic properties of coated products. Moreover, nutritional additives like minerals, vitamins, and probiotics, as well as sunflower oil, olive oil, n-hexanal, and D-limonene, have been used in edible films & coatings without compromising the goods' integrity. Lipid hydrolysates are applied to alginate films & coatings to enhance antibacterial, antioxidant, and visible light barrier characteristics. Alginates are also employed as a thickening agent in the printing and pharmaceutical sectors. In addition, it serves as an emulsifier, stabilizer, chelating agent, suspending agent, encapsulating agent, swelling, and gelling agent, as well as a film and membrane-forming material [56,57].

6.1.6 Chitosan
Chitosan, a biodegradable polymer, is derived from chitin. It can be present in the cell wall of fungi, yeast & insects as well as in the exoskeletons of crustacean shells, fungi, yeast, and insects. It is typically made by deacetylating α-chitin in a 40–50% aqueous alkali solution at 100–160 °C. It acquired the second position in terms of availability immediately after cellulose [20].

Chitosan, obtained from biological sources, can be used to package food instead of synthetic petroleum-based packaging [58]. Chitosan's functionalization and interaction with other compounds allow it to have a wide spectrum of biological actions, including antibacterial and antifungal properties [20]. Chitosan's molecular architecture is prepared by randomly dispersed β-(1, 4)-linked D-glucosamine & N-acetyl-D-glucosamine units [59]. Fig. 18.7 depicts the molecular structure of chitosan [60].

The crosslinked polymeric network of the chitosan chain involves amine and dicarboxylic functional groups that improve its mechanical properties thereby enhancing its transfection efficiency and gene transfer. The majority of the glucopyranose residues in chitosan are deacetylated to form 2-amino-2-deoxy-β-D-glucopyranose, whereas natural chitins are homopolymers of 2-acetamido-2-deoxy-β-D-glucopyranose residues that are arranged into crystalline microfibrils. Both chitin and chitosan are found naturally in combination with other polysaccharides, minerals, and proteins. Chitin is available in 3 polymorphic forms, viz. α, β & γ-forms, which might be all made of stacks of chains joined via way of means of amide bonds (CO−NH bond). The organization, packaging, and polarity of neighbouring chains in consecutive sheets determine the creation of their distinct types [61]. Due to its vast biological features such as biodegradability, biocompatibility, and edibility, chitosan has a diverse set of applications. Chitosan has antimicrobial and antibacterial properties and hence is used as an antimicrobial food packaging material.

Antibacterial properties of chitosan get up because of the electrostatic interconnection of undoubtedly charged polycationic chitosan with negatively charged additives in microbial cell membranes [62]. The molecular weight, series of acetamido and amine groups, and purity of the product, all affect the structural attributes of chitosan. Chitin is insoluble in water and most organic solvents. Chitin deacetylation creates a loose primary amine group, which makes chitosan incredibly soluble in dilute acid solutions (pH < 6.0).

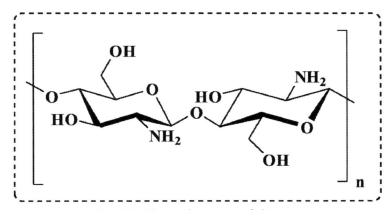

Fig. 18.7 Chemical structure of chitosan.

The charge and amine functional group of chitosan are responsible for its properties. Primary amines get protonated at low pH resulting in positively charged amine and thus become soluble in water. This cationic polyelectrolyte can interact with diverse molecules. At acidic pH, it shows high ionic conductivity and behaves as polyelectrolytes [61]. The degree of deacetylation influences many properties of chitosan including solubility, susceptibility to biodegradation, the extent of swelling in water, bioactivity, biocompatibility, etc. [63]. However, chitosonium salts are unsuitable for packing because of their hydrophilic nature and low water resistance [59].

Amino and hydroxyl groups of glucosamine residue of chitosan are susceptible to chemical modification and enhance their functional properties. The conformation of chitosan molecules in weakly acidic solutions depends on several factors, including solvent temperature, solvent ionic strength, pH, and dielectric constant. Chitosan is a linear amino polysaccharide containing excessive nitrogen content material, permitting it to shape sturdy intermolecular H-bonding ensuing in the creation of a polymeric network-like structure [61,63]. Chitosan is a versatile film-forming and coating material. It can be made from a variety of celluloses, including quaternized hemicelluloses, hydroxypropyl methylcellulose, methylcellulose, and micro fibrillated cellulose, as well as other biopolymers, to create composite films that increase its mechanical, biological, and physicochemical properties. These films exhibit properties like improved resistance, uniformity, elasticity, and transparency making them suitable for food packaging application [62,64,65].

Chitosan's traits may be altered with the aid of using blending it with extra components like plasticizers, proteins, and lipids [66]. Multilayer film formation is another advanced method in which several layers of component films are merged to enhance the mechanical, physical & transport attributes of the film for food safety [67]. Due to its diverse qualities, the chitosan-based film has created a new avenue in the food packaging business [64].

6.1.7 Agar

Agar is a gelatinous polymer derived from certain marine red algae of the Rhodophyceae family. Agar comes from two main commercial sources: Gelidium sp. and Gracilaria sp. It's the main ingredient in seed weed extract. The gelling fraction part is Agarose, and the non-gelling fraction part is agaropectin and both are combined to make agar. Agarose is a neutral linearly arranged unit of D-galactose & 3—6 repeating units of anhydro-L-galactose associated together by alternating α-(1, 3) & β (1, 4) glycosidic connections (Fig. 18.8 [68]).

Agaropectin is a byproduct of the commercial agar production process. The ability of agar to form gel is affected by temperature changes. A viscous fluid and a thermoreversible gel are formed in warm water and a temperature less than the gelling temperature (90—103 °C), respectively. Owing to its superior mechanical strength & mild water

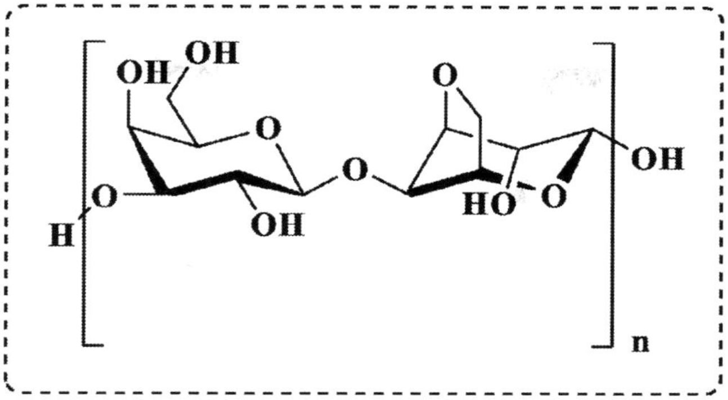

Fig. 18.8 Chemical structure of agar.

barrier characteristics, it is frequently employed in the development of packaging materials. Pure agar-based films suffer from brittleness, poor thermal stability, and high moisture permeability. As a result, agar-based films are combined with various biopolymers, as well as nanomaterials, plasticizers, hydrophobic components, and antibacterial agents, to improve their qualities and expand their applicability [69,70]. For food packaging applications, the employment of agarose coating on cellulose extract paper substrate can be done [71]. Both the food quality and shelf-life are increased with agar-based coatings [72].

6.2 Protein-based polymers

Protein-based polymers are nature-based, and capable of substituting synthetic non-biodegradable polymers. It is easily available, cheap, thermally accessible, and non-toxic, and is derived from both plant and animal sources. Gluten, soy proteins, maize germ, wheat glutenin, gladin, and zein are plant proteins, while casein, whey, egg, fish, and collagen are animal proteins [73].

Proteins are primarily made up of amino acids. Globular and fibrous proteins are the two forms of proteins found in nature. Globular proteins are water, aqueous acid, base, or salt soluble proteins that serve a variety of activities in living organisms. Fibrous proteins are water-insoluble and form hydrogen bonds with one another to form fibers [74]. To make protein-based edible films, caseinate, lactic acid, serine, and collagen proteins are employed [75]. Proteins have a 3D network structure composed of several amino acid units. During the formation of protein films & coatings, crosslinking of protein structure plays a vital role [76].

Proteins are good materials for edible food packaging because they effectively prevent moisture & flavour loss, control gas, and actively deliver substance. The strength of edible films is determined by intermolecular interactions between two neighbouring amino acid side chains, thus making them weak barriers against liquids, vapours, and gases. At low relative humidity, protein-based polymer films have good oxygen barrier qualities, but because they are hydrophilic, they have a poor barrier to water vapor. The molecular characteristics of proteins, such as their molecular weight, charge, conformation, flexibility & thermal stability, have a substantial impact on their ability to form films [74,77].

Protein films outperform polysaccharides in terms of mechanical characteristics [75]. The nature & distribution of both the amino acid sequence and the degree of elongation and the total number of interactions across protein chains, all have a significant impact on their mechanical properties [78]. Because of its superior barrier capabilities against gases like oxygen, whey protein coating has emerged as one of the most favourable applications in the packaging field. Whey protein coating provides a barrier to Aroma, humidity, fat, and oxygen. However, the inherent brittleness of whey protein coatings is a major drawback. The addition of plasticizing chemicals like sorbitol & glycerol to whey protein coatings to overcome the problem. Furthermore, in the food packaging industries, whey-based multilayer films reduce CO_2 emissions and resource use [79].

Among several protein sources, canola protein is isolated by aqueous extraction technology. It exhibits thermoplastic properties in the presence of plasticizing agents (water, polyethylene glycol, glycerol, and sorbitol) that allow it to attain glass transition ($T_g < -50\ °C$) and undergoes easy deformation and processability [80,81]. Peanuts represent a principal source of protein that exhibits good functional properties with high nutritional value. Consequently, the majority of recent research focuses on creating films using peanut proteins to improve their functional qualities [82]. Corn is a prolamin protein that is high in zein. Fig. 18.9 [83] shows the chemical structure of zein. Zein-based biofilms and coatings are prepared because of their excellent film-forming characteristics. Gelatin is readily available and inexpensive and it has been utilized to make edible films & coatings since the 1960s. Hydrolysis of a fibrous hydrophobic protein produces gelatin. Fig. 18.10 [84] illustrates the chemical structure of gelatin. Gelatin-based films or coatings provide good transparency, barrier, and mechanical properties [74]. However, gelatin-based films are not as effective as synthetic polymers (LDPE, HDPE, and others) in terms of characteristics. Plasticization, production of composite films, crosslinking, and use of nanotechnology have been investigated as ways to circumvent its limitations.

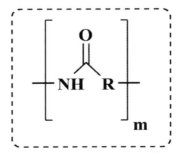

Fig. 18.9 Chemical structure of zein.

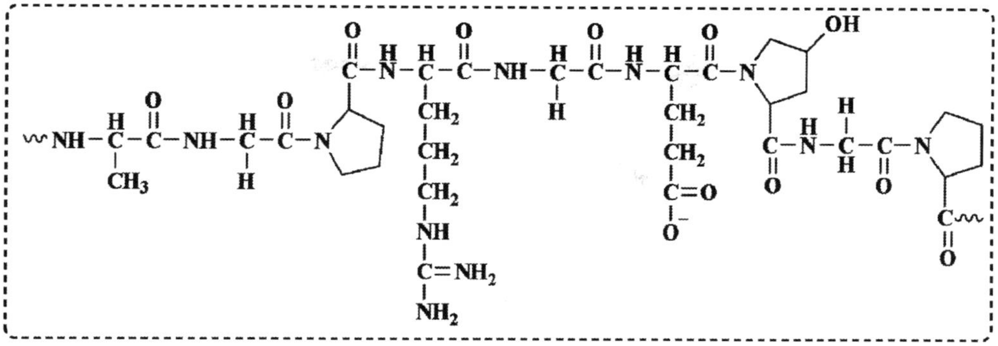

Fig. 18.10 Chemical structure of gelatin.

Additionally, antibacterial agents, bacteriocins, and acidulants (e.g citric acid, sodium lactate, tartaric acid, potassium sorbate, malic acid, lactic acid, etc.) are introduced into protein films to increase their antimicrobial, mechanical, and rheological properties [78].

Proteins are renewable raw resources frequently employed in the development of environment friendly bioplastics, particularly used for food packaging [76]. Whey protein coatings have been put to the test as edible films. Protein films made from peanuts, fish, fruits, or cereals, for example, can be employed because they are biodegradable and reusable. Protein films are used in foods such as cakes, pizza, and pastries to prevent moisture loss. It can also be used in pharmaceutical industries as coating or films to control the release of drugs [79].

6.3 Lipid-based polymers

Lipids are bio-based nonpolar materials like fatty acids, waxes, or fats. Waxes are derived from two different sources. Paraffin wax is a petroleum-based product whereas beeswax is extracted from the insect bumblebees.

Lipids are water insoluble and hence provide an excellent barrier to moisture. They contain long fatty acid chains having carbon 14–18 carbon atoms and are widely used for forming films & coatings. Other biopolymers, like polysaccharides or proteins combined with lipids, offer greater mechanical as well as barrier properties than pure lipid coatings. Waxes also prevent oxygen and moisture from entering the food, resulting in a smooth outer surface. The non-toxic nature of waxes makes them a popular choice for edible coatings on fresh fruits & vegetables [23]. Acetylated monoglyceride coatings have been employed as moisture retardants in the food industry (e.g., chicken and beef) during storage. Shellac resins, a blend of aliphatic and acyclic hydroxyl acid polymers, released by *Laccifer lacca* insects, are also utilized in packaging. This resin can be dissolved in both alcohols and basic alkaline solutions. The resins are typically used to cover citrus fruits to provide a high gloss and extra gas-permeable barriers [85].

Essential oils are complex lipid oils naturally produced from flowers, seeds, buds, wood, leaves, fruits, roots, barks, and twigs of plant materials. Monoterpenes, diterpenes, and triterpenes are the most common natural volatile molecules found in essential oils. They also contain antibacterial, antiviral, antioxidant, antimycotic, antiparasitic, antioxygenic, antiseptic, and antibiotic phenolic chemicals and oxygenated derivatives. Essential oils are commonly used as additives in food packaging materials to enhance the shelf-existence of food goods due to their superior biological performance. Essential oils are employed in conjunction with biodegradable polymers like chitosan, PLA, and gelatin [86].

6.4 Synthetically produced bio-polyesters

Synthesis of bio-based polyesters is highly demanded due to limited petroleum resources, ever-increasing oil prices, and emission of greenhouse gases [87]. As an alternative to petroleum-based products, biodegradable polyesters are commonly utilized for food packaging. Despite their high tensile strength, high modulus, and good processability, they possess considerable biodegradable and thermoplastic properties. Biopolyesters are categorized as aromatic & aliphatic polyesters and can be obtained from renewable as well as non-renewable sources [88].

6.4.1 Aliphatic polyesters

Aliphatic polyesters are biodegradable polymers derived from polycondensation reactions of glycol and aliphatic diacids of medium-sized monomers (C_6–C_{12}) prepared by fermentation of fungi as well as elastase. Aliphatic polyesters are odorless and lack of thermal and mechanical properties. The biodegradability of aliphatic polyesters is more than aromatic polyesters due to their balanced hydrophobicity and hydrophilicity. Procter and Gamble Co. manufactures commercially available aliphatic copolyester under the brand name Nodax [18,88]. The different types of aliphatic polyesters are discussed below.

6.4.1.1 Polylactic acids

Polylactic acids are synthetic biodegradable polymers obtained from both fossil and other renewable sources. They are made from starch and/or sugars by utilizing different polymerization processes including ring-opening, enzymatic, polycondensation, and direct methodologies like azeotropic dehydration. Among different polymerization techniques, the direct and ring-opening polymerization techniques are the most common [89,90]. The structure of the PLA is portrayed in Fig. 18.11 [91].

Because of their unique mechanical and biological properties, PLAs have a wide range of uses. Biodegradability, biocompatibility, renewability, thermoplastic processability, and antibacterial characteristics make them suitable as packaging materials [92]. Organic acids, extracts of plants like lemon, chelating agents such as EDTA, essential oils, enzymes like lysozyme & other substances can be added to boost their antibacterial activity. One of the

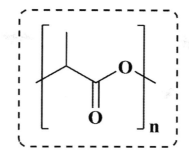

Fig. 18.11 Chemical structure of polylactic acid.

key features of PLA-based food packing applications is its capacity to serve as a good carrier for antibacterial agents without impairing the composting and degrading process [93]. PLAs are also non-toxic to humans. PLAs' tensile, thermal, and barrier characteristics are like the commonly used petroleum-based polymers (polystyrene, polycarbonate, polyethylene terephthalate). These characteristics are determined by the molecular weight & intrinsic stereochemistry architecture of the polymer. The molecular architecture of material controls the degree of crystallinity, processing temperature, and mechanical properties which ultimately rule on thermal behavior [94]. Moreover, the stereochemical composition, molecular weight & crystallinity also affect the degradation behavior. PLAs offer excellent mechanical characteristics, including high Young's modulus, & tensile and flexural strengths. Polylactides exist in various forms in nature because of the chiral character of lactic acid. PLAs are soluble in chlorinated solvents, tetrahydrofuran, hot benzene, and dioxane [95]. However, inherent brittleness (less than 10% extension at rupture), weak moisture barrier performance & low heat resistance limit their application. The production of PLA also has an adverse impact on humans as well as the environment because of deforestation and the requirement of excessive water for growing starch-based crops [93,96]. The blending or composites using PLA are expected to improve the characteristics while lowering the price of the final product without compromising biodegradability. To make composites, nature-based fibers like wool, hemp, ramie, flaxseed, kenaf, and other biodegradable polymers like starch, protein, PHB, and PCL are commonly used as fillers or reinforcing agents. Currently, nanofibers, nanocomposites, and nano additives are particularly used in food applications to enhance material performance, reduce cost, and widen the spectrum of PLA applications [97].

The use of naturally sourced fillers and additives can enhance the properties of the PLA [23]. PLA is widely employed as it is considered one of the most potent polymers for medical and pharmaceutical applications. Additionally, polylactic acids are frequently employed in food packing because of their distinct mechanical & physical characteristics. PLA is used in compostable cups, cutlery, bags, straws, and various other food packaging products. PLA is also employed for packaging carbonated water, yogurts, fresh juices, dairy drinks, frozen fries, pasta, herbs, bread, organic poultry, etc. Recently, the use of PLA for packing products with limited shelf-life foods like fruits & vegetables is also reported [90,98]. PLA is also being blended with proteins, starches, and other biopolymers to generate totally degradable and renewable packaging materials [93].

6.4.1.2 Polybutylene succinate (PBS)

PBS is an aliphatic polyester that is made using a combination of lipase-catalyzed synthesis, direct melt polymerization with butan-diol and succinic acid, and transesterification polycondensation. Furthermore, PBS can be synthesized via cyclic monomers' polymerization (ring-opening), tetrahydrofuran, and succinic anhydride. PBS is one of numerous compounds and polymers that are created using succinic acid as a starting ingredient. Low molecular weight polyesters associated with lower melt strength and melt viscosity create difficulties in film blowing. Direct melt polymerization is a straightforward technique that can produce high molecular weight PBS. In addition, condensation polymerization followed by chain extension produces high molecular weight PBS which comprises the coupling of two PBS chains. But the incorporation of chain extenders deteriorates the biosafety properties which ultimately affects the biodegradability of the synthesized PBS.

PBS exhibits a high degree of crystallinity and good thermal and mechanical characteristics. Its mechanical characteristics are quite like polypropylene. The mechanical characteristics of PBS like flexibility, tensile & impact strengths, transparency, and biodegradability, are dependent on their crystal structures & level of crystallinity. The thermal characteristics of PBS, influence the use & processing of semicrystalline polymer materials. Its molecular formula is represented in Fig. 18.12 [99].

PBS and its copolymers are biodegradable in a variety of environments, including river and sea beads, activated sludge, soil burial, and compost. Copolymerization is a simple way to modify the characteristics of PBS. Random copolymers possess varied properties with changing compositions of copolymers. Copolymerisation typically results in a decrease in crystallinity, a lower heat distortion temperature, and increased elongation. Copolymers' rheological properties are controlled by the polymer's branch content as well as its molecular weight. Branching enhances the polymer's melt strength, elongational flow & tension hardening feature that aids in uniform expansion in polymer processing with a strong orientation.

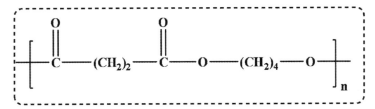

Fig. 18.12 Chemical structure of polybutylene succinate.

To generate PBS having longer branches, a branching agent such as glycerol, trimethylolpropane glycerol, or other monomers containing multifunctional groups are added. The amount of branching agent should be kept to a minimum to prevent gel formation, (e.g., less than 0.5—1 %). PBS products are frequently used as disposable grocery bags, mulch film, packing material, and other items [23,100].

6.4.1.3 Polycaprolactone (PCL)

PCL, a biodegradable thermoplastic polymer, is manufactured from non-renewable oil feedstock. It is an environment-friendly material that is resistant to water, solvents, oil, and chlorine. It degrades quickly, has a low viscosity, and is simple to manufacture with a melting point between 58 and 60 °C & a glass transition temperature of 60 °C [101]. The structure of PCL is portrayed in Fig. 18.13 [102].

It is water-insoluble & is semicrystalline in nature. At room temperature, PCL dissolves in dichloromethane, chloroform, toluene, cyclohexanone, carbon tetrachloride, benzene, and 2-nitropropane. It is sparingly soluble in 2-butanone, acetone, dimethylformamide, ethyl acetate, and acetonitrile while it's totally insoluble in alcohol, diethyl ether, and petroleum ether. PCL's typical molecular weight is normally between 3000 and 90,000 g/mol, and this weight also influences the crystallinity of the substance. PCL's number average molecular weight typically ranges from 3,000 to 90,000 g/mol, and the molecular weight also affects how crystallinity is [103]. With increasing molecular weight, crystallinity decreases. PCL as well as its copolymers have been applied in several biomaterials and biomedical fields due to its brilliant flexibility, biocompatibility, and thermoplasticity.

However, extensive commercialization of PCL is limited owing to its complex and expensive production with moderate mechanical properties. To get around these restrictions, PCL has been copolymerized or combined with additional chemicals. Extensive research is going on toward the development and characterization of PCL nanocomposites to attain improved desirable properties. Carbon nanotubes, cellulose-based nanofillers, and functional nanofillers are some of the most used nanofillers in bionanocomposite preparation.

In the case of food packaging, PCL and starch are combined to create a cost-effective, high-quality biodegradable material that is used to make trash bags [18]. The existing

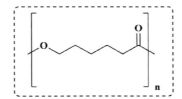

Fig. 18.13 Chemical structure of polycaprolactone.

petroleum-based packaging materials can be replaced with novel, high-performance, lightweight green nanocomposite materials. Another important factor in the packaging of food is to develop high barrier characteristics to restrict the dissemination of carbon dioxide, oxygen, water vapor, and flavoured compounds. Moreover, intelligent qualities of antibacterial activity, oxygen scavenging capabilities, enzyme immobilization, etc. can be added to food packaging by applying various nanostructures [103].

6.4.1.4 Polylactide aliphatic copolymer (CPLA)

CPLA is a polymer composed of lactide and aliphatic polyesters. The hardness and softness of the mixture vary according to the amount of polyester material present. CPLA is thermally stable up to 200 °C and easy to process. During combustion, the heating value, as well as carbon dioxide generation, is reduced to half compared to commercial polymers like polyethylene and polypropylene without producing any toxic substance. It naturally begins to deteriorate after 5—6 months and is finished after 12 months. The decomposition is faster (2 weeks) if it is composted with food garbage [18].

6.4.1.5 Polyvinyl alcohol (PVA)

PVA, a synthetic hydrophilic, biodegradable polymer, is created by hydrolysing polyvinyl acetate. It is non-toxic and commercially available at a low cost. PVA was deemed to be a safe biomaterial for food additives at the 61st meeting of the joint FAO/WHO expert committee in 2003, and the USDA has approved its usage when meat & poultry items are packaged. PVA is flexible and resistant to acid and alkali environments. It also has strong tensile strength and gas barrier qualities. PVA and other hydrophilic polymers, such as chitosan, are particularly miscible because of the intermolecular hydrogen bonding between the hydroxyl groups of the two molecules. The blending of PVA with chitosan shows high compatibility and their films exhibit homogeneous structure, reduce the production cost, increase the stability of films, and improve the mechanical properties. The molecular weight of PVA is 27 kDa [104]. Due to their potential for usage as food packaging materials, PVA-based products are cheap, easily available, biodegradable, non-toxic, and have strong film-forming ability & water processability. However, PVA possesses some disadvantages like high water solubility and low tensile strength. These restrictions can be removed by combining the polymer with crosslinking components to improve film characteristics. Food's nutritional value, appearance, and shelf life are significantly impacted by the interaction of packing material with food ingredients. Both organic and inorganic chemical crosslinking reagents are employed to improve the stability of PVA films. Organic reagents like diethyl carbonate, glutaraldehyde, & formaldehyde, as well as inorganic reagents like boric acid and phosphoric acids, react with the hydroxyl groups of PVA and modulate its physicochemical properties. Furthermore, crosslinkers, which include organic acids like citric acid and oxalic acid, are added to the PVA film to increase its thermomechanical stability and bactericidal activity. It has been proposed that cyclodextrin can also be used in food packaging. Interestingly, lactic acid, tartaric acid & malic acid, have important roles as a crosslinker as they enhance the mechanical as well as biological properties such as bactericidal activity, optical properties, hygroscopicity, and aqueous solubility, which are essential for food packaging material. Functional groups of organic acids actively participate in forming hydrogen bonds with PVA film.

PVA is used as a coating agent for food supplements besides serving as a material for food packing. Further, it is used for enzyme immobilization, protein purification, & membrane separation. Moreover, it is employed in cosmetics, paper, and textile industries as well as several medical applications [105]. The structural representation of PVA is shown in Fig. 18.14 [106].

6.4.2 Aromatic polyesters
6.4.2.1 Polyethylene furanoate (PEF)

PEF is a bio-sourced polyester prepared by polymerizing monoethylene glycol (MEG) & 2,5-furan-dicarboxylic acid (FDCA). FDCA is a monomeric unit prepared by biofermentation of sugars. FDCA can be used as an alternative to oil-based purified terephthalic acid (TA) which is the main component of polyethylene terephthalate (PET) to manufacture PEF. The structure of PEF is displayed in Fig. 18.15 [107]. Wheat or corn-based sugars such as wood, corn, etc. are considered as First-generation biofeedstock. Stover, wheat-straw, bagasse, etc. are considered second-generation biofeedstock to produce PEF [23,98].

Despite being environmentally benign, PEF has superior mechanical, thermal, and barrier qualities versus amorphous PET. When compared to amorphous PET, amorphous PEF has an 11-fold decrease in oxygen permeability, a 19-fold decrease in carbon dioxide permeability, a 5-fold decrease in water diffusion, and a 2.8-fold decrease in water permeability (at 35 °C) [108]. Its water vapor barrier property is two times that of PET. PEF and PET both have glass transition temperatures of 74 °C and 86 °C, respectively. PEF melts at 235 °C, which is lower than PET (365 °C). However, research conducted by the European Food Safety Authority (EFSA) investigated the safety of PEF [23].

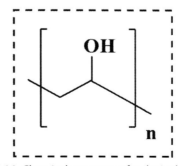

Fig. 18.14 Chemical structure of polyvinyl alcohol.

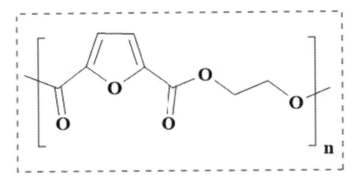

Fig. 18.15 Chemical structure of polyethylene furanoate.

7. Biodegradable polymers' limitations in food packaging

The development of biodegradable polymers for food packaging is the attention of many researchers due to a variety of motivating causes. In this context, environmental problems have emerged as a pressing concern for the entire international community. The chemicals that are used in the production of petroleum-based plastic packaging have adverse side effects on human health including hormone disruption, reduced fertility, genital malformations, and many more [109]. Therefore, nowadays, petroleum-based plastic packages are replaced with eco-friendly biodegradable materials. Biodegradable material includes both natural and synthetic bio-derived products. Natural biodegradable polymers like starch, proteins, lipids, and aliphatic polyesters, are broadly applied in the packing of vegetables, fruits, meats & seafood. These packaging materials preserve and enhance food quality, halt oxidation, and increase food product shelf life. Biopolymers are employed as food-grade coatings and films to reduce moisture loss, prevent oxidation, and so on. Each of them has already been discussed thoroughly [110]. Unfortunately, bio-based packaging materials have limited commercialization primarily due to cost factors, and hence their performance is still under par compared to plastic materials. The mechanical, functional, and biological properties of biomass-derived products are inferior to those of fossil-based plastics. Compared to polymers made from oil, bio-based packaging materials had a reduced shelf life because of these limitations. Because of the high hygroscopicity of biodegradable coatings and films, they become unstable in humid conditions and when in touch with high water content food items [111]. Thus, weak barrier properties booted interest in developing new strategies for improving barrier properties. Polymer blending, application of high barrier materials on biopolymer-based coating, and multi-layered films are the most frequently used strategies to promote barrier properties [112]. Biopolymer coatings suffer from poor processability due to their foaming behavior, high viscosity, and mainly higher brittleness of the final coating layers which lead to inhomogeneities and defects [113]. To overcome the challenges regarding films and coatings made of biopolymers, several advanced techniques are under study. Currently, the mechanical, structural & functional characteristics of polymers are enhanced by plasticization, chemical, thermal, and enzymatic crosslinking, ionizing radiation, pH modification, and the addition of antioxidant or antimicrobial compounds, nanomaterials, and lipids during or after film production [111].

Moreover, the blending of two or more biopolymers is an emerging trend to induce desired functionality into food packaging material [109]. These trends can also generate some drawbacks like a reduction in tensile strength and elasticity compared to single-component film [114].

8. Role or impact of different ingredients in biopolymer modification
8.1 Plasticization

Plasticizers are non-volatile low molecular weighted compounds that are commonly employed as additives in the polymer industries. Plasticizers are substances or materials that are added to another substance or material, usually plastic or an elastomer, to improve the flexibility, tensile strength, workability, and processability of polymers [115]. Polymer plasticity is mostly reliant on chemical composition, molecular weight & the effect of the plasticizer's functional groups. A precise description of the interaction between biopolymers and plasticizers is still unexplored.

Glycerol and sorbitol are plasticizers that enhance the physical, chemical & functional qualities of biodegradable and edible films while balancing integrity and preventing cavities and cracks in the polymeric matrix [116]. External and internal plasticizers are two types of plasticizers known in polymer science. External plasticizers can be easily eliminated through extraction, mitigation, or evaporation even when they do not chemically interact with the polymer chain. Internal plasticizers soften the polymers, which lowers the glass transition temperature & elastic modulus. They also have bulky structures and remain as the polymer molecules' components. By copolymerization or reacting with the original polymer, they become a part of the final product. Plasticizers are also classified as primary and secondary plasticizers. Primary plasticizers are water-soluble and rapidly form gels at a normal temperature range. Secondary plasticizers have lower compatibility and gel formation capability.

Plasticizers are linear or cyclic hydrocarbons with a molecular weight of 300—600. Since they are small molecules, they may readily fit between molecules in polymer chains, which usually change the polymer network's conformation and boost mobility. The degree of elasticity of polymers is greatly affected by molecular weight, chemical makeup & functional group of a plasticizer [117]. Plasticizers for bio-based polymers are classified as either hydrophilic or hydrophobic. The amount of plasticizer applied to the polymers has a big impact on the way the film forms. Hydrophilic plasticizers have high permeability to water, whereas hydrophobic or water-insoluble plasticizers lead to film drying. Dehydration of polymer matrix produces strong cohesive films and the addition of plasticizer reduces intermolecular pressures among the polymer chains thereby enhancing flexibility & chain mobility. Plasticizers also decrease brittleness and avoid shrinking during storage and processing. However, numerous studies have documented the negative effects of plasticizers on consumable films. Plasticizers reduce T_g and water barrier properties. Water itself is one of the most powerful natural plasticizers commonly used in hydrocolloid-based films. Polyols (glycerol and sorbitol) are polysaccharide-based hydrophilic compounds commonly employed in starch-based films to render film toughness, flexibility, and transparency, but they provide poor water barrier characteristics. Consequently, some sugar, amino acids, surfactant, and fatty acids are added to increase their mechanical and barrier qualities [115]. In comparison to unplasticized starch films, glycerol and sorbitol plasticized starch films are uniform, smooth, clear, and contain fewer insoluble particles. According to reports, glycerol-plasticized films made from jackfruit seed starch exhibit qualities like low opacity, average water vapour permeability, & good mechanical characteristics [118]. The pectin fraction of pineapple peels are also used as natural plasticizer to produce edible films for food packaging [119]. To make sugar palm (arenga pinnata) starch biodegradable films less brittle and fragile and more acceptable for use in food packaging, suitable plasticizers such as glycerol, sorbitol, or a combination of the two are utilized [120].

PLA is blended with PHB polymer for food packaging to enhance the crystallinity of the film. Although this blending process increases the oxygen barrier but decreases wettability. Thus, to establish a balance between performance and ductility, two ecologically friendly plasticizers—poly (ethylene glycol) (PEG) and acetyl (tributyl citrate) (ATBC)—have been introduced to the polymer matrix [121].

8.2 Chemical processes

Two significant problems that are frequently seen in food packaging applications are the excessive absorption of moisture and macromolecular reformations. In addition to innovative approaches like plasticization and incorporation of nanofillers, chemical processes involving etherification, esterification, crosslinking, and grafting are also employed. Among these processes, crosslinking is a suitable method for biopolymer materials to overcome the inherent deficiencies regarding barriers as well as mechanical properties. The crosslinking technique incorporates the creation of chemical bonds between several polymeric chains to create a more robust 3-D network. Based on their types of bond formation, crosslinking agents are classified as covalent crosslinking, ionic bond crosslinking, and physical crosslinking [122]. In the case of polysaccharide-based films, carboxylic groups of crosslinking agents form an intermolecular covalent di-ester linkage with hydroxyl groups of polysaccharides. Physical, chemical, and enzymatic crosslinkers are the three basic types of crosslinkers. Furthermore, crosslinking & post-crosslinking techniques have been applied to polymers. While in the latter, crosslinking takes place after the polymer is generated, in the former, the crosslinker is applied directly during polymer synthesis, allowing the end material to be received in one step. The interaction between crosslinking agents and polymers is influenced by the chemical composition, the existence of active functional groups, the molecular weight of biopolymers, and crosslinker-polymer compatibility. Water solubility, the major parameter of food packaging, can be enhanced by replacing the hydroxyl group of polysaccharide units with hydrophobic ester groups. Internal bonds of biopolymeric films become strong because of the addition of crosslinking agents into the polymer film structure. Crosslinking agents have been shown to enhance mechanical qualities such as tensile strength and reduce their elongation at break. In food packaging, crosslinking agents like glutaraldehyde, boric acid, citric acid, etc. are frequently employed with polysaccharide-sourced films. Citric acid, a biobased carboxylic acid found in fruits, is frequently utilized as a crosslinker due to its nontoxicity, affordability, and ability to stabilize polysaccharide materials.

Crosslinking creates a three-dimensional network that is closely connected and acts as a barrier for water from migrating through food packing materials. Since pure starch-based films are brittle, they are chemically treated using crosslinking agents such as citric acid and carboxymethyl cellulose (CMC), etc. to decrease water vapor permeability, moisture absorption, solubility, and improve tensile strength. Citric acid-crosslinked peanut protein films reportedly have decreased compatibility but enhanced dry and wet film strength. Citric acid crosslinking of potato starch composite films improves their mechanical, antibacterial, and tensile strength (by 29%) [122–124].

8.3 Antimicrobial and antioxidants

Presently, the novel preservation strategies of foodstuffs include the development of packaging material by adding antioxidant and antimicrobial agents. Foodstuffs like fresh fish, red meat, etc. are difficult to store in absence of oxygen. The rate of the degradation process increases with the direct addition of antioxidant material [125]. However, applying antioxidants to packaging materials has several benefits over adding them directly to food. For example, lowering the number of chemical additions, controlling their release into the food matrix, eliminating extra processes used while adding antioxidants, etc. [88]. Therefore, antioxidant packaging material has been developed to protect oxygen-sensitive foodstuff. Natural antioxidants like essential oils (coriander oil, thyme oil, etc.) are widely used in packaging. The lipophilic nature of essential oils helps reduce water absorption capabilities and improve the mechanical (tensile strength, optical structure, elasticity), antimicrobial and antioxidant properties. However, strong flavour limits its utilization. The antioxidant capability of chitosan-based polymeric films has also been increased by adding green tea extract, another natural antioxidant [125]. Antioxidants with high thermal stability, such as α-tocopherol & olive leaf extract, are used in the production of Ecoflex and

Ecoflex/PLA films [88,126]. Literature reports reveal that the extracts of grape seeds are a great alternative to artificial antioxidants containing phenolic compounds. Red grape seed extract added to soy protein concentrate films created by compression molding and casting techniques exhibits increased antioxidant capacity [126].

Antimicrobial packaging integrates antimicrobial substances in bio-based packaging materials to stop the growth of microorganisms in food. Antimicrobial packaging provides food protection, improves, and prolongs shelf-life by preventing pathogenic microbial growth. Universal food-borne diseases decrease by the utilization of antimicrobial agents. Antimicrobial compounds for food preservation are synthesized chemically or isolated from the plant, animal, and microbial biomass. Essential oils are naturally available sources of bioactive agents that can also function as antibacterial substances. These have a range of defence mechanisms against microbial growth, including altering the structure of enzymes, deteriorating the phospholipid layer of cell membranes, and endangering the genetic makeup of bacteria [125]. Phenolic compounds, aldehydes, and oxygenated terpenoids, present in essential oils, are mainly responsible for antimicrobial activity. Due to their hydrophobicity, essential oils can interrelate with the lipids found in the mitochondria & microorganisms' cell membranes. For instance, an essential oil having thymol and carvacrol can disrupt *Escherichia coli*'s cell membrane, allowing its component eugenol to interact with the appropriate protein part of the microorganism [16]. Garlic essential oil is produced by steam distillation of garlic bulbs. It contains diallyl disulphide (60%), allyl propyl disulphide (16%), and diallyl trisulphide (20%), with a tiny quantity of disulphide & presumably polysulfide. In a study, garlic oil was added to alginate-sourced edible coatings or films, and the physical & mechanical characteristics of the alginate film as well as the antibacterial activity of garlic oil against harmful bacteria like *E. coli*, *Salmonella typhi.*, *Staphylococcus aureus* & *Bacillus cereus* were studied. To prevent bacteria from growing on edible film, garlic oil can be added to packing materials [127]. Anthocyanins, organic acids, and phenolic compounds also act as strong antimicrobial agents, particularly against *E. coli*. Organic acids, in their protonated form and at high pH conditions, can inhibit the growth of both fungal and bacterial cells by entering their plasma membrane. Based on the different quantities of chitosan, the positively charged glucosamine residue of chitosan rapidly interacts with the membrane of a microbe that is negatively charged, causing leakage of proteins and other internal components [16]. *Thymus moroderi/Thymus piperella* essential oil was recently used in a study to test the antibacterial effectiveness of chitosan edible films. Additionally, it lessens lipid oxidation and water vapor permeability while improving the antibacterial and antioxidant characteristics of chitosan films [128]. To produce food packaging with better antibacterial action than pure chitosan-based films, rosemary essential oil is typically added to chitosan films. Active biodegradable food packaging has been developed using thyme essential oil-activated sweet potato starch/MMT nanoclay nanocomposite sheets [86].

8.4 Scavengers and absorbers

Absorbers, scavengers, and inactive packaging technologies have taken up a significant portion of the market. They consist of ethylene removers, odor absorbers, scent emitters, and carbon dioxide absorber emitters. Of them, barrier packing, oxygen scavengers, and moisture absorbers account for more than 80% of the market. These cutting-edge technologies protect the food products' freshness and quality while extending their shelf lives. In packaged food goods, oxygen can speed up oxidation reactions that degrade food and have negative effects on it, such as reduced nutritional quality, microbial growth, color changes, and development of off-flavours, which may lead to the production of ethylene in fruits & vegetables. Therefore, foods that are prone to oxygen have been packaged in multilayer structures made of high-barrier packaging materials like aluminium foil or ethylene-vinyl alcohol copolymers [128]. To minimize the concentration of oxygen in the packed material that oxidizes, oxygen scavenging chemicals are externally introduced to the material. This eliminates oxygen and reduces oxidative processes. These agents are incorporated in various forms such as sachets in the headspace & labels, and directly reinforced into packaged material. Some oxygen scavengers are metals like Pd, Ti, Fe, Co, Zn, etc., tocopherol, ascorbic acid, gallic acid, unsaturated hydrocarbons, photosensitive dyes, enzymes, etc. [129]. Iron-based oxygen scavenging agents are used commercially for oxygen-sensitive food in recent times [130]. Naturally occurring oxygen scavengers include phenolic compounds and some minerals (e.g., hydrotalcite). Gallic acid-loaded PEO/hydroxypropyl methylcellulose fibers have recently been created via electrospinning or mixing. When used as a packaging medium, the electrospun nanofibers exhibit good antioxidant activity and delay the oxidation of walnuts during storage. Similarly, the presence of ethylene degrades the shelf-life of fresh and leafy products accelerating the ripening or aging of harvested vegetables & fruits. Recent research has focused on the employment of ethylene-scavenging packing materials in the creation of active packing systems. Commercial ethylene scavengers are primarily based on potassium permanganate, but their application is constrained due to their high toxicity and color. To address their shortcomings, novel scavengers like nanoparticles, clays, activated alumina, and activated carbon are added to the packing materials, leading to the creation of a promising substitute for ethylene removal. The development of the chitosan-TiO_2 nanocomposite film exhibits superior barrier qualities and higher tensile strength when exposed to UV radiation, delaying the ripening of cherry tomatoes compared to pure chitosan and oxidized ethylene. Sothornvit and Sampoompuang have developed rice straw composite reinforced with activated carbon as ethylene scavengers and found a maximum level of ethylene scavenging capacity (77%). This composite is eco-friendly because it is reusable and can provide protection against mechanical damage. The most favorable conditions for the growth of microorganisms inside the packaging are both vapor and liquid forms of water. To eliminate surplus water in the headspace, moisture absorbers like calcium oxide, cellulose fibers, etc. have been used [131,132]. Moisture scavengers monitor the water content present in the food products. To prevent moisture absorption by food goods, moisture-sensitive foods must be sealed in a material that resists excessive humidity. In the absence of active barrier packaging, moisture is absorbed, resulting in microbial spoilage. Dry crispy items, such as cakes and biscuits, soften when exposed to too much moisture. Excess moisture is removed through the packing material, but this may cause dehydration or lipid oxidation in packaged foods. Therefore, the development of humidity control packaging film is essential [15]. Desiccants are extensively used in food products to manage humidity in the packaging headspace like cheeses, chips, meats, popcorns, gums, sweets, nuts, & spices. Moisture scavengers' primary objective is to stop the growth of microorganisms like yeast, mold, and bacteria. Moisture scavengers are some moisture-absorbent sachets that include silica gel, cellulose fibers, zeolite, and sodium chloride. These sachets balance the desired humidity level in the foodstuff that inhibits microbial growth [133]. Moreover, other packaged foods such as fresh fish, meat, vegetables, and fruits are also needed to be kept under controlled humidity conditions [134].

Carbon dioxide absorbers, in addition to oxygen and moisture absorbers, are frequently employed in the food business to extend shelf-life and safeguard food quality. CO_2 addition to food items has several advantages. CO_2 inhibits microbial development in specific foods, like fresh meat, cheese, poultry, and baked goods, and lowers the rate of fresh food respiration [129]. Concomitant acidification of carbon dioxide forms carbonic acid which acts as an antimicrobial agent that changes the cell membrane, enzyme activity of bacteria, and cytoplasmic pH. The commonly used CO_2 emitters consist of sodium bicarbonate and organic acids [135]. Commercially, they are available in the form of absorbent pads and sachets containing moisture-activated sodium bicarbonate chemicals. On the other side, excessive levels of carbon dioxide may deteriorate food quality through oxidation. Therefore, to preserve the stability of the food, both CO_2 absorbers and emitters are necessary for active packaging systems. Calcium oxide or calcium hydroxide contained in silica gel can be used to remove CO_2 from packaging sachets. Usually, CO_2 emitters are applied in combination with oxygen scavengers in bifunctional sachets where the absorbed volume of oxygen is equal to the volume of produced carbon dioxide. CO_2 absorber in the form of freshpad is used as one of the active elements for packaging poultry, meat, and seafood [133].

8.5 pH alteration

Intelligent packaging provides food protection and can monitor food quality by sensing any change in the environment during initial packing conditions. This packing system detects, tracks, records, and shows signals to aid decision-making, improve quality, strengthen security, and provide information about potential concerns. Packaging often uses pH dyes or indicators, which are formed of a solid phase and a dye that is sensitive to pH fluctuations. The colors are derived from several natural sources, including fruits & vegetables. As food starts degrading, pH changes take place indicating food quality at the time of purchase and providing real-time quality information to consumers. There are two types of indicators: direct and indirect. Direct indicators provide information regarding freshness, damage, time & temperature. Traceability or tracking indicators are examples of indirect or passive indicators, and this is a growing area of research today. Only limited reports are available and hence have high future research potential. Due to its sensitivity, speed of response, safety, compact size, low cost, and non-invasive nature, intelligent pH indicator packaging is in increasing demand. There are many synthetic colors available, including chlorophenol red, bromophenol blue, and bromocresol purple. However, Synthetic chemicals should be avoided due to potential health risks to consumers. Researchers are looking for natural pH indicators to help them overcome these constraints. Recently, natural dyes containing anthocyanin, curcumin, and flavonoids extracted mainly from flowers and fruits, have shown great potential for the development of pH dyes as they change color at different pH values [136]. These dyes can be integrated into packaging structures to monitor color changes arising due to the changes in acidic and basic components (organic acids, ammonia, CO_2, etc.) present in food products. Prietto et al. developed pH-sensitive sheets, derived from cornstarch & anthocyanin, extracted from red cabbage and the black bean seed coat, as pH indicators and evaluated the impact of anthocyanin on the color spectra of the films and reported that it can be used in intelligent packaging [137]. By using a casting technique, Rhim and Ezati created pH-responsive pectin-sourced films integrating silver nanoparticles (SNP) and curcumin. They then assessed the films' thermal, barrier, antibacterial, antioxidant, and color-changing capabilities in response to ammonia vapor. The results indicate that it has a strong potential for applications in intelligent & active packing [138,139].

8.6 Indicators

Intelligent or smart packaging materials, some indicators, or biosensors are used to monitor or communicate information about food quality. Various indicators like time-temperature indicator (TTI), radiofrequency identification (RFID), biosensor, gas indicator, microwave doneness indicator, ripeness indicator, and many others can be incorporated either inside or outside the packaging material. However, the commercial application of these advanced technologies in food packing has been limited [129]. A lot of food products are susceptible to temperature changes, which shortens their shelf lives. CO_2 is a significant ingredient that is utilized in food packaging as a scavenger or indicator. Therefore, TTIs have been created and are being used in the food business to achieve the sustainability of food items. Compared to other temperature monitoring systems, TTIs are tiny, user-friendly, and cost-effective. When food products are subjected to high temperatures, TTIs produce several changes such as chemical, electrical, biological, etc. A radiofrequency identification system (RFID) is a wireless communication system that tracks supplier information and delivers real-time data regarding the temperature, nutrition, and relative humidity of foodstuffs. This ensures food quality and safety while also enhancing supply chain effectiveness. RFID tags can be embedded with items of packaging materials, but their usage is limited due to high costs. To address this issue, fast-growing research is essential to receive a low-cost RFID as soon as possible [139,140]. CO_2 level rises as food starts degrading because of the growth of microorganisms. Additionally, CO_2 is employed as an indication in a changing environment as a calorimetric indicator by combining an amino acid (L-lysine), a polypeptide (ε-poly-L-lysine, EPL), and a naturally occurring dye anthocyanin and evaluating its effectiveness [141].

8.7 Ion radiation

To address the limitations of biopolymers in food packaging, the radiation method is becoming a typical treatment applied to biopolymer composites. It's a physical treatment for biopolymer-based food packaging materials that improve their functional qualities. It uses high-energy electron beams of UV– and γ-radiation to irradiate biopolymeric materials [142]. γ-rays are penetrated and deposited into the thin layers of polymeric materials thereby sterilizing the foods as well as packaging materials [143]. γ-ray can promote crosslinking and degrade the polymer chains. γ-radiation has been used to explore the characteristics and structure of PVA and bacterial cellulose composite films. The films provide good resistance to γ-radiation making them suitable for their use as packaging materials for irradiated food products [144]. The requirement of costly instruments, disposal of nuclear waste, and strict access to γ-ray generators are some of the serious disadvantages of using γ-ray on biopolymers. UV rays are superior to other sources (γ-ray, X-rays, electron beam) of ionizing radiation for crosslinking purposes due to their several advantages like low cost, easy accessibility, and low penetration power. UV radiation is reported to provide better crosslinking of gelatin and collagen and can readily polymerize a range of monomers. The effects of UV light on several biopolymers, including chitosan, whey protein, starch & carboxymethylcellulose, have been investigated. Crosslinks occur in an aqueous solution because of UV radiation's ability to produce free radicals and damage starch polymer chains. The ability of UV radiation to alter the mechanical and physicochemical characteristics of protein-based films is significant. In a recent investigation, it has been observed that the exposure of fish gelatin films to UV radiation in addition to crosslinking agents (ribose and lactose sugar) increased the tensile strength & percentage elongation at break. It is an environmentally benign technique to enhance the packaging qualities of biopolymers. In a study, starch-based biopolymer film has been prepared using photochemical reactions by exposing the films to different UV regions and various functional properties of modified starch solutions. Results revealed that UV irradiation can be considered a suitable process for the improvement of starch-based films [142,143,145].

9. Advances in food packaging

Recently, sustainability is one of the concerning factors. Different smart and sustainable approaches have been introduced into the packaging system. Synthetic polymers are replaced by biodegradable polymers. Films based on chitosan and other protein-derived bio-composite polymers are reportedly used [146]. Active packaging is used for safety and sensory properties, e.g., CO_2 scavengers/emitters, ethylene scavengers, aroma emitters, etc. are being used. Food condition is indicated by intelligent packaging. Devices for intelligent packaging might be microbiological, self-heating, or self-cooling containers, growth indicators, physical shock indicators, freshness indicators, sensors, time-temperature indicators, gas sensing dyes, etc. [147].

Nanotechnology allows us to reform the characteristics of the packaging materials on a very minute scale. Recent research has highlighted the potential use of nanotechnology within the food sector. Various nanosensors have been developed to get information about the internal packaging conditions and improve the properties like the gas barrier, temperature control, and moisture stability. Nanofillers in combination with polymers are also developed [148]. There has also been advancement in the packaging system by making it edible which has been discussed earlier in this chapter.

9.1 Tetra pack

In 1987, a USA-based company named Tetra Pak received great recognition for developing induction apparatus for the sealing of thermo-coated material. They invented a packaging material named tetra pack by combining three packaging materials: paper, polyethylene, and aluminium. It consists of six layers as shown in Fig. 18.19 [149]. Each layer has a specific function. It comprises 75% paper, 20% polyethylene, and 5% aluminium.

9.2 Nanotechnology

Nanotechnology is the process of fabricating polymeric materials at the nanometric scale to attain particular & desired material properties. This technique has greatly improved the growth of active & intelligent food packaging. Nanotechnology can give various packaging materials the functionality they need to safeguard food better throughout storage and transit. Nanotechnology has been used to improve packaing materials in two main ways.
1. Application of nanocoating on polymer surface and
2. Nanophase dispersion inside the polymer matrix.

These two methods are essential for improving the barrier, mechanical, physicochemical, and functional qualities of packaging materials [150]. Nanomaterials like nanofillers (<100 nm), nanoparticles, nanofibrils, nanotubes, and nanorods are used to improve the functionality of biopolymers [151]. The fundamental properties of polymer materials like flexibility, durability, barrier properties, recyclability, antimicrobial activity, oxygen scavenging capability, and heat resistance can also be modified by incorporating different nanomaterials like montmorillonite clay, kaolinite clay, zinc-oxide nanoparticles, coated silicate, silver nanoparticles, & titanium dioxide. They all function as barriers to block out oxygen, carbon dioxide, and other undesirable substances [152]. A review of the literature suggests that the generation of reactive oxygen species causes metal oxide nanoparticles such as zinc oxide, titanium dioxide, copper oxide, etc. to have potent antibacterial properties. The size, shape, and surface charge of particles and groups of nanostructures have a significant influence on the generation of ROS which ultimately provides antimicrobial activity [153]. The quantity of additional nanofillers has a substantial effect on the mechanical characteristics of bio nanocomposites. The tensile strength, modulus, rigidity, and affinity at the interface of biopolymer increase with the increase in the number of nanofillers. Due to the high surface-to-volume ratio of nano clay polymer composites (like MMT), they are said to have good gas barrier characteristics. Processed meat, candy, cheese, cereal, boil-in-the-bag items, and extrusion-coated packaging materials for dairy products, fruit juices, etc. are all packed using nano clay-based polymer materials. Polymer materials such as thermoformed containers, PET bottles for carbonated beverages, and multilayer film packaging can all be altered by nano clays. They can also improve the gas & moisture barrier qualities of bio-based polymer products. Graphene, a nanoplate material, is gaining attention because of its use as a graphene chemical compound for the development of barrier characteristics and as graphene sheets in the compound composite. Despite their barrier capabilities, nano clays have also been shown to enhance the mechanical characteristics of several thermosets, thermoplastic, and elastomeric polymers [150]. Exfoliated nano clays on dispersion into polymer matrix create a maze structure providing a twisting route to moving gases and thus significantly lowering the permeation rate [112]. Nanofillers like MMT, silver, ZnO, etc. also provide antimicrobial properties in food packaging applications. Among others, silver metal is commonly used to produce bio nanocomposite material due to its stability, low volatility, and antimicrobial characteristics. It is frequently employed as an antifungal agent due to its great toxicity against a variety of pathogens. Silver nanoparticles coupled with bio nanocomposite films made of cellulose, agar, and gelatin have demonstrated effective antibacterial action. Biopolymers act as biosensors to detect environmental changes like temperature, oxygen level, degradability, humidity, and contamination of food products. These responses are to be monitored particularly during storage & processing to ensure the freshness of the food [151]. ZnO nanoparticles are non-toxic compounds with antibacterial action due to their larger surface area. Transparent ZnO/polycarbonate nanocomposite films are prepared, using ZnO nanoparticles having diameters in the 15–20 nm range, for use in applications involving food packaging [154]. Chitosan-based nanocomposite fabricated films & coatings prolong the shelf-life of a variety of food and agricultural items [153]. PVA/Nanocellulose/Ag-NPs nanocomposite films have effective antibacterial properties against *E. coli* and *S. aureus* [155].

Biocomposites are derived from biodegradable & renewable materials such as coffee grounds, cellulose fibers, etc. Nanosized components synthesized from biopolymers (PHA, PBTA, PLA) are termed bio nanocomposites. Nanomaterials or antimicrobial nanoparticles are reinforced into packaging materials to fulfil biopolymeric properties and to compete with the global market needs [156]. Starch and its derivatives are commonly used as edible, safe, and completely degradable polymers used to produce bio nanocomposites.

In recent times, bio-based nanomaterials like nanofibrillated cellulose and cellulose nanocrystals have received rapid attention in food packaging. The mechanical properties, abundance, reinforcing abilities, low density, and biodegradability of nanocellulose make cellulose materials ideal candidate for polymer nanocomposites in food packaging. Despite being extremely sensitive to moisture, these bio-based nanoparticles are made from natural fibers and are used to increase strength and barrier qualities [152,156].

10. Characterization of biodegradable packaging material

Typically, polymers are divided into three categories—thermoplastics, thermosets, and elastomers—based on their thermal and mechanical characteristics. The thermal properties of biodegradable packaging are determined by different analytical techniques like TGA, DSC, FTIR, NMR, DTA etc.

The most widely used methods are:

10.1 Differential scanning calorimetry

It is a physical characterization methodology that's accustomed to studying the thermal behavior of pure polymers, polymer blends, copolymers, and compounds. DSC is a non-isothermal technique giving us detail on how temperature change can affect the properties of the packaging material i.e., energy changes with continuous heating and cooling. This facilitates the determination of transitional temperatures, like the melting point (T_m), crystalline temperature (T_c) & glass transition temperature (T_g), For instance, starch-based polymers (found in wheat) may be easily plasticized and have a low glass transition temperature of 170 °C [157].

10.2 Thermogravimetric analysis

It provides insight into the relationship between weight change or gain in relation to temperature. The amount of mass shrivelled at controlled atmospheric condition for given time or temperature delivers details about the thermal as well as oxidative stability of materials. A TGA thermogram may help to recognize the composition of the materials. TGA helps in studying weight loss/gain due to loss of volatiles, oxidation, or decomposition of materials. Thermoplastics, films, fibers, and other macromolecular materials can all be examined using this analytical technique. TGA measurement can be used in the industrial sector to choose materials for end-use applications, whether in terms of product quality or product attributes [157].

10.3 Fourier transform infrared spectroscopy

It is among the most significant technologies for assessing food packaging sheets because of its advantage in sensitivity, rapidity, and affordability. It is used for analyzing biodegradable polymer films produced from polymers of carbohydrates and protein sources [158]. It had been used to investigate the linkage of the biopolymers in packaging material. e.g., animal protein and antioxidants [159], application and effect of heat treatments on films [160], determination of chemical interaction of polysaccharides and functional groups [161], assessment of water content in packaging films, etc.

10.4 Nuclear magnetic resonance

It is most known as the NMR spectroscopy technique used to identify protein and other complex materials. NMR is preferred because it produces distinguishable signals for different functional groups. NMR is used for accessing the safety of food material in contact with packaging film [162]. NMR is employed for the accessibility of faster degradation of packaging material faster within the environment, leading to entire mineralization or bio-assimilation of the film [163].

The biodegradability of the packaging material can be tested by different methods like soil burial test, test with river water, aerobic test, composting UV degradation, etc.

11. Polymer fabrication technology for packaging food

The food packaging sector often employs a variety of fabrication techniques. A brief discussion of some of the techniques is jotted below.

11.1 Compression molding

This procedure involves adding a precise quantity of polymers to a heated mold cavity, followed by simultaneous application of pressure and heat. In the mold, the polymer melt flows and takes on the desired shape. To release the product from the mold cavity, the mold is cooled and opened. It is generally used for making trays, cups, boxes, etc. [164].

11.2 Injection molding

It is a process where molten polymers are injected into one or more mold cavities under high pressure, cooled, and opened the mold to remove the product. Injection molding equipment comprises of an injection unit and a clamp unit. The clamping device holds the mold while the injection unit melts and forces the molten polymer to inject inside it. The opening and closing of the mold can be done with the help of hydraulic or mechanical toggle systems. The thermoplastic starch-based materials, TPS, PHA, PLA, PBS, and PVA are biodegradable polymers used in injection moldings [165].

11.3 Blow molding

In this procedure, a parison is created by heating and inflating a thermoplastic polymeric material. Compressed air is utilized to blow the molten polymer into the cavity of the female mold, within the cavity of the partner, where the polymer takes the shape of the mold. The compressed air is used up as the shape cools until it sets. The mold is opened to collect the blown article. There are various blow molding techniques such as injection, free extrusion, forced extrusion, stretch blow molding, and coextrusion blow molding. These techniques differ in the formation of the parison and the process of imposing orientation on the product. This process is used for making carbonated-beverage containers, pouches, films, etc. Coextrusion blow molding is used for making multi-layer packaging systems [165].

11.4 Extrusion molding

In this procedure, pressure is used to drive the molten polymer to pass through a shaped die. The melting is done in an extruder consisting of screws, closely fitted inside a barrel with sufficient clearance to rotate the screw. Polymer is entered at one end and molten polymer extrudates from the other end. It then enters the die and takes the shape of the die. It can be used for making films, sheets, etc. [165].

11.5 Extrusion coating

This method is frequently employed in the food packaging sector, particularly when a moisture barrier and an oxygen barrier are needed. In this process, a highly viscous polymer melt is allowed to force out onto a substrate where it is cooled and forms a continuous coating. The heating and melting of polymers are done by using an extruder. The extrudate film's width and thickness may reduce before contacting the substrate. Therefore, pressure and proper polymeric melt management are essential for coating performance. Sometimes adhesion promoters are applied to the substrate for improving the adhesion between polymer and substrate. It can be used for obtaining single or multiple layers. In the case of multiple layers, several extruders containing different polymers are fed into one die to get the desired product [165]. A rough sketch of the process is portrayed in Fig. 18.16.

11.6 Thermoforming

In this procedure, a thermoplastic film is heated, made pliable, and then molded under pressure. The sheet is generally produced by extrusion. It is used mainly for making ice cream containers, disposable dishes, and packaging purposes. There are different processes to get thermoformed products. The simplest process is to use a female mold cavity having several holes drilled through it and the bottom part of the mold is attached with a vacuum pump for drawing the heated sheet material into the mold (Fig. 18.17).

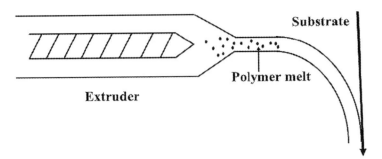

Fig. 18.16 Typical diagram of extrusion coating.

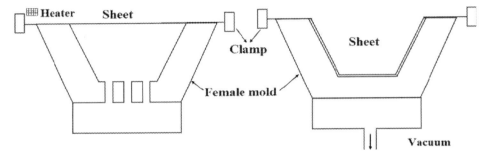

Fig. 18.17 Typical diagram of thermoforming process.

There are two major types of thermoformed materials: thin gauge thermoforming where the thickness is generally kept below 2.5 mm and thick gauge where the thickness is maintained above 2.5 mm. Thin gauge sheets are generally used in packaging sectors. The thermoformed products should have characteristics like higher self-life, heat stability for processes like microwave cooking, sterilization reseal ability, and low-temperature performance for frozen foods, etc. Nowadays multilayer co-extended films are also used for thermoformed packaging purposes. Every layer of the film has some specific polymers that act as a barrier to moisture, oxygen, heat, and so on.

Various biodegradable polymers based on renewable resources are emerging as alternatives to conventional polymers. PLA is widely used due to its good clarity and mechanical properties. Thermoplastic starch or starch-based polymers in combination with polymers derived from petroleum are also being used. These are not fully biodegradable, but this can allow for increasing the share of renewable polymers in the thermoformed product.

Dynamic Mechanical Thermal Analysis (DMTA), in which the changes of modules with temperature fluctuation are recorded, is one of the most significant techniques for characterizing thermoformed products. Besides this, biaxial deformation at elevated temperatures has also been an important parameter used for checking the thermoforming behaviour of polymer [165].

11.7 Electrospinning

Conventional food packaging may not be able to eliminate the contamination occurring in the food industry chain consisting of cultivation, processing, transportation, storage, and retail displaying stages. When designing active food packaging systems, various active agents like antimicrobials, antioxidants, oxygen scavengers, ethylene scavengers, and carbon dioxide emitters can be added. This can be extremely helpful in controlling the resistance of the barrier between packed food and external environments. These agents can be assembled into the packaging materials by coating, encapsulation, surface modification, or immobilization. To make films from polymers for packaging purposes, many fabrication techniques such as extrusion, thermoforming, casting, and so on are commonly used [165].

Electrospinning is a new fabrication technique for creating polymer fibers with a variety of morphologies and structures that are controlled by adjusting process parameters. The continuous polymer fiber is subsequently used as a collector, resulting in a fibrous film. This process involves applying a high voltage electric field to a syringe that serves as the reservoir for the polymer melt. The charged polymer jet is extruded, stretched, and finally stored on a collector that is placed on the ground when the supplied electric force is greater than the droplet's surface tension created at the syringe tip. The structure and composition of fibers can be done by different electrospinning techniques like blending, coaxial, and emulsion. Similarly, the orientation of fibers can be changed by rotating mandrel, moving spinneret, magnetically, etc. The electrospinning process is sketched in Fig. 18.18.

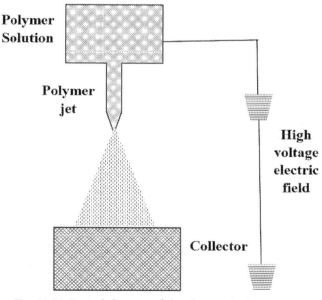

Fig. 18.18 Typical diagram of the electrospinning process.

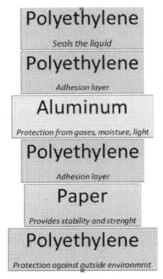

Fig. 18.19 Tetra packaging.

Electrospun fibers have some unique properties which include high porosity, large surface-to-volume ratio & ultrafine structures. These characteristics enhance the encapsulation efficiencies of incorporating bioactive compounds. Thermosensitive compounds can be encapsulated as this process doesn't require any heat. Another advantage of this process is the rapid removal of solvent which may also pose a threat to packaged foods. The major limitation of the process is the low productivity. This technique can use a variety of biopolymers, like polysaccharides and proteins, however, the main disadvantages are their poor mechanical and moisture barrier qualities [165] (Fig. 18.19).

12. Conclusion and future scope

According to Wise Guy Reports, the market for food packaging is anticipated to reach US$14.3 billion in 2022, growing at a CAGR of 17%. India's food packaging market is predicted to expand to US$73 billion in 2022 at a CAGR of 18%, reflecting the rising need for biopolymers as biodegradable food packaging material, according to a report by the Economic Times [152,166]. In the 20th century, several cutting-edge, inventive food packaging techniques emerged to provide consumers with wholesome, long-lasting, high-quality food without having an adverse effect on the environment. The most recent advancements in food packaging include active, intelligent, sustainable packaging, & packaging that makes use of cutting-edge technology. Active packaging provides long-life security and high food goods quality by adding additives like antioxidant & antimicrobial compounds and scavengers or absorbers, but it is incapable of warning real-time visual information regarding food goods quality, durability, and safety. Hence, researchers nowadays, incline toward smart or intelligent packaging as it can perform diverse functions like identifying, monitoring, processing, tracing, recording, and communicating information by using sensors or indicators. These functions foster decision-making efficiency, increase security, improve quality, etc. Radiofrequency identification, gas indications, microwave indicators, time-temperature indicators, and many other features are all part of intelligent packaging. The further development of these inventions may offer improved knowledge for preserving food quality and shelf life in various environmental circumstances. Intelligent packaging does not necessarily result in a longer shelf life for food or an increase in its quality. Therefore, combining active packaging with intelligent packaging can advance food packaging technology soon [140]. Smart packaging also refers to the fusion of active and intelligent packaging. Smart packaging may follow the food product and offer more precise information. Employing cutting-edge active packaging techniques can help mitigate the downsides of using antioxidants such as essential oils, α-tocopherol, etc. The development of active packaging materials having extra protection facilities to prevent the loss of activity will be of great concern to researchers in the coming days [131].

More emphasis should be given to the use of nanomaterials like clays, cellulose fibers, carbon nanotubes, etc. to get improved properties of the packaging materials at an affordable cost. In one report, nanobiosensor consisting of carbon nanotubes as transmitters and strands of DNA as sensors have been deployed to determine the odor and taste of food. Several reports have been published on the use of nanobiosensors in the packaging sector. This area can be explored further in the future. Although polymers used for packaging materials such as paper, metal, etc. are safe, they may pose some risk due to the migration of additives. Therefore, emphasis should be given to reaching a consensus in respect of food additives and testing conditions. To guarantee the calibre of packaging materials as well as customer safety, extensive revision lists comprising of hazardous monomers that are used to create polymers, additives, etc. must be prepared and thoroughly revised.

Modern technology approaches could lower food product quality and have a negative impact on people's health. However, most remedies that are offered in the literature cannot be effectively applied realistically to the packaged foods available in the market. Hence, research is going on focusing on the degree of food protection and its safe implementation to consumers [133]. The use of more computerized controlled technology for designing new cost-effective materials/processes and managing the quality of production in future may further widen the area of application. Future growth in the food packaging sector is predicted to be rapid due to rising consumer demand, the development of packaging technology and equipment, and the introduction of need-based packaging materials made of biodegradable polymers.

References

[1] G. Robertson (Ed.), Food Packaging:Principles and Practice, Taylor and Francis, CRC Press, 2013.

[2] R. Coles, D. McDowell, M.J. Kirwan, Food Packaging Technology, vol. 5, Taylor and Francis, CRC Press, 2003.

[3] M.P. Hekkert, L.A. Joosten, E. Worrell, W.C. Turkenburg, Reduction of CO2 emissions by improved management of material and product use: the case of primary packaging, Resour. Conserv. Recycl. 29 (1−2) (2000) 33−64.

[4] K. Marsh, B. Bugusu, Food packaging-roles, materials, and environmental issues, J. Food Sci. 72 (2007) 39−55.

[5] D. Twede, History of packaging, in: The Routledge Companion to Marketing History, Taylor and Francis, 2016, pp. 115−130.

[6] G. Porter, Cultural forces and commercial constraints: designing packaging in the twentieth-century United States, J. Des. Hist. 12 (1999) 25−43.

[7] S. Chaudhary, The role of packaging in consumer's perception of product quality, Int. J. Soc. Sci. Res. 3 (2014) 17−21.

[8] K. Mulder, M. Knot, PVC plastic: a history of systems development and entrenchment, Technol. Soc. 23 (2001) 265−286.

[9] K. Lee, Quality and safety aspects of meat products as affected by various physical manipulations of packaging materials, Meat Sci 86 (2010) 138−150.

[10] Tetra packaging (Mishra, M. L., amp; Shukla, U. N. Chapter-4 food packaging. And storage. Home Sci. Exten., 89, 47).

[11] G.L. Robertson, Food Packaging: Principles and Practice, Taylor and Francis, CRC Press, 2016.

[12] Y.T. Kim, B. Min, K.W. Kim, General characteristics of packaging materials for food system, in: Innovations in Food Packaging, 2014, pp. 13−35.

[13] S.A. Cichello, Oxygen absorbers in food preservation: a review, Food Sci. Technol. 52 (2015) 1889−1895.

[14] K. Gaikwad, S. Singh, Y.S. Negi, Ethylene scavengers for active packaging of fresh food produce, Environ. Chem. Lett. 18 (2020) 269−284.

[15] L. Vermeiren, F. Devlieghere, M. van Beest, N. de Kruijf, J. Debevere, Developments in the active packaging of foods, Trends Food Sci. Technol. 10 (1999) 77−86.

[16] A.M. Khaneghah, S.M.B. Hashemi, S. Limbo, Antimicrobial agents and packaging systems in antimicrobial active food packaging: an overview of approaches and interactions, Food Bioprod. Process. 111 (2018) 1−19.

[17] S. Kamel, N. Ali, K. Jahangir, S.M. Shah, A.A. El-Gendy, Pharmaceutical significance of cellulose: a review, Express Polym. Lett. 2 (2008) 758−778.

[18] V. Siracusa, M. Dalla, Biodegradable polymers for food packaging: a review, Trends Food Sci. Technol. 19 (2008) 634−643.

[19] G. Davis, J.H. Song, Biodegradable packaging based on raw materials from crops and their impact on waste management, Ind. Crop. Prod. 23 (2006) 147−161.

[20] A.D. Tripathi, S.K. Srivastava, A.J.A.Y. Yadav, Biopolymers potential biodegradable packaging material for food industry, in: Polymers for Packaging Applications, vol 153, Apple Academic press, 2014.

[21] J. Rydz, M. Musioł, B. Zawidlak-Wegrzyńska, W. Sikorska, Present and future of biodegradable polymers for food packaging applications, in: Biopolymers for Food Design, Elsevier Science, 2018, pp. 431−467.

[22] R. Grujić, D. Vujadinović, D. Savanović, Biopolymers as food packaging materials, in: Advances in Applications of Industrial Biomaterials, Springer International Publishing, 2017, pp. 139−160.

[23] Y. Liu, S. Ahmed, D.E. Sameen, Y. Wang, R. Lu, J. Dai, W. Qin, A review of cellulose and its derivatives in biopolymer-based for food packaging application, Trends Food Sci. Technol. 112 (2021) 532−546.

[24] C.L. Reichert, E. Bugnicourt, M.B. Coltelli, P. Cinelli, A. Lazzeri, I. Canesi, M. Schmid, Bio-based packaging: materials, modifications, industrial applications and sustainability, Polymers 12 (2020) 1558.
[25] D.K.P.K. Lavanya, P.K. Kulkarni, M. Dixit, P.K. Raavi, L.N.V. Krishna, Sources of cellulose and their applications-A review, Int. J. Drug Dev. Res. 2 (2018) 19−38.
[26] J. George, S.N. Sabapathi, Cellulose nanocrystals: synthesis, functional properties, and applications, Nanotechol. Sci. Appl. 8 (45) (2015).
[27] H. Azeredo, H. Barud, C.S. Farinas, V.M. Vasconcellos, A.M. Claro, Bacterial cellulose as a raw material for food and food packaging applications, Front. Sustain. Food Syst. 3 (7) (2019).
[28] P. Cazón, M. Vázquez, Bacterial cellulose as a biodegradable food packaging material: a review, Food Hydrocoll. 113 (2021) 106530.
[29] P.K. Gupta, S.S. Raghunath, D.V. Prasanna, P. Venkat, V. Shree, C. Chithananthan, K. Geetha, An update on overview of cellulose, its structure and applications, Cellulose (2019) 846−1297.
[30] K. Wasilewska, K. Winnicka, Ethylcellulose-a pharmaceutical excipient with multidirectional application in drug dosage forms development, Materials 12 (2019) 3386.
[31] B. Tajeddin, Cellulose-based polymers for packaging applications, in: V. Thakur (Ed.), Lignocellulosic Polymer Composites, Wiley, 2014, pp. 477−498.
[32] G. Fotie, S. Limbo, L. Piergiovanni, Manufacturing of food packaging based on nanocellulose: current advances and challenges, Nanomaterials 10 (2020) 1726.
[33] I. Gan, W.S. Chow, Antimicrobial poly (lactic acid)/cellulose bionanocomposite for food packaging application: a review, Food Packag. Shelf Life. 17 (2018) 150−161.
[34] S. Pokhrel, A review on introduction and applications of starch and its biodegradable polymers, Int. J. Environ. 4 (2015) 114−125.
[35] M.K. Lauer, R.C. Smith, Recent advances in starch-based films toward food packaging applications: physicochemical, mechanical, and functional properties, Compr. Rev. Food Sci. Food Saf. 19 (2020) 3031−3083.
[36] S. Jha, P. Rohilla, K. Singh, Starch based packaging materials: a review, Int. J. Res. Anal. Rev. 4 (2017) 209−211.
[37] U. Shah, A. Gani, B.A. Ashwar, A. Shah, M. Ahmad, A. Gani, F.A. Masoodi, A review of the recent advances in starch as active and nanocomposite packaging films, Cogent Food Agric 1 (2015) 1115640.
[38] B. Khan, M. Bilal Khan Niazi, G. Samin, Z. Jahan, Thermoplastic starch: a possible biodegradable food packaging material-a review, J. Food Process Eng. 40 (2017) e12447.
[39] M. Vilar, Starch-based Materials in Food Packaging: Processing, Characterization and Applications, Elsevier Science, 2017.
[40] S. Philip, T. Keshavarz, I. Roy, Polyhydroxyalkanoates: biodegradable polymers with a range of applications, J. Chem. Technol. Biotechnol. 82 (2007) 233−247.
[41] P.M. Visakh, Polyhydroxyalkanoates (PHAs), their blends, composites and nanocomposites: state of the art, new challenges and opportunities, in: Polyhydroxyalkanoate (PHA) Based Blends, Composites and Nanocomposites, Royal Society of chemistry, 2014, pp. 1−17.
[42] F. Masood, Polyhydroxyalkanoates in the food packaging industry, in: Nanotechnology Applications in Food, Elsevier Science, 2017, pp. 153−177.
[43] K. Khosravi-Darani, D.Z. Bucci, Application of poly (hydroxyalkanoate) in food packaging: improvements by nanotechnology, Chem. Biochem. Eng. Q. 29 (2015) 275−285.
[44] S. Taguchi, K.I. Matsumoto, Evolution of polyhydroxyalkanoate synthesizing systems toward a sustainable plastic industry, Polym. J. 53 (2021) 67−79.
[45] P. Ragaert, M. Buntinx, C. Maes, C. Vanheusden, R. Peeters, S. Wang, L. Cardon, Polyhydroxyalkanoates for food packaging applications, in: Reference Module in Food Science, Elsevier, 2019.
[46] Z.A. Raza, S. Abid, I.M. Banat, Polyhydroxyalkanoates: characteristics, production, recent developments, and applications, Int. Biodeterior. Biodegrad. 126 (2018) 45−56.
[47] T. Vanitha, M. Khan, Role of pectin in food processing and food packaging, in: Pectins-extraction, Purification, Characterization and Applications, Intech Open, 2019.

[48] P.J.P. Espitia, W.X. Du, R. de Jesús Avena-Bustillos, N.D.F.F. Soares, T.H. McHugh, Edible films from pectin: physical-mechanical and antimicrobial properties-A review, Food Hydrocoll. 35 (2014) 287–296.

[49] M.R. Saboktakin, Biomedical properties study of modified chitosan nanoparticles for drug delivery systems, in: Nanopharmaceutics: The Potential Application of Nanomaterials, vols. 129–172, World Scientific, 2013.

[50] H.J. Kang, C. Jo, J.H. Kwon, J.H. Kim, H.J. Chung, M.W. Byun, Effect of a pectin-based edible coating containing green tea powder on the quality of irradiated pork patty, Food Control 18 (2007) 430–435.

[51] C. Mellinas, M. Ramos, A. Jiménez, M.C. Garrigós, Recent trends in the use of pectin from agro-waste residues as a natural-based biopolymer for food packaging applications, Materials 13 (2020) 673.

[52] E. Ghorbani, A. Dabbagh Moghaddam, A. Sharifan, H. Kiani, Emergency food product packaging by pectin-based antimicrobial coatings functionalized by pomegranate peel extracts, J. Food Qual. 10 (2021) 2021.

[53] P. Ishwarya S, P. Nisha, Advances and prospects in the food applications of pectin hydrogels, Crit. Rev. Food Sci. Nutr. 0 (2021) 1–25.

[54] G.A. Martău, M. Mihai, D.C. Vodnar, The use of chitosan, alginate, and pectin in the biomedical and food sector-biocompatibility, bioadhesiveness, and biodegradability, Polymers 11 (2019) 1837.

[55] D.R. Sahoo, T. Biswal, Alginate and its application to tissue engineering, SN Appl. Sci. 3 (2021) 1–19.

[56] T. Senturk Parreidt, K. Müller, M. Schmid, Alginate-based edible films and coatings for food packaging applications, Foods 7 (2018) 170.

[57] M.S.A. Aziz, H.E. Salama, M.W. Sabaa, Biobased alginate/castor oil edible films for active food packaging, LWT 96 (2018) 455–460.

[58] R.B. Nambiar, P.S. Sellamuthu, A.B. Perumal, E.R. Sadiku, O.A. Adeyeye, The use of Chitosan in food packaging applications, in: Green Biopolymers and Their Nanocomposites, Springer Singapore, 2019, pp. 125–136.

[59] P. Fernandez-Saiz, Chitosan polysaccharide in food packaging applications, in: Multifunctional and Nanoreinforced Polymers for Food Packaging, Springer Singapore, 2011, pp. 571–593.

[60] J. López-García, M. Lehocký, P. Humpolíček, P. Sáha, HaCaT keratinocytes response on antimicrobial atelocollagen substrates: extent of cytotoxicity, cell viability and proliferation, J. Funct. Biomater. 5 (2014) 43–57.

[61] V. Zargar, M. Asghari, A. Dashti, A review on chitin and chitosan polymers: structure, chemistry, solubility, derivatives, and applications, ChemBioEng Rev 2 (2015) 204–226.

[62] S. Zivanovic, R.H. Davis, D.A. Golden, Chitosan as an antimicrobial in food products, in: Handbook of Natural Antimicrobials for Food Safety and Quality, vol 2, Elsevier Science, 2015, pp. 153–181.

[63] S. Bautista-Baños, G. Romanazzi, A. Jiménez-Aparicio (Eds.), Chitosan in the Preservation of Agricultural Commodities, Elsevier Science, 2016.

[64] H. Wang, J. Qian, F. Ding, Emerging chitosan-based films for food packaging applications, J. Agric. Food Chem. 66 (2018) 395–413.

[65] M. Taylor (Ed.), Handbook of Natural Antimicrobials for Food Safety and Quality, Elsevier Science, 2014.

[66] P. Cazón, M. Vázquez, Applications of chitosan as food packaging materials, in: Sustainable Agriculture Reviews, Springer International publishing, 2019, pp. 81–123.

[67] A. Muxika, I. Zugasti, P. Guerrero, K. de la Caba, Applications of Chitosan in Food Packaging, Springer Singapore, 2017.

[68] F. Shahidi, M.J. Rahman, Bioactives in seaweeds, algae, and fungi and their role in health promotion, J. Food Bioact 2 (2018) 58–81.

[69] F.S. Mostafavi, D. Zaeim, Agar-based edible films for food packaging applications-A review, Int. J. Biol. Macromol. 159 (2020) 1165–1176.

[70] P. Kanmani, J.W. Rhim, Antimicrobial and physical-mechanical properties of agar-based films incorporated with grapefruit seed extract, Carbohydr. Polym. 102 (2014) 708–716.

[71] T.R.K. Dora, S. Ghosh, R. Damodar, Synthesis and evaluation of physical properties of Agar biopolymer film coating-an alternative for food packaging industry, Mater. Res. Express 7 (2020) 095307.
[72] L.F. Wang, J.W. Rhim, Preparation and application of agar/alginate/collagen ternary blend functional food packaging films, Int. J. Biol. Macromol. 80 (2015) 460–468.
[73] M.Z. Elsabee, E.S. Abdou, Chitosan based edible films and coatings: a review, Mater. Sci. Eng. C 33 (2013) 1819–1841.
[74] B. Hassan, S.A.S. Chatha, A.I. Hussain, K.M. Zia, N. Akhtar, Recent advances on polysaccharides, lipids and protein based edible films and coatings: a review, Int. J. Biol. Macromol. 109 (2018) 1095–1107.
[75] S.A. Mohamed, M. El-Sakhawy, M.A.M. El-Sakhawy, Polysaccharides, protein and lipid-based natural edible films in food packaging: a review, Carbohydr. Polym. 238 (2020) 116178.
[76] M. Zubair, A. Ullah, Recent advances in protein derived bio-nanocomposites for food packaging applications, Crit. Rev. Food Sci. Nutr. 60 (2020) 406–434.
[77] S. Pirsa, K. Aghbolagh Sharifi, A review of the applications of bioproteins in the preparation of biodegradable films and polymers, Chem. Lett. 1 (2020) 47–58.
[78] S.J. Calva-Estrada, M. Jiménez-Fernández, E. Lugo-Cervantes, Protein-based films: advances in the development of biomaterials applicable to food packaging, Food Eng. Rev. 11 (2019) 78–92.
[79] M. Schmid, K. Dallmann, E. Bugnicourt, D. Cordoni, F. Wild, A. Lazzeri, K. Noller, Properties of whey-protein-coated films and laminates as novel recyclable food packaging materials with excellent barrier properties, Int. J. Polym. Sci. 7 (2012) 2012.
[80] Y. Zhang, Q. Liu, C. Rempel, Processing and characteristics of canola protein-based biodegradable packaging: a review, Crit. Rev. Food Sci. Nutr. 58 (2018) 475–485.
[81] H. Chen, J. Wang, Y. Cheng, C. Wang, H. Liu, H. Bian, W. Han, Application of protein-based films and coatings for food packaging: a review, Polymers 11 (2019) 2039.
[82] C.G. Riveros, M.P. Martin, A. Aguirre, N.R. Grosso, Film preparation with high protein defatted peanut flour: characterisation and potential use as food packaging, Int. J. Food Sci. Technol. 53 (2018) 969–975.
[83] S. Teilaghi, J. Movaffagh, Z. Bayat, Preparation as well as evaluation of the nanofiber membrane loaded with nigella sativa extract using the electrospinning method, J. Polym. Environ. 28 (2020) 1614–1625.
[84] S. Kommareddy, D.B. Shenoy, M.M. Amiji, Gelatin nanoparticles and their biofunctionalization, in: Nanotechnologies for the Life Sciences: Online, Wiley Interscience, 2007.
[85] J. Wróblewska-Krepsztul, T. Rydzkowski, G. Borowski, M. Szczypiński, T. Klepka, V.K. Thakur, Recent progress in biodegradable polymers and nanocomposite-based packaging materials for sustainable environment, Int. J. Polym. Anal. Charact. 23 (2018) 383–395.
[86] T. Huang, Y. Qian, J. Wei, C. Zhou, Polymeric antimicrobial food packaging and its applications, Polymers 11 (2019) 560.
[87] K.M. Zia, A. Noreen, M. Zuber, S. Tabasum, M. Mujahid, Recent developments and future prospects on bio-based polyesters derived from renewable resources: a review, Int. J. Biol. Macromol. 82 (2016) 1028–1040.
[88] B. Marcos, C. Sárraga, M. Castellari, F. Kappen, G. Schennink, J. Arnau, Development of biodegradable films with antioxidant properties based on polyesters containing α-tocopherol and olive leaf extract for food packaging applications, Food Packag. Shelf Life 1 (2014) 140–150.
[89] Ö. Süfer, Y. Celebi Sezer, Poly (lactic acid) films in food packaging systems, Food Sci. Nutr. Technol. 2 (2017) 000131.
[90] I.G. de Moura, A.V. de Sá, A.S.L.M. Abreu, A.V.A. Machado, Bioplastics from agro-wastes for food packaging applications, in: Food Packaging, Elsevier, Academic Press, 2017, pp. 223–263.
[91] M.A. Souza, J.E. Oliveira, E.S. Medeiros, G.M. Glenn, L.H. Mattoso, Controlled release of linalool using nanofibrous membranes of poly (lactic acid) obtained by electrospinning and solution blow spinning: a comparative study, J. Nanosci. Nanotechnol. 15 (2015) 5628–5636.

[92] N.C. Rusmiati, N. Rahmawati, L.E. Fitri, T. Nurseta, E.N. Sutrisno, A.T.E. Awaludin, The ethanol extract of red banana peel (M. Acuminata colla) induce cell death and inhibit Metastatic of breast cancer, Ann. Romanian Soc. Cell Biol. 25 (2021) 17184—17194.

[93] I.S. Tawakkal, M.J. Cran, J. Miltz, S.W. Bigger, A review of poly (lactic acid) -based materials for antimicrobial packaging, J. Food Sci. Technol. 79 (2014) 1477—1490.

[94] K. Hamad, M. Kaseem, H.W. Yang, F. Deri, Y.G. Ko, Properties and medical applications of polylactic acid: a review, Express Polym. Lett. 9 (2015) 435—455.

[95] R.A. Abd Alsaheb, A. Aladdin, N.Z. Othman, R. Abd Malek, O.M. Leng, R. Aziz, H.A. El Enshasy, Recent applications of polylactic acid in pharmaceutical and medical industries, J. Chem. Pharm. Res. 7 (2015) 51—63.

[96] C. González-Martínez, A. Chiralt, Combination of poly (lactic) acid and starch for biodegradable food packaging, Materials 10 (2017) 952.

[97] S.A. Malomo, R. He, R.E. Aluko, Structural and functional properties of hemp seed protein products, J. Food Sci. 79 (2014) 1512—1521.

[98] A.J.J.E. Eerhart, A.P.C. Faaij, M.K. Patel, Replacing fossil based PET with biobased PEF; process analysis, energy and GHG balance, Energy Environ. Sci. 5 (2012) 6407—6422.

[99] H. Storz, K.D. Vorlop, Bio-based plastics: status, challenges and trends, Appl. Agric. For. Res. 63 (2013) 321—332.

[100] J. Xu, B.H. Guo, Microbial succinic acid, its polymer poly (butylene succinate), and applications, in: Plastics from Bacteria, Springer Berlin Heidelberg, 2010, pp. 347—388.

[101] V. Siracusa, I. Blanco, S. Romani, U. Tylewicz, P. Rocculi, M.D. Rosa, Poly (lactic acid)-modified films for food packaging application: physical, mechanical, and barrier behavior, J. Appl. Polym. Sci. 125 (2012) 390—401.

[102] A. Mahapatro, D.K. Singh, Biodegradable nanoparticles are excellent vehicle for site directed in-vivo delivery of drugs and vaccines, J. Nanobiotechnology 9 (2011) 1—11.

[103] R.M. Mohamed, K. Yusoh, A review on the recent research of polycaprolactone (PCL), in: Advanced Materials Research, Trans Tech Publications Ltd, 2016, pp. 249—255.

[104] H. Haghighi, S.K. Leugoue, F. Pfeifer, H.W. Siesler, F. Licciardello, P. Fava, A. Pulvirenti, Development of antimicrobial films based on chitosan-polyvinyl alcohol blend enriched with ethyl lauroyl arginate (LAE) for food packaging applications, Food Hydrocoll. 100 (2020) 105419.

[105] S. Suganthi, S. Vignesh, J.K. Sundar, V. Raj, Fabrication of PVA polymer films with improved antibacterial activity by fine-tuning via organic acids for food packaging applications, Appl. Water Sci. 10 (2020) 1—11.

[106] L.A. Majewski, Alternative Gate Insulators for Organic Field-Effect Transistors, (Doctoral dissertation, University of Sheffield, Department of Physics and Astronomy), 2006.

[107] J.G. Rosenboom, D.K. Hohl, P. Fleckenstein, G. Storti, M. Morbidelli, Bottle-grade polyethylene furanoate from ring-opening polymerisation of cyclic oligomers, Nat. Commun. 9 (2018) 1—7.

[108] N. Poulopoulou, N. Kasmi, M. Siampani, Z.N. Terzopoulou, D.N. Bikiaris, D.S. Achilias, G.Z. Papageorgiou, Exploring Next-Generation Engineering Bioplastics: poly (alkylene furanoate)/poly (alkylene terephthalate) (PAF/PAT) Blends, Polymers 11 (2019) 556.

[109] O.A. Adeyeye, E.R. Sadiku, A.B. Reddy, A.S. Ndamase, G. Makgatho, P.S. Sellamuthu, T. Jamiru, The use of biopolymers in food packaging, in: Green Biopolymers and Their Nanocomposites, Springer Singapore, 2019, pp. 137—158.

[110] R. Grujic, M. Vukic, V. Gojkovic, Application of biopolymers in the food industry, in: Advances in Applications of Industrial Biomaterials, Springer International Publishing, 2017, pp. 103—119.

[111] J. Nilsen-Nygaard, E.N. Fernández, T. Radusin, B.T. Rotabakk, J. Sarfraz, N. Sharmin, M.K. Pettersen, Current status of biobased and biodegradable food packaging materials: impact on food quality and effect of innovative processing technologies, Compr. Rev. Food Sci. Food Saf. 20 (2021) 1333—1380.

[112] A. Arora, G.W. Padua, Nanocomposites in food packaging, J. Food Sci. 75 (2010) R43—R49.

[113] V. Jost, C. Stramm, Influence of plasticizers on the mechanical and barrier properties of cast biopolymer films, J. Appl. Polym. Sci. 133 (2016).

[114] S.P.S. Aung, H.H.H. Shein, K.N. Aye, N. Nwe, Environment-friendly biopolymers for food packaging: starch, protein, and poly-lactic acid (PLA), in: Bio-based Materials for Food Packaging, Springer Singapore, 2018, pp. 173−195.

[115] M.G.A. Vieira, M.A. da Silva, L.O. dos Santos, M.M. Beppu, Natural-based plasticizers and biopolymer films: a review, Eur. Polym. J. 47 (2011) 254−263.

[116] P. Bergo, P.J.A. Sobral, Effects of plasticizer on physical properties of pigskin gelatin films, Food Hydrocoll 21 (2007) 1285−1289.

[117] V. Tyagi, B. Bhattacharya, Role of plasticizers in bioplastics, MOJ Food Process. Technol. 7 (2019) 128−130.

[118] D.Z. Šuput, V.L. Lazić, A. Jelić, L.B. Lević, L.L. Pezo, N.M. Hromiš, S. Popović, The effect of sorbitol content on the characteristics of starch based edible films, J. Process. Energy Agric. 17 (2013) 106−109.

[119] P. Rodsamran, R. Sothornvit, Preparation and characterization of pectin fraction from pineapple peel as a natural plasticizer and material for biopolymer film, Food Bioprod. Process. 118 (2019) 198−206.

[120] M.L. Sanyang, S.M. Sapuan, M. Jawaid, M.R. Ishak, J. Sahari, Effect of plasticizer type and concentration on physical properties of biodegradable films based on sugar palm (Arenga pinnata) starch for food packaging, J. Food Sci. Technol. 53 (2016) 326−336.

[121] M.P. Arrieta, M.D. Samper, J. López, A. Jiménez, Combined effect of poly (hydroxybutyrate) and plasticizers on polylactic acid properties for film intended for food packaging, J. Polym. Environ. 22 (2014) 460−470.

[122] F. Garavand, M. Rouhi, S.H. Razavi, I. Cacciotti, R. Mohammadi, Improving the integrity of natural biopolymer films used in food packaging by crosslinking approach: a review, Int. J. Biol. Macromol. 104 (2017) 687−707.

[123] H. Wu, Y. Lei, J. Lu, R. Zhu, D. Xiao, C. Jiao, M. Li, Effect of citric acid induced crosslinking on the structure and properties of potato starch/chitosan composite films, Food Hydrocoll. 97 (2019) 105208.

[124] N. Reddy, Q. Jiang, Y. Yang, Preparation and properties of peanut protein films crosslinked with citric acid, Ind. Crops Prod. 39 (2012) 26−30.

[125] M. Asgher, S.A. Qamar, M. Bilal, H.M. Iqbal, Bio-based active food packaging materials: sustainable alternative to conventional petrochemical-based packaging materials, Food Res. Int. 137 (2020) 109625.

[126] E.M. Ciannamea, P.M. Stefani, R.A. Ruseckaite, Properties and antioxidant activity of soy protein concentrate films incorporated with red grape extract processed by casting and compression molding, LWT 74 (2016) 353−362.

[127] Y. Pranoto, V.M. Salokhe, S.K. Rakshit, Physical and antibacterial properties of alginate-based edible film incorporated with garlic oil, Food Res. Int. 38 (2005) 267−272.

[128] Y. Ruiz-Navajas, M. Viuda-Martos, E. Sendra, J.A. Perez-Alvarez, J. Fernández-López, In vitro antibacterial and antioxidant properties of chitosan edible films incorporated with Thymus moroderi or Thymus piperella essential oils, Food Control 30 (2013) 386−392.

[129] A.L. Brody, B. Bugusu, J.H. Han, C.K. Sand, T.H. Mchugh, Innovative food packaging solutions, J. Food Sci. 73 (2008) 107−116.

[130] K.K. Gaikwad, S. Singh, A. Ajji, Moisture absorbers for food packaging applications, Environ. Chem. Lett. 17 (2019) 609−628.

[131] C. Zhang, Y. Li, P. Wang, H. Zhang, Electrospinning of nanofibers: potentials and perspectives for active food packaging, Compr. Rev. Food Sci. Food Saf. 19 (2020) 479−502.

[132] A. Valdés, A.C. Mellinas, M. Ramos, M.C. Garrigós, A. Jiménez, Natural additives and agricultural wastes in biopolymer formulations for food packaging, Front. Chem. 2 (6) (2014).

[133] J. Wyrwa, A. Barska, Innovations in the food packaging market: active packaging, Eur. Food Res. Technol. 243 (2017) 1681−1692.

[134] S. Yildirim, B. Röcker, M.K. Pettersen, J. Nilsen-Nygaard, Z. Ayhan, R. Rutkaite, V. Coma, Active packaging applications for food, Compr. Rev. Food Sci. Food Saf. 17 (2018) 165−199.

[135] A.K. Haghi (Ed.), Electrospinning of Nanofibers in Textiles, Taylor and Francis Group, CRC Press, 2019.
[136] E. Balbinot-Alfaro, D.V. Craveiro, K.O. Lima, H.L.G. Costa, D.R. Lopes, C. Prentice, Intelligent packaging with pH indicator potential, Food Eng. Rev. 11 (2019) 235–244.
[137] L. Prietto, T.C. Mirapalhete, V.Z. Pinto, J.F. Hoffmann, N.L. Vanier, L.T. Lim, E. da Rosa Zavareze, pH-sensitive films containing anthocyanins extracted from black bean seed coat and red cabbage, LWT 80 (2017) 492–500.
[138] P. Ezati, J.W. Rhim, pH-responsive pectin-based multifunctional films incorporated with curcumin and sulfur nanoparticles, Carbohydr. Polym. 230 (2020) 115638.
[139] R. Priyadarshi, P. Ezati, J.W. Rhim, Recent advances in intelligent food packaging applications using natural food colorants, ACS Food Sci. Technol. 1 (2021) 124–138.
[140] J.W. Han, L. Ruiz-Garcia, J.P. Qian, X.T. Yang, Food packaging: a comprehensive review and future trends, Compr. Rev. Food Sci. Food Saf. 17 (2018) 860–877.
[141] F. Saliu, R. Della Pergola, Carbon dioxide colorimetric indicators for food packaging application: applicability of anthocyanin and poly-lysine mixtures, Sens. Actuators B Chem. 258 (2018) 1117–1124.
[142] V. Goudarzi, I. Shahabi-Ghahfarrokhi, Photo-producible and photo-degradable starch/TiO_2 bio-nanocomposite as a food packaging material: development and characterization, Int. J. Biol. Macromol. 106 (2018) 661–669.
[143] M. Haji-Saeid, M.H.O. Sampa, A.G. Chmielewski, Radiation treatment for sterilization of packaging materials, Radiat. Phys. Chem. 76 (2017) 1535–1541.
[144] I.M. Jipa, M. Stroescu, A. Stoica-Guzun, T. Dobre, S. Jinga, T. Zaharescu, Effect of gamma irradiation on biopolymer composite films of poly (vinyl alcohol) and bacterial cellulose, Nucl. Instrum. Methods Phys. Res. B: Beam Interact. Mater. B 278 (2012) 82–87.
[145] I. Shahabi-Ghahfarrokhi, V. Goudarzi, A. Babaei-Ghazvini, Production of starch based biopolymer by green photochemical reaction at different UV region as a food packaging material: physicochemical characterization, Int. J. Biol. Macromol. 122 (2019) 201–209.
[146] H. Haghighi, F. Licciardello, P. Fava, H.W. Siesler, A. Pulvirenti, Recent advances on chitosan-based films for sustainable food packaging applications, Food Packag. Shelf Life. 26 (2020) 100551.
[147] A. Nura, Advances in food packaging technology-a review, Int. J. Postharvest Technol. Innov. 6 (2018) 55–64.
[148] L. Angiolillo, S. Spinelli, A. Conte, M. Alessandro Del Nobile, Recent advances in food packaging with a focus on nanotechnology, Recent Pat. Eng. 11 (2017) 174–187.
[149] G. Robertson, Food packaging: principles and practice, in: Taylor and Francis Group, third ed., CRC Press, 2013.
[150] Y. Wyser, M. Adams, M. Avella, D. Carlander, L. Garcia, G. Pieper, J. Weiss, Outlook and challenges of nanotechnologies for food packaging, Packag. Technol. Sci. 29 (2016) 615–648.
[151] S.H. Othman, Bio-nanocomposite materials for food packaging applications: types of biopolymers and nano-sized filler, Agric. Agric. Sci. Procedia. 2 (2014) 296–303.
[152] P. Chaudhary, F. Fatima, A. Kumar, Relevance of nanomaterials in food packaging and its advanced future prospects, J. Inorg. Organomet. Polym. Mater. 30 (2020) 5180–5192.
[153] S. Kumar, A. Mukherjee, J. Dutta, Chitosan based nanocomposite films and coatings: emerging antimicrobial food packaging alternatives, Trends Food Sci. Technol. 97 (2020) 196–209.
[154] V. Dhapte, N. Gaikwad, P.V. More, S. Banerjee, V.V. Dhapte, S. Kadam, P.K. Khanna, Transparent ZnO/polycarbonate nanocomposite for food packaging application, Nanocomposites 1 (2015) 106–112.
[155] M.S. Sarwar, M.B.K. Niazi, Z. Jahan, T. Ahmad, A. Hussain, Preparation and characterization of PVA/nanocellulose/Ag nanocomposite films for antimicrobial food packaging, Carbohydr. Polym. 184 (2018) 453–464.
[156] H. Moustafa, A.M. Youssef, N.A. Darwish, A.I. Abou-Kandil, Eco-friendly polymer composites for green packaging: future vision and challenges, Compos. B: Eng. 172 (2019) 16–25.
[157] P. Ye, K.S. Fiorini, Thermal analysis of biodegradable material: from modulated temperature DSC to fast scan DSC, Am. Lab. 39 (2007) 25–26.

[158] Z.N. Hanani, Y. Roos, J. Kerry, Fourier Transform Infrared (FTIR) spectroscopic analysis of biodegradable gelatin films immersed in water, in: 11th International Congress on Engineering and Food, ICEF11, Elsevier, Athens, Greece, 2011.
[159] A. Jongjareonrak, S. Benjakul, W. Visessanguan, M. Tanaka, Antioxidative activity and properties of fish skin gelatin films incorporated with BHT and α-tocopherol, Food Hydrocoll 22 (2008) 449–458.
[160] M.S. Hoque, S. Benjakul, T. Prodpran, Effect of heat treatment of film-forming solution on the properties of film from cuttlefish (Sepia pharaonis) skin gelatin, J. Food Eng. 96 (2010) 66–73.
[161] Y. Pranoto, C.M. Lee, H.J. Park, Characterizations of fish gelatin films added with gellan and κcarrageenan, LWT 40 (2007) 766–774.
[162] A. Gratia, D. Merlet, V. Ducruet, C. Lyathaud, A comprehensive NMR methodology to assess the composition of biobased and biodegradable polymers in contact with food, Anal. Chim. Acta 853 (2015) 477–485.
[163] M. Avella, J.J. De Vlieger, M.E. Errico, S. Fischer, P. Vacca, M.G. Volpe, Biodegradable starch/clay nanocomposite films for food packaging applications, Food Chem. 93 (2005) 467–474.
[164] R.A. Tatara, Compression molding, in: Applied Plastics Engineering Handbook, Elsevier Science, 2017, pp. 291–320.
[165] L.T. Lim, R. Auras, M. Rubino, Processing technologies for poly (lactic acid), Prog. Polym. Sci. 33 (2008) 820–852.
[166] S. Amith, H.A. Kalody, B.R. Menon, Review on developments and future scope in using biodegradable materials for food packaging, Int. J. Eng. Res. 8 (2278–0181) (2019).

CHAPTER 19

Application of environmentally benign biodegradable composite in intelligent and active packaging

Mira chares Subash and Muthiah Perumalsamy

Department of Chemical Engineering, National Institute of Technology, Tiruchirappalli, Tamilnadu, India

1. Introduction

Plastic production is about three hundred and twenty-two million tons in the year 2015 throughout the world. India produces nearly eight million tons of plastic annually and contributes nearly 43% of plastic in the packaging industry [1]. The production and processing of plastic lead to high emission of greenhouse gases, and global warming, and threatens public health by venomous emissions. The adverse effect of non-degradable plastic necessitates research to develop biodegradable composite. The biodegradable composite synthesized with eco-friendly raw materials and biomaterials is more welcomed [2]. The advantage of biodegradable plastics is sustainability, eco-friendliness, degradability, biobased and less excretion of greenhouse gases. The biodegradable composites manufactured from renewable sources can mimic the features of conventional materials. Research study states that bioplastic production increased to 2.0 million tons from 1.6 million tons during the 2013−5 period. Biodegradable plastic and composites contribute about 0.6−0.7 tons in food packaging applications. Most biodegradable plastics are fabricated of Polylactic acid, Polyesters, biodegradable starch blends and polyhydroxy alkanoates. The subsequent bioplastics focuses on the material durability for consumer and industrial applications. The biological degradable materials are employed in composite to satisfy the physical, thermal and optical barriers of biomaterials. Current research study focusses on biodegradable materials and biodegradable composites in smart packaging applications. Smart packaging involves extended features with incorporation of sensors to predict the quality, life-span and products safety in packaging.

1.1 Composite

The admixture of materials with variant properties is called Composite material. The increased interest in the synthesis of multifunctional materials is due to the evolution of new structural materials which perform (a) multiple constitutional functions, (b) combined non-structural and structural functions, or (c) both simultaneously [3].

The superiority of material preference is toward traditional material for improvised features and application of composite material. The composite substance is capable of specialized strength, resistance to electricity, and stiffness [4]. A composite material can replace the conventional materials as composite material possesses advantages such as lightweight, flexibility, electric insulation, durability, and resistance to chemical agents. The composites can be categorized based on the nature of matrix composites such as ceramic, metal, glass, translucent concrete, cement, wood-plastic, carbon, papier mache and synthetic foams. The hunt for composite materials has led to the development of modeling the mechanical behavior, structural element analysis of composite substance either as laminate or sandwich beams, shells and plates [5].

1.2 Types of composites

- Ceramic matrix composite: The ceramic matrix composite is made of ceramic spread out in a ceramic matrix. The properties of ceramic matrix composites are high-temperature shock and resistance from fracture.
- Metal composite matrix: The ceramic matrix composite is made by the metal spread out in a ceramic matrix.
- Reinforced concrete: The reinforced concrete is prepared with a material with high tensile strength
- Glass fiber concrete: The concrete is added into a glass fiber with higher zirconia content to make the glass fiber reinforced content
- Translucent concrete: Translucent concrete is prepared by the concrete which encases optic fibers
- Modified wood: The combination of wood with speciality materials like veneer or cheaper materials to synthesize the engineered wood
- Plywood: The plywood is prepared by the combination of thin layers at diverse angles
- Modified bamboo: The bamboo strips are combined to form engineered bamboo. The properties of engineered bamboo are highly compressive, tensile strength, and flexural strength
- Linoleum/Parquetry: A square of many hardwood cubes is joined together to make parquetry. It is often sold as an ornamental piece
- Wood-plastic mixture: The casting of wood fiber in plastic to form the matrix
- Cement wood fiber matrix: The casting of mineralized wood pieces in cement
- Fiberglass: The combination of glass fiber and plastic to form a flexible fibreglass
- Carbon Fiber reinforced polymer: The plastic with the high tensile strength to weight ratio is reinforced with carbon fiber
- Sandwich Panel: The sandwich panel are prepared by the layering of composites
- Composite honeycomb: The honeycomb shape is made by the selection of composites in many hexagons

- Papier-mache: Bounding of paper with adhesive
- Plastic coated paper: The coating of plastic with paper such as playing cards
- Synthetic foams: The synthesis of light-weight materials by the filling of metals, ceramics, or plastic with glass micro balloon, carbon, or plastic micro balloon

1.3 Need for biodegradable composite

The raw material cost, fabrication and assembly cost, and susceptibility to impact damage are the disadvantages of synthetic polymer composite materials. These drawbacks urge the need for biodegradable composite materials. Biodegradable composites are usually made by a combination of plant fiber products and bioplastics. A biocomposite polymer is an admixture of the biologically degradable polysaccharide as its matrix form and a biodegradable filler as the reinforcement [6]. Biodegradable composites are used to lessen the amount of disposal in packaging materials by replacing petrochemicals with biodegradable components. The biodegradable composites are easily degradable as they possess the capacity to break down into carbon-di-oxide, organic material, and water vapor. Biodegradable composites cause less environmental pollution, and minimum greenhouse gas emissions compared to the other composites. The biodegradable composites and plant fiber composites could satisfy the drawbacks of synthetic polymer composites. Biodegradable composites possess advantages such as renewable sources, biodegradability, environmental benefits and sustainability [7].

2. Biodegradable composite

Biodegradable composites were synthesized using natural fibers and polyhydroxy butyrate (PHB) as a matrix [8]. Green composites are commonly known as biocomposites formed by the combination of natural fibers with degradable resins which can be disposed of without polluting the environment. The natural fibers used for the synthesis of biodegradable composites are coir, flax fiber, hemp fiber, Jute fiber, Kenaf fiber, and sisal fiber [4]. Apart from the natural fiber, mixed reinforced fibers are also used in the process of biodegradable composites. The matrix reinforced fibers employed in the manufacture of biocomposites are cellulosic compounds, hydroxy alkanoates, polylactide material, polyester amide, polyvinyl alcohol compound, and thermal plastic starch. The surface moderation of natural fibers intensifies the mechanical and sustainable properties [9]. The biodegradable composites have manifold applications such as surgical sutures in the medical field, first-aid dressings, tissue engineering, immobilization of enzymes, controlled and released drug delivery, genetic engineering, organelle preservation, medical devices and nanotechnology [10].

2.1 Characteristics of biodegradable composite

The degradable composites can be classified ground on the nature of the material used for the synthesis of biodegradable composites. They can be either matrix-based composites or fiber-reinforced matrix composites [10]. The matrix reinforced polymers can be cellulose, polyhydroxyalkanoates, polylactide compound, polyester, polyvinyl, thermal plastic starch, chitosan, gelatin and methylcellulose. Similarly, natural fibers namely hemp, coir, flax, jute, kenaf and sisal can be used for biodegradable composite applications. The succeeding table illustrates the properties of biomaterials used in composite applications (Table 19.1).

2.2 Synthesis of biodegradable composite

Biodegradable composites can be synthesized by casting solution method, Mix-Melt method, electrospinning, thermal pressing, extruding and blowing film method. These methods are explained as follows.

2.2.1 Extrusion

The commercial method of processing biopolymer is called extrusion. The extrusion can be carried out in two ways either as Single-screw extruding or Twin-screw extruding. The single screw extruding method is desirable as it is low cost and simpler. For example, starch can be modified into an amorphous form by a single screw extrusion method. The aliphatic esters can be only processed by extrusion method as they possess a lower melting point.

2.2.2 Film blowing

Film blowing is the commercial strategy to manufacture film from thermoplastics. In this process, the film will be extruded through a slit die by blowing air to the center position of the die. Further, the film is cooled, modified to a flat shape and subjected to extrusion over an isolated bubble of air. The standard thickness of the film is from 0.007 mm to 0.125 mm. The blowing film process requires elevated viscosity resin to pull the film

Table 19.1 Properties of biomaterials.

Material variety	Material density (kg/m^3)	Material strength (Mpa)	Young modulus of the material (Gpa)	Elongation (%)	Moisture content (wt %)
Coir	1150	140—593	4—6	25—30	8
Flax	1500	500—1500	27.6	2.7—3.2	8—12
Hemp	1480	550—900	30—70	1.6—1.8	6.2—12
Jute	1460	393—773	26.5	1.5—1.8	12.5—13.7
Kenaf	1400	930	53		
Sisal	1450	511—680	9.4—22	2.0—4	10—22
Polyester	1250	40—90	2—5	2.0—4.5	—
Polyhydroxy alkanoates	1140	35—100	3—6	1—6	—
Chitosan	1456	200	7.06	3.9	4.10
Gelatin	1720	40.8		6—7	8—13
Methylcellulose	1319	51—55		2.82	12.5

from the die in an upward direction. The film blowing method contributes excellent mechanical property to film formation.

2.2.3 Casting
The casting method is preferred to make soft and stretchy films. In this process, the film preparation is carried out by depicting a resin web from the die to the rollers for cooling. The film obtained by the casting method possesses excellent optical properties and lower mechanical properties compared to the film blowing method.

2.2.4 Thermopressing
Thermopressing is a technical method in which a plastic sheet is heated to a bearable temperature, molded into a specific shape and trimmed further to create the desired product. Thermopressing can be either vacuum-based or pressure based. The vacuum-based thermopressing employs heat and pressure to create the final configuration of plastic sheets. The sheets are heated further and placed over a mold to obtain the desired shape.

2.2.5 Electrospinning
The electrospinning process is a process of drawing continuous polymeric fiber from either polymeric solutions for polymer melt based on the electrohydrodynamics phenomenon. The electrospinning technique is the combination of electrical spraying and traditional solution dry fiber spinning. The electro-spinning method generates a nonwoven web of nanofibers when maximum voltage electricity is enforced to the liquid solution and collector tank. The nanofibers are extruded from the nozzle column which liberates a jet. The synthesized jet fibers are further dried and unloaded on the collector tank to fabricate polymer composite.

2.2.6 Melt-mix method
The melt mixing process involves the dispersion of nanoparticles by shear stress polymer matrix in the molten state configuration. The working principle of the melt mix method is similar to the long-established methods of polymer processing. Table 19.2 explains the production of biocomposites by different techniques and their advantages.

2.3 Environmental impacts of biodegradable composite
Biodegradable composite has earned interest in the rearmost decades owing to environmental concerns and the exhaustion of remnant resources. The synthesis of biodegradable composite makes use of biopolymer matrixes and lignocellulosic reinforcements thereby extending the application in biomedical, packaging, and structural applications. Biodegradable composites can replace synthetic polymers and synthetic fibers reinforced composites. The major advantage of biodegradable composites is renewable raw materials as resources, low eco-toxicity, sustainability, and biodegradability.

Table 19.2 Production of biodegradable composite by various techniques.

Biomaterial combination	Technique	Properties	Reference
Starch/chitosan/gelatin	Solution casting method	High density, fewer agglomerates, low moisture uptake	[1] [11] [12] [13]
Wheat-starch/lauric acid/chitosan	Solution casting method	Antimicrobial activity	
Corn starch/PVOH	Solution casting method	Enzyme rich, high biodegradability	
Starch/chitosan	Melt mix method	Higher tensile strength	
PVA/starch fibers	Electrospinning	Low weight ratio, irregular shape	
Degradable films from cassava	Thermopressing and casting	Transparent, mechanical resistance	
Starch from rice/flour	Casting	A new alternative, cheaper	
Starch from rice/Polyvinyl alcohol	Casting	Increased tensile strength	
Poly lactic acid/starch	Melt mix twin screw extruder	Superior mechanical behavior and toughness	
Poly lactic acid/corn	Twin-screw method	High biodegradability, compatibility	

2.4 Need for biodegradable composite in smart packaging

Food packages are practised to improvise the quality, safety, life-span, transport, storage and preservation. Polymer films are used in conventional packaging. The acrylic side-chain polymer can change phase reversibly at various temperatures. Once the side chain component melts, a dramatic increase of gas permeation occurs which helps to maintain the ratio of permeation of carbon-di-oxide to oxygen for particular products. Conventional packaging has the drawback of earlier product deterioration as it cannot preserve at elevated respiration rates pre-cut vegetables, fruits, lettuces, and salads [14]. Traditional packaging methods also make use of synthetic polymers such as polyamide, polypropylene, polyethene, polyethene terephthalate, ethylene, vinyl alcohol, polystyrene and polyvinylchloride as they possess mechanical and hurdle properties [15]. A drawback of synthetic polymers is undesirable environmental sequences, conversion of synthetic polymer to micro or nano-plastics in the environment and non-degradability. Owing to these drawbacks, passion for natural polymers such as lipids, phospholipids, surfactants and natural nanoparticles has increased. Except for natural polymers, biopolymers such as cellulosic compounds, plant chitin, shellfish chitosan, pectin, sodium or calcium alginate, agar, carrageenan, gelatin, zein, whey protein have been explored [16]. The major challenge in the exploitation of biopolymers in packaging is a synthesis of biofilm that can match the mechanical, optical and barrier properties of synthetic polymers. Researchers are therefore focused on the combination of biopolymers or biodegradable composite in packaging materials. The synthesis of biopolymer-based packaging with efficient functional properties has been done in the laboratory. Biji et al. [17] stated that attempts are done to extend the functional performance of biodegradable composite by either smart packaging or active packaging. Smart packaging is designed in such a way to respond to a specific trigger, pH variation, temperature fluctuations, gas level modification, exposure to light, chemical substance composition, enzyme activity and color. The resulting nanocomposites can successfully meet the optical, mechanical and techno-functional characteristics. The advantage of biodegradable composite in smart packaging is the incorporation of sensing materials into easily degradable films to provide technical information [18].

3. Smart packaging

Smart packaging is one of the packaging systems with embedded sensor technology. Smart packaging technology furnishes information about a specific product or brand by sensing them [19]. Smart packages involve a huge grouping of packages that leavers technology with distinct features such as shelf-life extension, monitor freshness, display of quality information, and mechanical, electronic, and responsive ink features. Smart packaging technology is embedded with barcodes and QR codes for the improvement of product and customer safety. The basic principle of smart packages is label scanning which reveals the package information and progression temperature. Further, smart packaging has been incorporated into integrated circuits to reveal functions such as screening vibration, acid value, slope, shock, moisture content, light, temperature, chemicals, bacteria, or viruses [20].

Smart packaging can regulate oxygen ingression and carbon-di-oxide egress by transpiration corresponding to the temperature. Therefore, the optimum atmosphere can be maintained throughout the product for storage and distribution, prolonged freshness, and high quality. Intelligent packaging mediates the customer by creating color change based on factors such as temperature modification during preservation, leaked packages, food spoilage, and the quality of the food [17]. The major profit of smart packages is that they can predict the presence of specific pathogens, and bacteria, moisture control, and prevents food from food spoilage. The process of smart packaging is explained by the pictorial representation of smart packaging (Fig. 19.1).

3.1 Process of smart packaging

Various monitoring devices are involved in the process of smart packaging and are described below in detail.

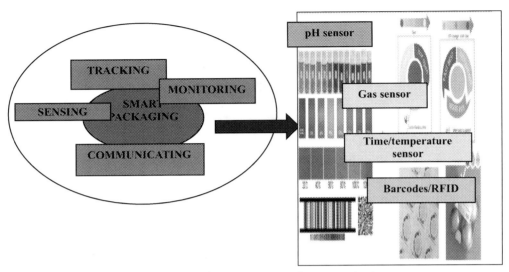

Fig. 19.1 Pictorial representation of smart packaging.

3.1.1 pH indicators

The pH indicator is employed to identify the significant alteration in the pH of the food by visible color change. The reason for the change in pH of the packed food is microbial growth, chemical reactions, and enzymatic activity [21]. In this process, natural pigment is usually preferred over synthetic dyes. The natural pigment undergoes color changes concerning moderation in the pH of the food. The principle of the calorimetric pH sensor is based on proton exchange between pigments and the medium around the food [22]. The specific volatile compounds are also used in some pH sensors to indicate the food modifications. Research studies reported using anthocyanins derived from variant botanical sources such as hibiscus, blueberry, soybean, purple corn, red cabbage and sweet potato. The factors affecting the response of natural pigment to the pH of the food are temperature, oxygen levels, light exposure and environmental conditions. The color change can be red (strongly acidic), purple (mildly acidic), violet (neutral), blue (mildly alkaline), green (moderately alkaline) and yellow (strongly alkaline) concerning change in pH [23]. The conversion of anthocyanin into various forms determines the color change. The possible conversion of anthocyanin into cationic oxonium, neutral carbinol pseudo-base, and anionic quinoidal base. Therefore, anthocyanins can be significantly incorporated into the biodegradable polymer for smart packaging applications.

3.1.2 Gas indicators

The gas indicator aims to determine the freshness of the food product. The concept behind these indicators is the freshness of fruits and vegetables, The fresh products produce ethylene gasses after natural respiration which acts as a measure of freshness and quality. The oxygen permeates the inner and outer surface of food packages which provokes the chemical degradation and oxidation of food. In addition to oxygen permeation, microbes can also contaminate foods. The probability of microbial contamination, oxidation and degradation of food demands the need for smart biodegradable films to sense and indicate the presence of specific gases, quality of food and safety [24]. Most smart packaging materials are designed with the incorporation of O_2, CO_2, H_2S, ethylene and other volatile compound indicators. The gas sensors are developed based on the tendency of chemical degradation of natural pigments. An increase in gas concentration corresponds to the color change of natural pigments. The anthocyanins, gelatins and carotenoids degrade in presence of ammonia which produces a color change from a purple color to green color and then to yellow [25]. The gas sensors can be incorporated into biodegradable films by adhesive labels, printed layers and thin films. From the research studies, it is inferred that smart packaging with gas sensors was able to detect ethylene gas, ammonia compounds, hydrogen sulfide, and carbon-di-oxide levels. The smart biodegradable packaging consisting of tara gum/polyvinyl alcohol/curcumin was able to screen the modifications in the freshness nature of the shrimp by predicting the formation of ammonia gas. The increase in ammonia was inferred by the color variation of the smart film from yellow to brown color.

3.1.3 Time temperature indicators (TTI's)

Time Temperature Indicators (TTI's) are employed to indicate the irreversible changes such as color and shape concerning temperature history. These integrators imitate the certain quality parameter variation of the food product under similar temperature exposure. The parameters that can be monitored by TTI are changes in biological activity, acid-base reaction, melting point, polymerization, and spore formation. Generally, time-temperature indicators are employed to track food quality, safety, and elevated temperature exposure using natural pigment as indicators. The time-temperature profile can be understood by the color change of the natural pigments.

In addition, TTIs can also be used to predict the likelihood of food deterioration during storage [26]. The features of TTI are inexpensive design, customer-friendly and simple. They can be also categorized based on underlying principles of temperature prediction such as diffusion, polymerization, the growth rate of microbes, presence of microbes, thermal chromic reaction, photochromic reaction, electronic resonance, and surface plasmon resonance. The critical temperature indicator (CTI) is used to represent whether the food has been heated above or cooled below during its lifetime. The critical temperature-time indicator (CTTI) is used to indicate whether the food has been heated above or cooled below during the specified time. The thermal-time indicator (TTI) is used to determine the product history of heating and cooling. The temperature sensor obeys the Arrhenius temperature dependency and activation energy ranging from 10 to 40 kcal/mol. Research studies suggest the efficient sustainability of natural pigments as temperature sensors. The biodegradable film (chitosan/cellulose matrix) smart packaging with temperature-sensitive indicators produced irreversible color change from violet to yellow concerning an increase in temperature. The significant color change was visualized using time-temperature indicators based on sensor mechanisms such as microbial-based (green to red), polymer-based (colorless to blue), diffusion (from yellow to pink), and enzymatic reactions (from green to yellow and to red).

3.1.4 Fresh-check lifeline integrator
Fresh-Check Lifeline integrator is used as own-adhesive labels for packing perishable products. These integrators provide information such as the fresh condition of the product at the event of purchase and at home. Fresh-check Lifeline integrators can also be referred to as Bull's eye configuration or historical integrators. The basic mechanism behind these integrators was the color change of formulated polymer from diacetylene monomers. The polymer color is directly proportional to the exposure temperature and rate of quality loss of food. The increase in temperature corresponds to the rapid color change of the polymer.

3.1.5 Vistab indicator
Vistab indicator is used to determine the full history of the product based on enzyme reaction. Vistab indicator device contains bubble-like dot which includes two-component (a) enzyme solution (lipase with pH indicating dye compound) (b) substrate solution (triglycerides). The basic principle of the vistab indicator is that when pressure is applied to the plastic bubble, dot activation occurs at the beginning of the monitoring period which results in the breaking of the seal between the two components. Once the seal breaks, the ingredients of the two components are mixed, and pH change occurs which is indicated by color change. The green color of the dot indicates the initial shelf life of the product. Similarly, the product approaching end of shelf life is indicated by a yellow dot. The rate of the reaction is directly proportional to the temperature. An increase in temperature corresponds to a rapid reaction rate and a decrease in temperature slows the reaction rate. Vistab indicator with adhesives is available either as a single dot indicator or triple or three-dot indicators. Single or one dot integrators are used to monitor the cartons, and pallet labels of the product under transit temperature.

3.1.6 RipeSense indicator
RipeSense indicator is the pioneer intelligent ripeness indicator label globally. The working principle of the Ripe-Sense indicator is that design a label that senses the ripe stage of the product apart from branding the product. The ripe-Sense indicator has produced a revolution in marketing.

3.1.7 3 M monitor-mark

The 3 M Monitor Mark is based on the ester compound and phthalate mix colored with blue dye (under the required melting point). The change in melting point creates diffusion along with a wick which permits a signal for the duration of the indicator in the liquid stage. The polyester strip performs as a contact barrier between the liquid and the wick until it is monitored and switched on by the operator to start the device.

3.1.8 Radio frequency identification (RFID) tags

Radiofrequency identification tags are the modern form of data conveyor that can diagnose and track the product. Current smart packages are furnished with minute electronic materials and specifications such as LED lights, Bluetooth chips, RFID chips, alarms, and speakerphones These components are employed to communicate, monitor and trace the packaging materials. RFID tags have advanced features such as in the case of medicine packaging, monitoring of each pill consumption by the customer, and alarming sound if the wrong dosage is taken by the customers. The biodegradable packaging materials possess advantages such as increased food quality, shelf life and safety.

3.2 Current applications of biodegradable composite in smart packaging

Biodegradable composite plays a significant role in smart packaging. Various materials with different properties are used for suitable smart packaging applications. The material used in the biocomposite preparation and its properties and application is listed in the following Table 19.3.

4. Benefits of smart packaging

Smart packaging reduces the post-effect of packaging disposal on the environment by opting for sustainable packaging options for brands. In addition, smart packaging is identified to be ecofriendly, cost-effective, renewable sources, and has fewer carbon footprints. Smart packaging helps to track the product status.

4.1 Customer empowerment

Smart packaging involves the technique namely QR (Quick response), RFID (radio frequency identification) and NFC (Near field communication) which impresses the customer interface to learn about the purchased product. The customer will be aware of the ingredient research, nutritional value, and dietary scans by scanning the QR code. In addition, proprietary technology leads customers through user manuals and smartphones.

4.2 Quality control

To identify the condition of the product, smart packages are designed with advanced sensors and indicators. By removing the unwanted particulates from the compound, the life

Table 19.3 Biodegradable composite in smart packaging.

The material used for biocomposite preparation	Properties	Application	References
Chitosan and polyvinyl alcohol (40:60)	Tensile-11.02Mpa, Thickness-0.15 mm Produces pH change upon detoriation	Smart Packaging Applications	[18]
Poly (ester-urethane) incorporated with catechin	Transparency, UV-light transmission, water resistance, effective antioxidant activity, appropriate disintegration in compost		[27]
Methylcellulose chitosan/ nanofibres incorporated with saffron petal anthocyanins	Increased tensile strength, antimicrobial and antioxidant activity		[28]
Cassava starch with blueberry residue	Excellent barrier properties prevent food deterioration, Aromatic compounds in biopolymers are advantageous		[29]
Chitosan with black soybean seed coat extract	Antioxidant and antimicrobial activity, UV resistance		[30]
Gelatin with red cabbage extracts	Improved mechanical behavior, prevents deterioration with pH changes		[25]
Chitosan integrated with extract of purple-fleshed sweet potato extract	UV light barrier film, thermal stability, thickness and water solubility		[31]
Agar combined with root extracts of *Arnebia euchroma*	Extracts produce color change upon fish spoilage		[32]
Gelatin with curcumin	Sense pH change, antioxidant and hydrophobicity of films		[33]
K-carragenan with curcumin	pH change sense, redness shift under alkaline condition		[34]
Chitosan with blueberry/ blackberry pomace extracts	Increased elongation, surface hydrophobicity, thermal stability		[35]

Table 19.3 Biodegradable composite in smart packaging.—cont'd

The material used for biocomposite preparation	Properties	Application	References
PVA/nanocellulose/Ag composite films	Mechanical and antibacterial properties		[36]
Chitin nanofibre/methylcellulose/Red barberry anthocyanins	Films exhibited excellent antioxidant and antimicrobial activity and the ability to detect quality changes.	Smart packaging of fish	[37]
Chitosan/corn starch/*Hibiscus rosasinensis*	Films exhibited high optical and surface morphological properties and are sensitive to pH changes.	Smart packaging of chicken	[38]
Agar/Tapioca starch/Red cabbage anthocyanin	Anthocyanins differ in color in response to quality changes in sausage during storage.	Smart packaging of sausage	[39]
Cassava starch/blueberry residue anthocyanin	Anthocyanins vary in color in response to pH (quality) changes in chicken during storage.	Smart packaging of sausage	[40]
Chitosan/anthocyanin	Anthocyanins modify color in response to quality changes in pork and fish during storage.	Smart packaging of Pork/Fish	[41]
Methylcellulose/chitosan nanofibre/Barberry anthocyanin	Chitosan provides antimicrobial activity while anthocyanins change and modify in response to changes in meat quality during storage.	Smart packaging of red meat	[21]
PVA/glucomannan/sapan wood extracts	The wood extract changed color in response to quality changes in a banana during storage	Smart packaging of banana	[42]
Starch coupled with polyvinyl alcohol/purple sweet potato anthocyanin	The anthocyanins gave a color change in response to alterations in milk quality. The films also exhibited antimicrobial activity	Smart packaging of Milk	[43]

span of the packed product can be improved. Smart packaging also helps in counterfeit detection, easier accessibility of genuine quality products from the manufacturer, and long-lasting.

4.3 Customer connectivity

Smart Packaging offers ideal practices, an advanced customer interface, a better shopping experience and improves interaction with customers. Smart packaging guides customers to invest in the business with advanced technologies. The customers experience better technologies that impact the business positively.

4.4 Prediction analysis

The availability of the product with a similar brand and name which creates substantial profits in the false product can be predicted by smart packaging. The product quality can be ensured without the interference of any counterfeit-proof products, substitution, or from creating theft. The smart packaging will allow customers to track their products and communicate legitimacy. The public reputation of a particular brand can be maintained with smart packaging.

4.5 Predictive planning

The most predominant step is to gain customers' orders whenever they require a specific product or replacement. Smart packaging integrations allow manufacturers to have a better understanding of the product and create communication platforms such as email or messages for the customers in need of help or replacement. The advanced feature of smart packaging incorporates the automation process, integration of QR codes, and cloud-based solutions which brings customers to repurchase through websites.

4.6 Brand transparency

Smart packaging helps customers track the source of the product materials, manufacturing company details, and relevant information integral to the customer. Through smart packaging, people are made aware of their product details, show interest to purchase eco-friendly products and spend more on their favorite products. Apart from these information transparencies, the holding corporation accountable for their social, economic, and environmental impacts has been given importance to the everyday customer.

4.7 Increasing likelihood of positive experience

Smart packaging mediates the customer through web-based information, tutorial video links, usage information, and the positive impact of the specific product. Moreover, customers can interact, express their concerns, and clarify their doubts through available customer support lines in a time-friendly manner. The customers will be able to identify the new product, ingredients, or materials and screen out the best product among other products. Therefore, smart packaging creates an engaging opportunity for the customer to learn about the purchased product and its specific details.

5. Advanced techniques in smart packaging

5.1 Active packaging

Active packages mediate with the package contents for condition modification, especially to freshness extension or lifespan extension. These packages are foremost employed in food and beverage applications. Active packaging aims to maintain the lifespan and standard quality of the vegetables, and livestock products by incorporating the additives into the packaging system. In addition to these, active packaging makes use of nanomaterials, to interact directly with food or the environment for the protection of the product. The active packaging improves product lifespan and the standard quality of the packed product during storage. The integration device employed in active packaging contains filters for light, absorbers for oxygen and ethylene, and surface coatings for antimicrobial and humidity regulation. For example, sometimes active packages may liberate antimicrobial agents which preclude the growth of microorganisms by sensing the freshness of the product. Similarly, plastic bottle beer packages contain absorbers for oxygen in the screw caps to prolong the lifespan of the beer bottle by up to six months. Further, ethylene

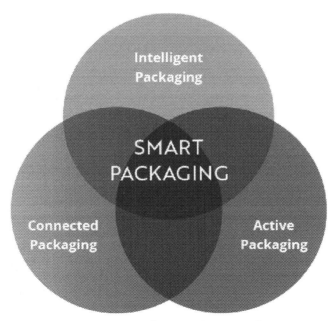

Fig. 19.2 Advanced features of smart packaging.

absorbers are used in film packaging systems to absorb the ripening hormone ethylene for extensive lifespan assurance.

5.2 Connected packaging

The connected packages incorporate the special code printed inside the food package. The customers shall activate those special codes with help of a mobile phone to get their content exclusively. For example, Smash-box eyeshadow packaging utilized a Touch code link to lodge an unnoticeable code on a card that is placed inside the packaging. The invisible code on the card invites the customer to a specific website, where the customers can unlock the series of exclusive online professional makeup tutorial videos by touching their mobile device to the card. The code cannot be copied or interfered with graphic design. The customers can unlock the content by simply typing the short URL on their mobile device and touching the code using a mobile device. Connected packaging breaks the limit on food and beverage products. Connected packaging can be practised in almost all retail product categories such as toys, cosmetics, gaming and apparel. targets to practice digital experiences in packaging using a mobile phone tab or QR code. Connected packaging creates a pathway for the customer to learn about the particular brand and product information, exclusive content offers and delivers tangible outcomes. Branded companies make use of connected packaging to offer exclusive content, deliver coupons, and collectable items and allow customers to buy the particular authenticated product. The benefits of connected packaging technology are popularity attention with wholesale marketers, generating deeper relationships with customers for elevating brand loyalty and elevating the lifetime value of the customer. Further, Connected packages interact with consumers through a code on the package that can be activated with a mobile device to deliver a specific product with exclusive content (Fig. 19.2).

6. Future scope and challenges of smart packaging

Smart packaging fabricated with biodegradable materials or composites in food industries has created a positive impact on improvised quality, lifespan and welfare of the foods. Biodegradable smart packaging can reduce waste in the environment. The natural pigments incorporated in smart packaging can detect freshness, quality, safety, pH modifications, gas levels and temperature change. Recent smart packaging has advanced features of antimicrobial and antioxidant incorporation. These features can be monitored and screened without damaging the package. Moreover, research studies with the prediction of lifetime history of exposure to light, oxygen, and pH have been attempted to incorporate into future smart packaging applications. Food packaging industries approach smart packaging with a moderated atmosphere for high-value product packages and niche merchandise namely organic foods. The biodegradable composite can meet the economic cost and environmental sustainability. Apart from the positive impacts, rigorous optical, mechanical, barrier and stability requirements act as hurdles in the commercial usage of smart packaging in the food industry. Future smart packaging depends on the generation of biodegradable smart packaging with new natural materials, innovative structural designs and a processing approach. The proper field study of biomaterial sources is essential to understand the food components interaction and the interaction of the biodegradable polymer during food preservation and food processing.

7. Conclusion

Biodegradable polymers are useful in reducing environmental waste, plastic production and creating a green economy. Similarly, the biodegradable films synthesized from natural fibers, renewable feedstocks, and agricultural waste are eco-friendly, cost-effective and sustainable. The utilization of bioplastics is increasing at a rapid rate in food packaging applications. However, at present biodegradable polymers only replaces about 2% of the synthetic plastic all over the world. The hurdles in the successful replacement of synthetic plastic with biodegradable plastic are complicated downstream operations, stability and durability of bioplastics. Therefore, forthcoming research has to be focused on the improvisation of thermal and physical properties of bioplastic in smart packaging applications.

References

[1] S. Mangaraj, A. Yadav, M.B. Lalit, S.K. Dash, N.K. Mahanti, Application of biodegradable polymers in food packaging industry : a comprehensive review, J. Packag. Technol. Res. 3 (77–96) (2019), https://doi.org/10.1007/s41783-018-0049-y.
[2] R.C.F. Cheung, T.B. Ng, J.H. Wong, Y.W. Chan, Chitosan: an update on potential biomedical and pharmaceutical applications, Mar. Drugs 13 (8) (2015) 5156–5186, https://doi.org/10.3390/md13085156.
[3] R.F. Gibson, A review of recent research on mechanics of multifunctional composite materials and structures, Compos. Struct. 92 (2010) 2793–2810, https://doi.org/10.1016/j.compstruct.2010.05.003.
[4] I. Uddin, S. Thomas, R.K. Mishra, A.M. Asiri, Sustainable polymer composites and nanocomposites, Sustain. Polym. Compos. Nanocomposites. 1–1440 (2019), https://doi.org/10.1007/978-3-030-05399-4.
[5] K.W.H. Altenbach, J. Altenbach, Classification of composite materials. In: mechanics of composite structural elements, Mech. Compos. Struct. Elem. 1–14 (2004), https://doi.org/10.1007/978-3-662-08589-9_1.
[6] K. Rohit, S. Dixit, A review—future aspect of natural fiber reinforced composite, Polym. from Renew. Resour. 7 (43–60) (2016), https://doi.org/10.1177/204124791600700202.
[7] A.K. Mohanty, M. Misra, G. Hinrichsen, Biofibres, biodegradable polymers and biocomposites: an overview, Macromol. Mater. Eng. 276–277 (2000) 1–24, https://doi.org/10.1002/(SICI)1439-2054(20000301)276:1<1::AID-MAME1>3.0.CO;2-W.
[8] N.R. Paluvai, S. Mohanty, S.K. Nayak, Synthesis and modifications of epoxy resins and their composites: a review, Polym.—Plast. Technol. Eng. 53 (2014) 1723–1758, https://doi.org/10.1080/03602559.2014.919658.

[9] M. Bahrami, J. Abenojar, M.A. Martínez, Recent progress in hybrid biocomposites: mechanical properties, water absorption, and flame retardancy, Materials 13 (22) (2020) 5145, https://doi.org/10.3390/ma13225145.

[10] H.T. Sreenivas, N. Krishnamurthy, G.R. Arpitha, A comprehensive review on light weight kenaf fiber for automobiles, Int. J. Light. Mater. Manuf. 3 (2020) 328—337, https://doi.org/10.1016/j.ijlmm.2020.05.003.

[11] F.D.S. Larotonda, K.N. Matsui, V. Soldi, J.B. Laurindo, Biodegradable films made from raw and acetylated cassava starch, Brazilian Arch. Biol. Technol. 47 (2004) 477—484.

[12] N.A. Azahari, N. Othman, H. Ismail, Biodegradation studies of polyvinyl alcohol/corn starch blend films in solid and solution media, J. Phys. Sci. 22 (15—31) (2011).

[13] F. Parvin, M.A. Rahman, J.M.M. Islam, M.A. Khan, A.H.M. Saadat, Preparation and characterization of starch/PVA blend for biodegradable packaging material, Adv. Mater. Res. 123—125 (2010) 351—354, https://doi.org/10.4028/www.scientific.net/AMR.123-125.351.

[14] L.G. Angelini, M. Scalabrelli, S. Tavarini, P. Cinelli, I. Anguillesi, A. Lazzeri, Ramie fibers in a comparison between chemical and microbiological retting proposed for application in biocomposites, Ind. Crops Prod. 75 (2015) 178—184, https://doi.org/10.1016/j.indcrop.2015.05.004.

[15] K.J. Groh, T. Backhaus, B. Carney-Almroth, B. Geueke, P.A. Inostroza, A. Lennquist, H.A. Leslie, M. Maffini, D. Slunge, L. Trasande, A.M. Warhurst, J. Muncke, Overview of known plastic packaging-associated chemicals and their hazards, Sci. Total Environ. 651 (2019) 3253—3268, https://doi.org/10.1016/j.scitotenv.2018.10.015.

[16] L. Motelica, D. Ficai, A. Ficai, O.C. Oprea, D.A. Kaya, E. Andronescu, Biodegradable antimicrobial food packaging: trends and perspectives, Foods 9 (2020) 1—36, https://doi.org/10.3390/foods9101438.

[17] K.B. Biji, C.N. Ravishankar, C.O. Mohan, T.K. Srinivasa Gopal, Smart packaging systems for food applications: a review, J. Food Sci. Technol. 52 (2015) 6125—6135, https://doi.org/10.1007/s13197-015-1766-7.

[18] Y. Hidayato, N.A. Wijaya, M.W. Bintoro, V.P. Mulyani, S. Pratama, Development of biodegredable smart packaging from chitosan, polyvinyl alcohol(PVA) and butterfly pea flowers (Clitoria ternatea L.) anthocyanin extract, Food Res 5 (2021) 307—314.

[19] E. Psochia, L. Papadopoulos, D.J. Gkiliopoulos, A. Francone, M.E. Grigora, D. Tzetzis, J.V. de Castro, N.M. Neves, K.S. Triantafyllidis, C.M.S. Torres, D.N. Kehagias, D.N. Bikiaris, Bottom-up development of nanoimprinted PLLA composite films with enhanced antibacterial properties for smart packaging applications, Macromol 1 (2021) 49—63, https://doi.org/10.3390/macromol1010005.

[20] H. Moustafa, A.M. Youssef, N.A. Darwish, A.I. Abou-Kandil, Eco-friendly polymer composites for green packaging: future vision and challenges, Compos. Part B Eng. 172 (2019) 16—25, https://doi.org/10.1016/j.compositesb.2019.05.048.

[21] M. Alizadeh-sani, M. Tavassoli, E. Mohammadian, A. Ehsani, G.J. Khaniki, R. Priyadarshi, J. Rhim, pH-responsive color indicator films based on methylcellulose/chitosan nanofiber and barberry anthocyanins for real-time monitoring of meat freshness, Int. J. Biol. Macromol. 166 (2020) 741—750, https://doi.org/10.1016/j.ijbiomac.2020.10.231.

[22] K.S. Suslick, N.A. Rakow, A. Sen, Colorimetric sensor arrays for molecular recognition, Tetrahedron 60 (2004) 11133—11138, https://doi.org/10.1016/j.tet.2004.09.007.

[23] M. Alizadeh-sani, E. Mohammadian, J. Rhim, S. Mahdi, pH-sensitive (halochromic) smart packaging films based on natural food colorants for the monitoring of food quality and safety, Trends Food Sci. Technol. 105 (2020) 93—144.

[24] M. Sohail, D. Sun, Z. Zhu, Recent developments in intelligent packaging for enhancing food quality and safety, Crit. Rev. Food Sci. Nutr. 58 (2018) 2650—2662, https://doi.org/10.1080/10408398.2018.1449731.

[25] Y.S. Musso, P.R. Salgado, A.N. Mauri, Smart gelatin films prepared using red cabbage (*Brassica oleracea L*.) extracts as solvent, Food Hydrocoll 89 (2019) 674—681, https://doi.org/10.1016/j.foodhyd.2018.11.036.

[26] P. Müller, M. Schmid, Intelligent packaging in the food sector: a brief overview, Foods 8 (2019) 16, https://doi.org/10.3390/foods8010016.

[27] M.P. Arrieta, V. Sessini, L. Peponi, Biodegradable poly(ester-urethane) incorporated with catechin with shape memory and antioxidant activity for food packaging, Eur. Polym. J. 94 (2017) 111—124, https://doi.org/10.1016/j.eurpolymj.2017.06.047.

[28] M. Alizadeh-sani, M. Tavassoli, D. Julian, H. Hamishehkar, Multifunctional halochromic packaging materials : saffron petal anthocyanin loaded-chitosan nanofiber/methyl cellulose matrices, Food Hydrocoll 111 (2021) 106237, https://doi.org/10.1016/j.foodhyd.2020.106237.

[29] C. Leites, T. Garrido, J. Corralo, I. Cristina, K. De, Development and characterization of cassava starch films incorporated with blueberry pomace, Int. J. Biol. Macromol. 106 (2018) 834–839, https://doi.org/10.1016/j.ijbiomac.2017.08.083.

[30] S. Chen, X. Wei, Z. Sui, M. Guo, J. Geng, J. Xiao, D. Huang, Preparation of antioxidant and antibacterial chitosan film from *Periplaneta americana*, Insects 12 (2021) 53, https://doi.org/10.3390/insects12010053.

[31] H. Yong, X. Wang, R. Bai, Z. Miao, X. Zhang, J. Liu, Development of antioxidant and intelligent pH-sensing packaging films by incorporating purple-fleshed sweet potato extract into chitosan matrix, Food Hydrocoll 90 (2019) 216–224, https://doi.org/10.1016/j.foodhyd.2018.12.015.

[32] S. Huang, Y. Xiong, Y. Zou, Q. Dong, F. Ding, X. Liu, A novel colorimetric indicator based on agar incorporated with Arnebia euchroma root extracts for monitoring fish freshness, Food Hydrocoll 90 (2019) 198–205, https://doi.org/10.1016/j.foodhyd.2018.12.009.

[33] Y.S. Musso, P.R. Salgado, A.N. Mauri, Smart edible films based on gelatin and curcumin, Food Hydrocoll 66 (8–15) (2017), https://doi.org/10.1016/j.foodhyd.2016.11.007.

[34] J. Liu, H. Wang, P. Wang, M. Guo, S. Jiang, X. Li, S. Jiang, Films based on k-carrageenan incorporated with curcumin for freshness monitoring, Food Hydrocoll 83 (2018) 134–142, https://doi.org/10.1016/j.foodhyd.2018.05.012.

[35] P. Ezati, J. Rhim, pH-responsive chitosan-based film incorporated with alizarin for intelligent packaging applications, Food Hydrocoll 102 (2020) 105629, https://doi.org/10.1016/j.foodhyd.2019.105629.

[36] M. Salman, M. Bilal, K. Niazi, Z. Jahan, T. Ahmad, Preparation and characterization of PVA/nanocellulose/Ag nanocomposite films for antimicrobial food packaging, Carbohydr. Polym. 184 (2018) 453–464, https://doi.org/10.1016/j.carbpol.2017.12.068.

[37] M. Alizadeh, M. Tavassoli, H. Hamishehkar, D. Julian, Carbohydrate-based films containing pH-sensitive red barberry anthocyanins : application as biodegradable smart food packaging materials, Carbohydr. Polym. 255 (2021) 117488, https://doi.org/10.1016/j.carbpol.2020.117488.

[38] M. Othman, A.A. Yusup, N. Zakaria, Bio-polymer chitosan and corn starch with extract of Hibiscus rosa-sinensis (Hibiscus) as PH indicator for visually-smart food packaging, AIP Conf. Proc. (2018) 050004, https://doi.org/10.1063/1.5047198.

[39] T.D.W. Ata Aditya Wardana, Development of edible films from tapioca starch and agar , enriched with red cabbage (*Brassica oleracea*) as a sausage deterioration bio-indicator, Int. Conf. Eco Eng. Dev. 109 (2017) 012031, https://doi.org/10.1088/1755-1315/.

[40] L. Luchese, V.F. Abdalla, J.C. Spada, I.C. Tessaro, Evaluation of blueberry residue incorporated cassava starch film as pH indicator in different simulants and foodstuffs, Food Hydrocoll 82 (2018) 209–218, https://doi.org/10.1016/j.foodhyd.2018.04.010.

[41] H. Xiao-wei, Z. Xiao-bo, S. Ji-yong, G. Yanin, Z. Jie-wen, Z. Jianchun, H. Limin, Determination of pork spoilage by colorimetric gas sensor array based on natural pigments, Food Chem 145 (2014) 549–554, https://doi.org/10.1016/j.foodchem.2013.08.101.

[42] Ardiyansyah, M.F. Kurnianto, B. Poerwanto, A. Wahyono, M.W. Apriliyanti, I.P. Lestaro, Monitoring of banana deteriorations using intelligent-packaging containing brazilien extract (Caesalpina sappan L .), Int. Conf. Food Agric. 411 (2020) 012043, https://doi.org/10.1088/1755-1315/411/1/012043.

[43] B. Liu, H. Xu, H. Zhao, W. Liu, L. Zhao, Y. Li, Preparation and characterization of intelligent starch/PVA films for simultaneous colorimetric indication and antimicrobial activity for food packaging applications, Carbohydr. Polym. 157 (2017) 842–849, https://doi.org/10.1016/j.carbpol.2016.10.067.

Index

Note: Page numbers followed by "*f*" indicate figures and "*t*" indicate tables.

A

Absorbers, 349–350
Absorbing system, 179–180
Active packaging, 27–30, 28t, 29f, 66, 176–179, 322, 323t, 385–386
  home meals smart packaging, 144
  industrial application/market implementation, 77–82, 78f
    advantages of, 86
    antimicrobial and antioxidant packaging, 82
    carbon dioxide emitter, 79–80
    consumers, 89–90
    costs, 90
    ethanol emitters, 81–82
    ethylene scavengers, 80–81
    industrial barriers for, 86
    industries, prohibitive regulations for, 89
    intelligent packaging, 85, 85t
    lake of knowledge, 88
    moisture absorber/scavengers, 81
    move from lab to industrial scale, 88–89
    new manufacturing techniques, 90
    oxygen scavenger, 79
    safety concern, 88
    sustainability, 87
  market demand
    antioxidants, 122–123
    carbon dioxide emitter (CO_2), 122
    ethylene scavengers, 120–122
    flavor or odor absorbers, 122
    flavor or odor emitters, 122
    market demand, 118–123, 119f, 119t
    moisture scavengers, 119t, 120
    oxygen scavengers (OS), 120
  medicinal food supplements, 235–246
    carbon dioxide (CO_2) absorbers & emitters, 240
    commercial scavengers and emitters, 235, 236t–237t
    ethanol scavengers, 239
    ethylene scavengers, 238–239
    moisture scavengers, 239–240
    odor or taint scavengers, 239
    oxygen scavengers, 235–238
    preservative releasers, 241
  smart packaging technology, innovations in, 40–49, 42t
    antimicrobial packaging systems, 47–48
    antioxidant releaser, 46–47, 48f
    CO_2 emitters and absorbers, 45–46, 47f
    ethylene scavengers, 45, 46f
    flavor/odor releasers and absorbers, 48
    moisture absorbers, 41–43, 43f
    other active packaging, 48–49
    oxygen scavengers, 43–44, 44f
Agar, 334–335, 335f
Alginate, 331–332, 331f
Aliphatic polyesters, 338–343
  polybutylene succinate (PBS), 340–341, 340f
  polycaprolactone (PCL), 341–342, 341f
  polylactic acids, 338–340, 339f
  polylactide aliphatic copolymer (CPLA), 342
  polyvinyl alcohol (PVA), 342–343
Alloys of iron (steel), 130
Aluminum, 131–132, 313
  bottles, 133
  collapsible tubes, 133
  foil, 133
Anthocyanins (ANT), 217
Antimicrobial packaging (AP), 47–48, 158–159, 159f, 180–181
  antioxidants, 82, 347–348
  attributing polymer, 246
  medicinal food supplements, 241–246
Antimicrobials to polymers, surface fixation of, 245–246
Antioxidant packaging
  application of, 15, 16t
  endogenous antioxidant, 1–2
  exogenous antioxidant, 2–4
    ascorbic acid, 2
    carotenoid compounds, 2–3
    polyphenol compounds, 3–4
    tocopherol, 4
  food packaging material, 8–10, 9t
    glass, 9

Antioxidant packaging (*Continued*)
    metals, 10
    paper and paper, 9—10
    plastic, 8
  food products, oxidation in, 10—11
  packaging film to food products, 14—15
  peroxidation, mechanism of, 11
  production of films, 13—14
  smart packaging, types and features of, 12—13, 12f
  sources of, 5—8
    cereals, 7—8
    fruits and vegetables, 7
    herbs and spices, 5—6
    legumes, 7—8
    nut, 7—8
    tea, 6
  synthetic antioxidants, 5
  types of, 1—4
Antioxidant releaser, 46—47, 48f
Antioxidants, 122—123
Aromatic polyesters, 343
Article 3 of EC 1935/2004, 66
Asafoetida packaging, 63
Ascorbic acid, 2
Aseptic packaging, 322

B

Bacterial nanocellulose (BC), 217
Bakery based products, 63, 199—201
  confectionary products, 139
  ethanol emitters, 200
  oxygen scavengers, 200
Barcode-based technique, 250—251
Barcodes, 52—53
Beverage products, 139, 163, 198—199
  enzyme release packaging, 199
  flavor releasing packaging, 199
  gas releasing packaging, 198
  nutrient releasing packaging, 198—199
Bio-based plastics, 174
Biodegradable composite, 373
  casting, 375
  characteristics of, 373—374, 374t
  current applications of, 381
  electrospinning, 375
  environmental impacts of, 375
  extrusion, 374
  film blowing, 374—375
  melt-mix method, 375

  smart packaging, 376—377
  synthesis of, 374—375
  thermopressing, 375
Biodegradable polymers, greener approach food packaging, 323—324
  absorbers, 349—350
  active packaging, 322, 323t
  advances in, 353—355
  antimicrobial and antioxidants, 347—348
  aseptic packaging, 322
  biodegradable packaging, 323—324
  characteristics and criteria of, 319—322
  characterization of, 355—356
  chemical processes, 346—347
  coated unbleached kraft (CUK), 320
  differential scanning calorimetry, 355
  Fourier transform infrared spectroscopy, 356
  glass, 321
  history of, 318
  indicators, 351—352
  ion radiation, 352
  limitations, 344—345
  lipid-based polymers, 337—338
  metals, 320—321
  modified atmosphere packaging (MAP), 322, 324t
  nanotechnology, 353—355
  nuclear magnetic resonance, 356
  pH alteration, 350—351
  plasticization, 345—346
  plastics, 320
  polymer fabrication technology, 356—360
    blow molding, 357
    compression molding, 356
    electrospinning, 359—360
    extrusion coating, 357
    extrusion molding, 357
    injection molding, 356
    thermoforming, 357—358
  polysaccharide based polymers, 324—335
    agar, 334—335, 335f
    alginate, 331—332, 331f
    cellulose, 324—326, 325f
    chitosan, 332—334, 333f
    pectin, 329—331, 330f
    polyhydroxyalkanoates, 328—329, 329f
    starch, 326—328
  protein-based polymers, 335—337, 336f—337f
  scavengers, 349—350

solid bleached sulfate (SBS), 320
synthetically produced bio-polyesters, 338—343
   aliphatic polyesters, 338—343
   aromatic polyesters, 343
   polyethylene furanoate (PEF), 343, 343f—344f
Tetra pack, 353
thermogravimetric analysis, 355
uncoated recycled board (URB), 320
white-lined chipboard (WLCB), 320
Biosensors, 53, 54f, 252—257, 253t
BIS Standard 10146:1982, 65
BIS Standard 10171:1986, 65
Blended edible vegetable oils, 63
Blockchain systems, 219—220, 219f
   RFID systems, 219—220, 220t
Blow molding, 357
Brand transparency, 384
Brazil, regulatory aspects in, 109—111
Bureau of Indian standards (BIS), 105—107
Butylated hydroxyanisole, 29—30
Butylated hydroxytoluene, 29—30

C

Carbon dioxide emitter, 79—80, 122, 240, 279
Carotenoid compounds, 2—3
Carriers, 66
Carrot anthocyanins (CA), 217
Casting, 375
Cellulose, 324—326, 325f
Cereals, 7—8
Chemical sensors, 54
Chitosan, 332—334, 333f
Cholesterol and lactose, reduced content of, 202
Coated unbleached kraft (CUK), 320
Coating antimicrobial on polymer, 245
CO_2 emitters and absorbers, 45—46, 47f
Coffee and tea, 139
Cohesive peeling, 151
Commercial biosensors, 254—255
Commercial food products
   bakery based products, 199—201
      ethanol emitters, 200
      oxygen scavengers, 200
   beverage products, 198—199
      enzyme release packaging, 199
      flavor releasing packaging, 199
      gas releasing packaging, 198
      nutrient releasing packaging, 198—199
   comfort factors of, 205—207
   consumer advantages, 205—207
   consumer's value proposition, 205—206
   fish and seafood products, 202, 203t
   fruits and vegetable products, 201
   meat and poultry products, 203—204
      oxygen scavengers, 203—204, 205f
   milk-based products, 201—202
      cholesterol and lactose, reduced content of, 202
      oxygen scavenger, 202
   product use convenience, 206—207
Commercial scavengers and emitters, 235, 236t—237t
Composite, 371—372
   biodegradable composite, 373
      casting, 375
      characteristics of, 373—374, 374t
      electrospinning, 375
      environmental impacts of, 375
      extrusion, 374
      film blowing, 374—375
      melt-mix method, 375
      smart packaging, 376—377
      synthesis of, 374—375
      thermopressing, 375
   smart packaging, 377—381
      active packaging, 385—386
      biodegradable composite, current applications of, 381
      brand transparency, 384
      challenges of, 386—387
      connected packaging, 386
      customer connectivity, 384
      customer empowerment, 381
      fresh-check lifeline integrator, 380
      future scope, 386—387
      gas indicators, 378—379
      3 M monitor-mark, 381
      pH indicators, 378
      positive experience, increasing likelihood of, 384—385
      prediction analysis, 384
      predictive planning, 384
      process of, 377—381
      quality control, 381—384
      radio frequency identification (RFID) tags, 381
      RipeSense indicator, 380
      time temperature indicators (TTI), 379—380
      Vistab indicator, 380

Composite (*Continued*)
   types of, 372–373
Composite wood, 311–313
   aluminum, 313
   fiberboard, 312
   glass showcase materials, 313
   light alloy-based showcase materials, 313
   metal-based showcase materials, 312
   particleboard, 311
   plastic based showcase materials, 313
   plywood, 312
   PVC, 313
Compression molding, 356
Connected packaging, 172–174, 172f–173f, 386
Consumer, 89–90
   convenience, 66
   food waste reduction, smarter packaging for, 307–308
   packaging interface, 307
   perception, 124–125, 125f
   value proposition, 205–206
Container-based smart packaging technology, 276–278
   CO_2 non-dispersive infrared (NDIR) sensor, 276–278
   RFID sensors, 276
Cooked foods, sustainable metal packaging for food packaging materials, 286–296, 286f, 288t
   metal packing materials, 287–296, 289t–290t
      advantages, 292–296
      alternatives, 291–292
      disadvantages, 292–296
      normalized priorities, 299t
      results, 297–299, 297t
Curcumin (CR), 217
Customer
   connectivity, 384
   empowerment, 381
Customer's attraction, 135
Cutting edge advancement, 174–176
   sustainable packaging, innovative materials for, 175–176
Cybersecurity, 33–34

D

Data carriers, 99, 123–124
Date of manufacturing, 61
Date of packaging, 61

Di-ethylene polymerization through Microwave E Technology (DEMETO), 164
Directive 2002/46/EC, 65
Directive 90/496/EEC, the Health and Consumer Protection Directorate General Services, 65
Dried glucose syrup, 63
Durability, 134

E

Easy to exposed packing (EEP), 150–153
   easy peel sealant, 151–152
   easy to open for rigid packing, 151
   easy to peeling for flexible packing, 151
   laser perforation technology (LPT), 152–153
EC food law, Food law Act 1990, 60
Edible coatings, 181–182
Edible common salt, 63
Edible sensors, 55
Electrochemical biosensors, 254
Electronic sensing systems, 220–222, 221f
Electrospinning, 359–360, 375
Emerging technologies, 177–179
Endogenous antioxidant, 1–2
E-nose, 222
Environmental concerns of metal packaging, 137
Enzyme-based time-temperature indicator, 177
Enzyme release packaging, 199
Ethanol emitters, 81–82, 200
Ethanol scavengers, 239
Ethylene scavengers, 45, 46f, 80–81, 120–122, 238–239, 278
E-tongue systems, 222
European Union, regulatory aspects in, 100–104, 101f
   commission regulation, 103
   framework regulation, 101–103
   labeling requirement, 103–104
   specific legislations, 104
Exogenous antioxidant, 2–4
   ascorbic acid, 2
   carotenoid compounds, 2–3
   polyphenol compounds, 3–4
   tocopherol, 4
Extrusion, 374
   coating, 357
   molding, 357
   technique, 186, 186f

F

FDA regulations, 108
Ferrous oxide and photosensitive dyes, 97–98
Fiberboard, 312
Film-based technique, 241–242
Film blowing, 374–375
Fish and seafood packaging systems, 162, 202, 203t
    blockchain systems, 219–220, 219f
        RFID systems, 219–220, 220t
    electronic sensing systems, 220–222, 221f
    indicators, 216–218
        freshness indicators, 216–217, 216f
        leakage indicators, 217
    potential trends, 222–223
    sensors, 212–216
        biosensors, 214–216, 214f
        chemical sensors, 213–214, 214t–215t
    smart technologies, 211–212, 212f
    temperature-time indicators (TTI), 217–218, 218t
Flavor/odor absorbers, 48, 122
Flavor releasing packaging, 199
Flesh products, 139
Food Business Operators (FBO), 171
Food contact legislation, 62
Food grade titanium dioxide, 63
Food packaging materials, 8–10, 9t, 286–296, 286f, 288t
    glass, 9
    metals, 10
    paper and paper, 9–10
    plastic, 8
Food products, oxidation in, 10–11
Food safety and standards authority (FSSAI), 105–107
Food safety guidelines, food packaging
    different countries, packaging labeling regulations in, 65–68
        active packaging, 66
        carriers, 66
        food safety regulation, active packaging, 67
        food safety regulation, intelligent packaging, 67
        future trends, 67–68
        indicators, 66
        intelligent packaging, 66
        nanotechnology, 67
        sensors, 66
        smart packaging, legislation related to, 66–67
    food contact legislation, 62
    food safety laws, 61–62
        food packaging legislation, India, 61–62
    international legislation, 60
    introduction to, 59
    oils, packaging label requirements of, 64–65
    role of packaging, 59–60, 60f
    sale and license, condition for, 62–64
        packing and labeling of foods, 64, 65t
    WHO, 59
Food safety laws, 61–62
    food packaging legislation, India, 61–62
Food safety regulation
    active packaging, 67
    intelligent packaging, 67
Food supplements, 231t–232t
Food supply system, sustainability of, 177–179
Fourier transform infrared spectroscopy, 356
Fresh-check lifeline integrator, 380
Fresh cut vegetable (FCV) goods, 147, 147t–148t
Freshness indicators, 50–51, 177
Freshness sensors, 271–272
Fruit quality preservation
    fruit freshness, 269
    nutritional features and dietary implications, 268t
    smart packaging technologies, 269–279, 270f
        active technologies, 278–279
        container-based smart packaging technology, 276–278
        freshness sensors, 271–272
        intelligent packaging technologies, 272–276
Fruits and vegetable products, 7, 139, 162, 201

G

Gas indicators, 248–249, 378–379
Gas releasing packaging, 198
Gas sensors, 53–54
Glass, 9
    showcase materials, 313
GLOBAL G. A. P (Good Agricultural practices), 61
Global marketing companies, 111, 111t

H

HACCP, 61
Herbs and spices, 5–6
Home food replacement (HFR) technology
    antimicrobial packaging (AP), 158–159, 159f
    characteristics of, 149–150

Home food replacement (HFR) technology (*Continued*)
  easy to exposed packing (EEP), 150–153
  FCV product packaging, 155–157, 156f, 158t
  industry, necessary for, 149
  intelligent packaging (IP), 159–162, 160f–161f
  microwaveable packaging (MP), 153–155, 153f
Home meals smart packaging
  active packaging, 144
  consumer benefits of, 166
  fresh cut vegetable (FCV) goods, 147, 147t–148t
  home food replacement (HFR) goods, 146–149, 146f
  home food replacement (HFR) technology, 149–162
    antimicrobial packaging (AP), 158–159, 159f
    characteristics of, 149–150
    easy to exposed packing (EEP), 150–153
    FCV product packaging, 155–157, 156f, 158t
    industry, necessary for, 149
    intelligent packaging (IP), 159–162, 160f–161f
    microwaveable packaging (MP), 153–155, 153f
  innovative packaging
    beverage products, 163
    fish and sea-food food items, 162
    fruit and vegetable goods, 162
    meat & poultry goods, 162
  intelligent packaging, 144
  issues related to, 166–167
  need of, 145–146
  purposes of, 143, 144f
  ready to cook (RTC) goods, 148–149
  ready to heat (RTH) goods, 148
  ready to take (RTT) goods, 148
  trends for, 163–166, 165f
Humidity control, moisture absorbers for, 279
Humidity indicator-based technique, 249
Humidity sensors, 257

I

IBCSGate solution, 304–305
India and South East Asian countries, regulatory aspects in, 104–105
Indian Standards for Direct Food Contact Materials, 65
Indicators, 49–53, 66, 98–99, 123
  barcodes, 52–53
  fish and seafood packaging systems, 216–218
    freshness indicators, 216–217, 216f
    leakage indicators, 217
  freshness indicators, 50–51
  integrity indicators, 52, 52f
  RFID tags, 52–53, 53f
  supported with, 247–250
  supported without, 250–257
  time temperature indicators (TTIs), 50, 51f
Industrial application/market implementation
  active packaging, 77–82, 78f
    advantages of, 86
    antimicrobial and antioxidant packaging, 82
    carbon dioxide emitter, 79–80
    consumers, 89–90
    costs, 90
    ethanol emitters, 81–82
    ethylene scavengers, 80–81
    industrial barriers for, 86
    industries, prohibitive regulations for, 89
    intelligent packaging, 85, 85t
    lake of knowledge, 88
    moisture absorber/scavengers, 81
    move from lab to industrial scale, 88–89
    new manufacturing techniques, 90
    oxygen scavenger, 79
    safety concern, 88
    sustainability, 87
  advantages, 74, 74t–76t
  disadvantages of, 74, 74t–76t
  intelligent packaging, 83–85, 83f
    active packaging, 85, 85t
    advantages of, 86
    consumers, 89–90
    costs, 90
    indicators, 84–85
    industrial barriers for, 86–91
    industries, prohibitive regulations for, 89
    lake of knowledge, 88
    move from lab to industrial scale, 88–89
    new manufacturing techniques, 90
    radio frequency identification (RFID), 85
    safety concern, 88
    sensors, 84
    sustainability, 87
  packaging, 72
  smart packaging, 77
  traditional packaging, 73–74
Infant milk, 64

Inference, 181—182
Injection molding, 186—187, 187f, 356
Integrity indicators, 52, 52f
Intelligent packaging, 66, 123—124
 home meals smart packaging, 144
 industrial application/market implementation, 83—85, 83f
  active packaging, 85, 85t
  advantages of, 86
  consumers, 89—90
  costs, 90
  indicators, 84—85
  industrial barriers for, 86—91
  industries, prohibitive regulations for, 89
  lake of knowledge, 88
  move from lab to industrial scale, 88—89
  new manufacturing techniques, 90
  radio frequency identification (RFID), 85
  safety concern, 88
  sensors, 84
  sustainability, 87
 market demand
  data carrier, 123—124
  indicators, 123
  sensors, 124
 medicinal food supplements, 246—257
  barcode-based technique, 250—251
  biosensors-based technique, 252—257, 253t
  commercial biosensors, 254—255
  electrochemical biosensors, 254
  gas indicator-based technique, 248—249
  gas sensors, 256—257
  humidity indicator-based technique, 249
  humidity sensors, 257
  indicator, supported with, 247—250
  indicator, supported without, 250—257
  optical biosensors, 253—254
  RFID tags-based technique, 251—252
  time temperature indicator (TTI)-based technique, 247—248
  visual indicator-based technique, 249
 smart nanosensors for, 306
 smart packaging technology, innovations in, 49—55, 50t
  indicators. See Indicators
  legal aspects of, 55
  sensors. See Sensors
 technologies
  direct freshness, 272—275
   leak, 273—275, 274f
   microbial pathogens, 275
   ripeness, 273, 274f
   spoilage, 272, 273f
  indirect freshness, 275—276
   TTIs, 276
Intelligent to smart packaging
 absorbing system, 179—180
 active packaging, 176—179
 antimicrobial packaging, 180—181
 bio-based plastics, 174
 connected and smart packaging, 172—174, 172f—173f
 cutting edge advancement, 174—176
  sustainable packaging, innovative materials for, 175—176
 edible coatings, 181—182
 emerging technologies, 177—179
 enzyme-based time-temperature indicator, 177
 Food Business Operators (FBO), 171
 food supply system, sustainability of, 177—179
 freshness indicators, 177
 inference, 181—182
 radio frequency identification (RFID), 176—177
 releasing system, 180
International legislation, 60
International standard ISO 9000, 60
Ion radiation, 352
IS:20 specifications, 63
IS:1046 specifications, 63
IS:10142 specifications, 63
IS:10151 specifications, 63
IS:10910 specifications, 63
IS:11434 specifications, 63
IS:11704 specifications, 63
IS:12247 specifications, 63
IS:12252 specifications, 63
IS:13576 specifications, 63
IS:13601 specifications, 63

L

Label-based technique, 243—244
Laminated and metallized films, 133
Laser perforation technology (LPT), 152
Legislation
 absorption of gases, 97—98
 Brazil, regulatory aspects in, 109—111
 Bureau of Indian standards (BIS), 105—107

Legislation (*Continued*)
  data carriers, 99
  European Union, regulatory aspects in, 100–104, 101f
    commission regulation, 103
    framework regulation, 101–102
    framework regulation (Article 16), 103
    labeling requirement, 103–104
    specific legislations, 104
  FDA regulations, 108
  ferrous oxide and photosensitive dyes, 97–98
  food safety and standards authority (FSSAI), 105–107
  global marketing companies, 111, 111t
  India and South East Asian countries, regulatory aspects in, 104–105
  indicators, 98–99
  moisture absorbers, 97–98
  sensors, 99, 99t–100t
  South and Central America, regulatory aspects in, 109–111
  Threshold of regulation rule, 109
  US, regulatory aspects in, 107–108
Legumes, 7–8
Light alloy-based showcase materials, 313
Light weight metallic food packaging, 135
Lipid-based polymers, 337–338
Long shelf life, 134–135

M
Market demand
  active packaging, 118–123, 119f, 119t
    antioxidants, 122–123
    carbon dioxide emitter (CO_2), 122
    ethylene scavengers, 120–122
    flavor or odor absorbers, 122
    flavor or odor emitters, 122
    moisture scavengers, 119t, 120
    oxygen scavengers (OS), 120
  consumer perception, 124–125, 125f
  intelligent packaging (IP), 123–124
    data carrier, 123–124
    indicators, 123
    sensors, 124
  smart packaging, 117–118, 118f
Meat and poultry products, 162, 203–204
  oxygen scavengers, 203–204, 205f
Meat, temperature-sensitive smart packaging for, 307

Medicinal food supplements
  active packaging, 235–246
    carbon dioxide (CO_2) absorbers & emitters, 240
    commercial scavengers and emitters, 235, 236t–237t
    ethanol scavengers, 239
    ethylene scavengers, 238–239
    moisture scavengers, 239–240
    odor or taint scavengers, 239
    oxygen scavengers, 235–238
    preservative releasers, 241
  antimicrobial active packaging, 241–246
    antimicrobial attributing polymer, 246
    antimicrobials to polymers, surface fixation of, 245–246
    coating antimicrobial on polymer, 245
    film-based technique, 241–242
    label-based technique, 243–244
    polymers, 244–245
    sachet-based technique, 244
    tray-based technique, 243
  food supplements, 231t–232t
  future challenges, 257–261
  intelligent packaging, 246–257
    barcode-based technique, 250–251
    biosensors-based technique, 252–257, 253t
    commercial biosensors, 254–255
    electrochemical biosensors, 254
    gas indicator-based technique, 248–249
    gas sensors, 256–257
    humidity indicator-based technique, 249
    humidity sensors, 257
    indicator, supported with, 247–250
    indicator, supported without, 250–257
    optical biosensors, 253–254
    RFID tags-based technique, 251–252
    time temperature indicator (TTI)-based technique, 247–248
    visual indicator-based technique, 249
  pictorial presentation of, 233f
  smart packaging, 234–257, 234f
    necessity of, 230–233
    patented products of, 257, 258t–260t
  steps for, 232f
  traditional and smart packaging, 230
  types, 230
Melt-mix method, 375
Metal-based showcase materials, 312

Metal containers, 134
Metal corrosion, 135–136
Metal food packaging
  advantages of, 134–135
    customer's attraction, 135
    durability, 134
    light weight metallic food packaging, 135
    long shelf life, 134–135
    product protection, 134
    sustainability, 134–135
  alloys of iron (steel), 130
  aluminum, 131–132
  applications, 138–139, 138f
    bakery and confectionary products, 139
    beverages, 139
    coffee and tea, 139
    flesh products, 139
    fruits and vegetables, 139
    milk products, 138–139
  disadvantages of, 135–137
    environmental concerns of metal packaging, 137
    metal corrosion, 135–136
    metallic cans, protection and decoration of, 136–137
    metal packaging, health issues with, 137
    sightlessness of contents, 136
    storage issues, 136
  shapes of, 132–134, 132f
    aluminum bottles, 133
    aluminum collapsible tubes, 133
    aluminum foil, 133
    laminated and metallized films, 133
    metal containers, 134
    metal lids, 134
    retort pouches, 133–134
  Sn free steel (SFS), 131
  stainless steel, 131
  stannous/tin (Sn) plate, 130–131
  types of, 130–132
Metallic cans, protection and decoration of, 136–137
Metal lids, 134
Metal packaging, health issues with, 137
Metal packing materials, cooked foods, 287–296, 289t–290t
  advantages, 292–296
  alternatives, 291–292
  disadvantages, 292–296

  normalized priorities, 299t
  results, 297–299, 297t
Metals, 10
Microwaveable packaging (MP), 153–155, 153f
  MicroRite method, 154–155, 155f
  susceptor method, 154, 154f
Milk products, 64, 138–139, 201–202
  cholesterol and lactose, reduced content of, 202
  oxygen scavenger, 202
Mineral oil of food grade, 64
Ministerial Notification No. 92/2528, 106
Ministerial Notification No. 117/2532, 106
Ministerial notification No. 295/2548, 106
Ministry of Health of Malaysia drives Food Act, 106–107
3 M monitor-mark, 381
Modified atmosphere packaging (MAP), 29, 322, 324t
Moisture absorbers, 41–43, 43f, 97–98
Moisture scavengers, 119t, 120, 239–240
Multipiece package, 61

N

Nanotechnology, 34, 67, 353–355
Non-veg food, 62
Non-volatile tocopherols, 29–30
Nuclear magnetic resonance, 356
Nut, 7–8
Nutrient releasing packaging, 198–199
Nutrition protection, graphene-based nanosensors for, 306–307

O

Odor/taint scavengers, 239
Oils, packaging label requirements of, 64–65
Opportunities, 32–34
Optical biosensors, 253–254
Oxygen scavengers (OS), 28, 43–44, 44f, 79, 120, 200, 202, 235–238

P

Packaging film to food products, 14–15
Packaging systems, classification of, 26–30, 27f
  active packaging, 27–30, 28t, 29f
  definition, 30
Paper, 9–10
Particleboard, 311
Pectin, 329–331, 330f
Peroxidation, mechanism of, 11

PH alteration, 350–351
PH indicators, 378
Plasticization, 8, 313, 320, 345–346
Plywood, 312
Polybutylene succinate (PBS), 340–341, 340f
Polycaprolactone (PCL), 341–342, 341f
Polyethylene furanoate (PEF), 343, 343f–344f
Polyhydroxyalkanoates, 328–329, 329f
Polylactic acids, 338–340, 339f
Polylactide aliphatic copolymer (CPLA), 342
Polymer fabrication technology, 356–360
    blow molding, 357
    compression molding, 356
    electrospinning, 359–360
    extrusion coating, 357
    extrusion molding, 357
    injection molding, 356
    thermoforming, 357–358
Polymers, 244–245
Polyphenol compounds, 3–4
Polysaccharide based polymers, 324–335
    agar, 334–335, 335f
    alginate, 331–332, 331f
    cellulose, 324–326, 325f
    chitosan, 332–334, 333f
    pectin, 329–331, 330f
    polyhydroxyalkanoates, 328–329, 329f
    starch, 326–328
Polyvinyl alcohol (PVA), 342–343
Positive experience, increasing likelihood of, 384–385
Prediction analysis, 384
Predictive planning, 384
Prepackaged and packed foods, 62
Preservative releasers, 241
Principal display panel, 62
Product protection, 134
Product use convenience, 206–207
Protein-based polymers, 335–337, 336f–337f
Protein rich atta and Maida, 64
PVC, 313

Q

Quality checking, graphene-based nanosensors for, 306–307
Quality control, 381–384

R

Radio frequency identification (RFID), 52–53, 53f, 85, 176–177, 219–220, 220t, 251–252, 381
Ready to cook (RTC) goods, 148–149
Ready to heat (RTH) goods, 148
Ready to take (RTT) goods, 148
Regulation (EC) 178/20028 (General Food Law), 60
Releasing system, 180
Research, 31–34
    challenges, 32
    cybersecurity, 33–34
    opportunities, 32–34
Retort pouches, 133–134
RipeSense indicator, 380

S

Sachet-based technique, 244
Sale and license, condition for, 62–64
    packing and labeling of foods, 64, 65t
Sale and purchase of insecticides, 63
Scavengers, 349–350
Sensors, 53–55, 66, 84, 99, 99t–100t, 124
    biosensors, 53, 54f
    chemical sensors, 54
    edible sensors, 55
    fish and seafood packaging systems, 212–216
        biosensors, 214–216, 214f
        chemical sensors, 213–214, 214t–215t
    gas sensors, 53–54
Shelf life, 309
Sightlessness of contents, 136
Smart control systems, 308
Smart materials, 34
Smart nutrition packet method, 306–307
Smart packaging, 117–118, 118f, 376–377
    active packaging, 40–49, 42t
        antimicrobial packaging systems, 47–48
        antioxidant releaser, 46–47, 48f
        CO_2 emitters and absorbers, 45–46, 47f
        ethylene scavengers, 45, 46f
        flavor/odor releasers and absorbers, 48
        moisture absorbers, 41–43, 43f
        other active packaging, 48–49
        oxygen scavengers, 43–44, 44f
    applications, 30, 31f
    intelligent packaging, 49–55, 50t

indicators. *See* Indicators
legal aspects of, 55
sensors. *See* Sensors
legislation related to, 66—67
new technologies, 25—26
overview of, 40—55, 41f
packaging systems, classification of, 26—30, 27f
   active packaging, 27—30, 28t, 29f
   definition, 30
   research, 31—34
      challenges, 32
      cybersecurity, 33—34
      opportunities, 32—34
   types and features of, 12—13, 12f
   worldwide market prospect, 30—31
Smart showcase design, smart packaging
   accessories, 310
   color, 309
   composite wood, 311—313
      aluminum, 313
      fiberboard, 312
      glass showcase materials, 313
      light alloy-based showcase materials, 313
      metal-based showcase materials, 312
      particleboard, 311
      plastic based showcase materials, 313
      plywood, 312
      PVC, 313
   consumer food waste reduction, smarter packaging for, 307—308
   consumer/packaging interface, 307
   definition, 306—308
   IBCSGate solution, 304—305
   intelligent packaging, smart nanosensors for, 306
   materials, 310—313
   meat, temperature-sensitive smart packaging for, 307
   nutrition protection, graphene-based nanosensors for, 306—307
   principles, 308—310
   proposal, 313—314, 314t
   quality checking, graphene-based nanosensors for, 306—307
   shelf life, 309
   smart control systems, 308
   smart nutrition packet method, 306—307
   solid wood, 311
   use of light, 310
   wood-based showcase materials, 311

Sn free steel (SFS), 131
Solid bleached sulfate (SBS), 320
Solid wood, 311
South and Central America, regulatory aspects in, 109—111
Stainless steel, 131
Stannous/tin (Sn) plate, 130—131
Starch, 326—328
Storage issues, 136
Sugar boiled confectionery, 64
Sustainability, 87, 134—135
   agricultural residues, 189—190, 189f
   animal skin product, 190
   bamboo products, 187—188
   chicken feathers, 192
   eggshell product, 190—192
   wool products, 188—189
Sustainable materials, smart packaging from
   biochemical production of, 187—192
      sustainable agricultural residues, 189—190, 189f
      sustainable animal skin product, 190
      sustainable bamboo products, 187—188
      sustainable chicken feathers, 192
      sustainable eggshell product, 190—192
      sustainable wool products, 188—189
   mechanical production of, 186—187
      extrusion technique, 186, 186f
      injection molding technique, 186—187, 187f
Synthetically produced bio-polyesters, 338—343
   aliphatic polyesters, 338—343
   aromatic polyesters, 343
   polyethylene furanoate (PEF), 343, 343f—344f
Synthetic antioxidants, 5

T

Tea, 6
Temperature-time indicators (TTI), 217—218, 218t
Tetra pack, 353
Thermoforming, 357—358
Thermogravimetric analysis, 355
Thermopressing, 375
Thin-film electronics, 32, 34
Threshold of regulation rule, 109
Time temperature indicators (TTI), 50, 51f, 247—248, 379—380
Tocopherol, 4
Tray-based technique, 243

U
Uncoated recycled board (URB), 320

Use by date, 62

US, regulatory aspects in, 107–108

V
Vistab indicator, 249, 380

W
White-lined chipboard (WLCB), 320
WHO, 59
Wood-based showcase materials, 311

X
Xanthine oxide (XOD), 215

Z
Ziegler Natta, 151

Printed in the United States
by Baker & Taylor Publisher Services